Lyapunov Exponents
A Tool to Explore Complex Dynamics

Lyapunov exponents lie at the heart of chaos theory and are widely used in studies of complex dynamics. Utilising a pragmatic, physical approach, this self-contained book provides a comprehensive description of the concept. Beginning with the basic properties and numerical methods, it then guides readers through to the most recent advances in applications to complex systems. Practical algorithms are thoroughly reviewed and their performance is discussed, while a broad set of examples illustrates the wide range of potential applications. The description of various numerical and analytical techniques for the computation of Lyapunov exponents offers an extensive array of tools for the characterisation of phenomena, such as synchronisation, weak and global chaos in low- and high-dimensional setups, and localisation. This text equips readers with all of the investigative expertise needed to fully explore the dynamical properties of complex systems, making it ideal for both graduate students and experienced researchers.

Arkady Pikovsky is Professor of Theoretical Physics at the University of Potsdam. He is a member of the editorial board for *Physica D* and a Chaotic and Complex Systems Editor for *J. Physics A: Mathematical and Theoretical*. He is a Fellow of the American Physical Society and co-author of *Synchronization: A Universal Concept in Nonlinear Sciences*. His current research focusses on nonlinear physics of complex systems.

Antonio Politi is the 6th Century Chair in Physics of Life Sciences at the University of Aberdeen. He is Associate Editor of *Physical Review E*, a Fellow of the Institute of Physics and of the American Physical Society, and was awarded the Gutzwiller Prize by the Max-Planck Institute for Complex Systems in Dresden and the Humboldt Prize. He is co-author of *Complexity: Hierarchical Structures and Scaling in Physics*.

Lyapunov Exponents

A Tool to Explore Complex Dynamics

ARKADY PIKOVSKY

University of Potsdam

ANTONIO POLITI

University of Aberdeen

CAMBRIDGE
UNIVERSITY PRESS

CAMBRIDGE
UNIVERSITY PRESS

University Printing House, Cambridge CB2 8BS, United Kingdom

Cambridge University Press is part of the University of Cambridge.

It furthers the University's mission by disseminating knowledge in the pursuit of
education, learning and research at the highest international levels of excellence.

www.cambridge.org
Information on this title: www.cambridge.org/9781107030428

First published 2016

Printed in the United Kingdom by TJ International Ltd. Padstow Cornwall

A catalogue record for this publication is available from the British Library

Library of Congress Cataloguing in Publication data
Pikovsky, Arkady, 1956–
Lyapunov exponents : a tool to explore complex dynamics / Arkady Pikovsky,
University of Potsdam, Antonio Politi, University of Aberdeen.
pages   cm
Includes bibliographical references and index.
ISBN 978-1-107-03042-8 (Hardback : alk. paper)
1. Lyapunov exponents.   2. Differential equations.   I. Politi,   II. Title.
QA372.P655 2016
515′.352–dc23   2015032525

ISBN 978-1-107-03042-8 Hardback

Contents

Preface

With the advent of electronic computers, numerical simulations of dynamical models have become an increasingly appreciated way to study complex and nonlinear systems. This has been accompanied by an evolution of theoretical tools and concepts: some of them, more suitable for a pure mathematical analysis, happened to be less practical for applications; other techniques proved instead very powerful in numerical studies, and their popularity exploded. Lyapunov exponents is a perfect example of a tool that has flourished in the modern computer era, despite having been introduced at the end of the nineteenth century.

The rigorous proof of the existence of well-defined Lyapunov exponents requires subtle assumptions that are often impossible to verify in realistic contexts (analogously to other properties, e.g., ergodicity). On the other hand, the numerical evaluation of the Lyapunov exponents happens to be a relatively simple task; therefore they are widely used in many setups. Moreover, on the basis of the Lyapunov exponent analysis, one can develop novel approaches to explore concepts such as hyperbolicity that previously appeared to be of purely mathematical nature.

In this book we attempt to give a panoramic view of the world of Lyapunov exponents, from their very definition and numerical methods to the details of applications to various complex systems and phenomena. We adopt a pragmatic, physical point of view, avoiding the fine mathematical details. Readers interested in more formal mathematical aspects are encouraged to consult publications such as the recent books by Barreira and Pesin (2007) and Viana (2014).

An important goal for us was to assess the reliability of numerical estimates and to enable a proper interpretation of the results. In particular, it is not advisable to underestimate the numerical difficulties and thereby use the various subroutines as black boxes; it is important to be aware of the existing limits, especially in the application to complex systems.

Although there are very few cases where the Lyapunov exponents can be exactly determined, methods to derive analytic approximate expressions are always welcome, as they help to predict the degree of stability, without the need of actually performing possibly long simulations. That is why, throughout the book, we discuss analytic approaches as well as heuristic methods based more on direct numerical evidence, rather than on rigorous theoretical arguments. We hope that these methods will be used not only for a better understanding of specific dynamical problems, but also as a starting point for the development of more rigorous arguments.

The various techniques and results described in the book started accumulating in the scientific literature during the 1980s. Here we have made an effort to present the main (according to our taste) achievements in a coherent and systematic way, so as to make the understanding by potentially unskilled readers easier. An example is the perturbative approach of the weak-disorder limit that has already been discussed in other reviews; here we present the case of ellyptic, hyperbolic and marginal matrices in a systematic manner.

Although this is a book and, as such, mostly devoted to a coherent presentation of known results, we have also included novel elements, wherever we felt that some gaps had to be filled. This is for instance, the case of the finite-size effects in the Kuramoto model or the extension of the techniques developed by Sompolinsky et al. to a wider class of random processes.

As a result, we are confident that the book can be read at various levels, depending on the needs of the reader. Those interested in the bare application to some simple cases will find the key elements in the first three chapters; the following chapters contain various degrees of in-depth analysis. Cross references among the common points addressed in the various sections should help the reader to navigate across specific items.

The most important acknowledgement goes to the von Humboldt Foundation, which, supporting the visit of Antonio Politi to Potsdam with a generous fellowship, has allowed us to start and eventually complete this project. Otherwise, writing the book would have been simply impossible.

We happened to discuss with, ask and receive suggestions from various colleagues. We specifically wish to acknowledge V. N. Biktashev, M. Cencini, H. Chaté, A. Crisanti, F. Ginelli, H. Kantz, R. Livi, Ya. Pesin, G. Puccioni, K. A. Takeuchi, R. Tonjes and H.-L. Yang.

Antonio Politi wishes also to acknowledge A. Torcini and S. Lepri as long-term collaborators who contributed to the development of some of the results herein summarised.

Special thanks go to P. Grassberger, who, more than 10 years after the publication of a joint paper with G. D'Alessandro, S. Isola and Antonio Politi on the Hénon map, was able to dig out some data to determine the still most accurate estimate of the topological entropy of such a map. As laziness has prevented a dissemination of those results, we made an effort to include them in this book.

We also wish to thank E. Lyapunova, the grand-niece of A. M. Lyapunov, who provided a high-quality photograph of the scientist who originated all of the story.

We finally warmly thank S. Capelin of Cambridge University Press, who has been patient enough to wait for us to complete the work. We hope that the delay has been worthy of a much better product. Although surely far from perfect, at some point we had to stop.

1 Introduction

1.1 Historical considerations

1.1.1 Early results

The problem of determining the stability of a given regime (e.g. the motion of the solar system) is as old as the concept of the dynamical system itself. As soon as scientists realised that physical processes could be described in terms of mathematical equations, they also understood the importance of assessing the stability of various dynamical regimes. It is thus no surprise that many eminent scientists, such as Euler, Lagrange, Poincaré and Lyapunov (to name a few), engaged themselves in properly defining the concept of stability. Lyapunov exponents are one of the major tools used to assess the (in)stability of a given regime. Within hard sciences, where there is a long-standing tradition of quantitative studies, Lyapunov exponents are naturally used in a large number of fields, such as astronomy, fluid dynamics, control theory, laser physics and chemical reactions. More recently, they started to be used also in disciplines, such as biology and sociology, where nowadays processes can be accurately monitored (e.g. the propagation of electric signals in neural cells and population dynamics).

The reader interested in a fairly accurate historical account of how stability has been progressively defined and quantified can refer to Leine (2010). Here, we limit ourselves to the recapitulation of a few basic facts, starting from the Galilean times, when E. Torricelli (1644) investigated the stability of a mechanical system and conjectured (in the modern language) that a point of minimal potential energy is a point of equilibrium.

Besides mechanical systems, floating bodies provide another environment where stability is naturally important, especially to avoid roll instability of vessels. Unsurprisingly, the first results came from a Flemish (S. Stevin) and a Dutch (Ch. Huygens) scientist: at that time, the cutting-edge technology of ship-building had been developed in the Dutch Republic. In particular, Huygens' approach was quite modern in that he addressed the problem by explicitly comparing two different states. D. Bernoulli too dealt with the problem of roll-stability, emphasising the importance of the restoring forces, which make the body return towards the equilibrium state. L. Euler was the first to distinguish between stable, unstable, and indifferent equilibria and suggested also the possibility of considering infinitely small perturbations.

The concept of stability was further developed by J.-L. Lagrange, who formalised the ideas expressed by Torricelli (for conservative dynamical systems), clarifying that, in the presence of a vanishing kinetic energy, the minimum of the potential energy corresponds to a stable equilibrium. The corresponding theorem is nowadays referred to as "Lagrange-Dirichlet" because of further improvements introduced by J. P. G. L. Dirichlet.

In the nineteenth century, fluid dynamics provided many examples where the stability assessment was far from trivial. Some scientists (notably Lord Kelvin) were striving to unify physics under the paradigm of the motion of perfect liquids, and such an approach required the stability of various forms of motion. At a macroscopic level, in the attempt of predicting the Earth's shape, the problem of determining the stable shape of a rotating fluid, under the influence of the sole action of centrifugal and (internal) gravitational forces, was posed. The studies led to the conclusion that, in some conditions, ellipsoidal shapes are to be expected, but the problem was not fully solved (see Section 1.1.2 on Lyapunov's biography).

On a more microscopic level, hydrodynamics proved to be an extremely fertile field for the appearance of instabilities: concepts such as sensibility to infinitesimal and finite perturbations were present in the minds of esteemed scientists. G. Stokes was one of the pioneers: he stipulated that instabilities naturally occur in the presence of rapidly diverging flow lines, such as past a solid obstacle. Slightly later, H. Helmholtz and W. Thomson discovered that the surface separating two adjacent flows may lose its flatness. Contrary to the instability foreseen by Stokes, which was based only on conjectures, the latter one, nowadays referred to as the Kelvin-Helmholtz instability, was also derived directly from the hydrodynamics equations. Last but not least, Lord Kelvin strived to develop a vortex theory of matter, which, however, required the stability of the underlying dynamical regimes. Only at the end of his career did he convince himself that his ideas were severely undermined by the unavoidable presence of instabilities. The interested reader can look at the exhaustive review by Darrigol (2002).

Celestial mechanics proved to be another fruitful environment for the development of new ideas. In order to appreciate how relevant the subject was in those times, it is sufficient to mention that when P. S. Laplace studied perturbatively the behaviour of three gravitationally interacting particles (the so-called 3-body problem), he referred to it as to the "world system". Heavily relying on recent results by Lagrange, Laplace concluded that the semi-major axis of the orbits is characterised by periodic oscillations. Thus, he concluded in favour of stability, meaning that the fluctuations are bounded. A bit later, S. D. Poisson discovered that second- and third-order terms generate a secular contribution of the type $At \sin \alpha t$; however, as remarked by C. G. J. Jacobi, it was not clear whether such a contribution would survive a higher-order analysis. All in all, no clear answer had yet been given by the end of the nineteenth century. This is the reason why King Oscar II of Sweden decided to offer a prize for those who could find an explicit solution. H. Poincaré won the prize even though he did not actually solve the problem. On the contrary, his work established the existence of unavoidable high sensitivity to initial conditions: what was later called the 'butterfly effect' by the metereologist E. N. Lorenz.[1] Poincaré received the

---

[1] The expression 'butterfly effect' was arguably introduced by Lorenz in 1972, when he gave a talk at the American Association for the Advancement of Science entitled "Does the flap of a butterfly's wings in Brazil set off a tornado in Texas?"

prize for the revolutionary methods that he developed to gain insight about the behaviour of generic dynamical systems.

A last environment where stability turned out to be of primary importance is related to engineering applications. In the nineteenth century, with the advent of steam engines, it became necessary to regulate the internal pressure inside the boiler. This problem represented the starting point for the birth of a new discipline: automatic control theory. J. C. Maxwell analysed the stability of Watt's flyball regulator by linearising the equations of motion. Independently, I. A. Vyshnegradtsky used a similar approach to study the same problem in greater detail.

1.1.2 Biography of Aleksandr Lyapunov

Here, we briefly summarise some basic facts of the biography of Aleksandr Mikhailovich Lyapunov, mostly relying on Smirnov (1992) and Shcherbakov (1992).

Aleksandr Lyapunov was born in 1857 in Yaroslavl. After completing his gymnasium studies in Nizhny Novgorod, Lyapunov moved to the University of St. Petersburg, where the Mathematical Department was blooming under the direction of Pafnuty Chebyshev, who soon became the supervisor of his graduate studies. Chebyshev used to say that "every young scholar . . . should test his strength on some serious theoretical questions presenting known difficulties". As a matter of fact, Lyapunov got involved in a problem that had been earlier proposed to other students (he discovered this later in his career), namely that of determining the shape of a rotating fluid. As his efforts proved unsuccessful, Lyapunov

Fig. 1.1   A. M. Lyapunov in 1902, in Kharkov. Photo courtesy of Elena Alexeevna Lyapunova.

decided to refocus his work, preparing a dissertation entitled *On the stability of elliptic forms of equilibrium of rotating fluids*, which nevertheless allowed him to be awarded a Master's degree in applied mathematics (1884) and made him known in Europe. In 1885, Lyapunov was appointed Privatdozent in Kharkov, where he worked on the stability of mechanical systems. His main results were summarised in a remarkable thesis entitled *The general problem of the stability of motion*, which granted him a PhD at Moscow University (1892). The dissertation contains an extraordinarily deep and general analysis of systems with a finite number of degrees of freedom. Interestingly, Lyapunov mentioned H. Poincaré as one of his principal sources of inspiration.

In 1893, Lyapunov was promoted to ordinary professor in Kharkov. In the following years, he kept studying stability properties of dynamical systems, investigated the Dirchlet problem, and engaged himself in problems of probability theory, contributing to the central limit theorem and paving the way to the rigorous results obtained by his friend Andrei Markov. In 1901 he became head of the department of Applied Mathematics at the Russian Academy of Sciences in St. Petersburg (the position, without teaching duties, had been vacant since 1894, when Chebyshev died).

After having completed a cycle of papers on the stability of motion, Lyapunov came back to the question posed to him by Chebyshev about 20 years before and much related to the problem of determining the form of celestial bodies, earlier formulated by Laplace. While he was still struggling to find a solution, Lyapunov became aware of a book published by Poincaré in 1902 on the same problem and managed to acquire a copy. From a letter sent by Lyapunov to his disciple and close friend Steklov: "To my greatest surprise, I did not find anything significant in this book . . . Thus my work has not suffered and I apply myself to it afresh". The book by Poincaré essentially contained previous (known) concepts with little advancements.

Shortly after, the astronomer George Darwin (son of Charles Darwin) published some papers on the same subject, concluding that pear-shaped forms are to be expected. Lyapunov completed his studies in 1905: a treatise of about 1000 pages, with some mathematical calculations made up to 14 digits when necessary. He indeed discovered deviations from ellipsoids, but he also showed that pear-shaped forms are unstable. The controversy with Darwin went on for some years, until it was eventually settled in 1917, when another British astronomer, J. H. Jeans, confirmed that Lyapunov was right.

In 1917 Lyapunov left St. Petersburg for Odessa, so that his wife could receive treatment for tuberculosis. On the day of his wife's death, Aleksandr Lyapunov committed suicide.

1.1.3 Lyapunov's contribution

The first formal definition of stability was given by Lyapunov in his PhD thesis: a given trajectory is stable if, for an arbitrary ε, there always exists a δ such that all other trajectories starting in a δ-neighbourhood of the given one remain at most at a distance ε to it. He introduced also what was later called asymptotic stability, to refer to cases where sufficiently small perturbations eventually die.

Lyapunov introduced also two methods to assess the stability of a given solution. The first method was based on a "standard" perturbative analysis; he was very interested in identifying those cases where it is necessary to go beyond the first order to characterise correctly the perturbation dynamics. As a result, he introduced the "characteristic number" λ_L. It is basically the opposite of what is nowadays called the (characteristic) Lyapunov exponent. In fact, he defined λ_L as the exponential rate which has to be added to balance the growth rate of a given perturbation $\delta(t)$: in other words, assuming that $\delta(t) \approx e^{\lambda t}$, he would define the characteristic number λ_L as the value such that $\delta(t)e^{\lambda_L t}$ neither diverges nor converges exponentially.

The second, or direct method, deals with the introduction of a pseudo-energy function (nowadays called Lyapunov function) that vanishes in the equilibrium point and is otherwise positive, and decreases (or does not increase) along a generic trajectory. This is an extension of the ideas of Torricelli and Lagrange to a context where the potential energy is not defined a priori. Lyapunov's PhD thesis was translated into French in 1908, while one had to wait until 1992 to see the first English translation (Lyapunov, 1992).

Although Lyapunov himself attached more importance to the first method, he became famous for the second method. Nevertheless, even within Russia, the first practical applications of his stability theory were not made until the 1930s by N. G. Chetayev and I. G. Malkin at the Kazan Aviation Institute. The reader interested in the development of the second method is invited to consult Parks (1992).

1.1.4 The recent past

Although Lyapunov exponents (LEs) were formally introduced at the end of the nineteenth century, for a long time, they did not attract the attention of scientists. One reason is that most efforts were initially devoted to characterising the stability of either constant, or periodic dynamics, in which case the problem reduces to determining the eigenvalues of a suitable matrix (see Chapter 2).

A second reason why the application of the Lyapunov exponents was so much delayed with respect to the time of their definition was the difficulty of dealing with noncommuting entities. In fact, as shown in Chapter 2, the LEs are generated by multiplying (infinitely many) matrices. The first analytical result was obtained by Furstenberg and Kesten (1960), who basically proved the existence of the maximum and the minimum Lyapunov exponent. The full multiplicative ergodic theorem, ensuring the existence of as many Lyapunov exponents as the dimension of the space where the matrices operate, was proved later by Oseledets (1968) under fairly general conditions.

A further reason for the prolonged lack of specific studies of the Lyapunov exponents was the lack of workable instances of what was later defined as chaotic dynamics; in other words it was not clear which trajectories to consider for the underlying linearisation. It was only after the advent of the electronic computer that (approximate) trajectories could be generated and thereby characterised. The reception of the first physical model of a chaotic dynamical system, the Lorenz attractor (Lorenz, 1963), provides an enlightening

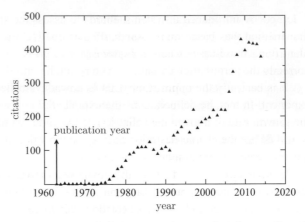

Fig. 1.2 Number of citations per year of the seminal paper (Lorenz, 1963), starting from the year of its publication. (Data from ISI Web of Science (www.wokinfo.com).)

view. In Fig. 1.2, one can see that the number of yearly citations of such a seminal paper remained pretty low until the late 1970s, when it started to explode. That was the time when computers became generally available to scientists, allowing them to work "experimentally" on chaotic dynamics. It is, in fact, in the same period that the relevant algorithms for the Lyapunov-exponent computation were developed (Shimada and Nagashima, 1979; Benettin et al., 1980a, b).

1.2 Outline of the book

Writing a book requires organizing a set of items in a sequential way, but in many cases, such as the present one, several interconnections are present, which make any ordering inconvenient. Therefore, we find it useful to summarise the book's content here, highlighting the connections among the various chapters.

Chapters 2 and 3 are devoted to the introduction of the main general properties of Lyapunov exponents and to the related numerical tools; the information contained therein will suffice for those interested in a basic Lyapunov analysis. The pseudocodes described in Appendix B provide further guidance for the implementation of the required algorithms. More precisely, the proper mathematical setups (continuous vs. discrete time) are introduced in Chapter 2, where they are followed by a discussion of the main properties, which include the effect of symmetries and the invariance of the LEs under changes of coordinates. In Chapter 3, various algorithms for the computation of LEs are introduced; they include the standard methods based on vector orthogonalisation at finite time intervals, as well as continuous methods. A specific section is devoted to models characterised by discontinuities in phase space where some care is required. The various sources of errors are also briefly discussed.

In nonlinear dynamical systems, there exists a relationship between LEs and various properties of their invariant measures, such as fractal dimensions and dynamical entropies. This connection is discussed in Chapter 6, where the Kaplan-Yorke and Pesin formulas are reviewed.

When computed over finite time, LEs naturally exhibit fluctuations that can be described via two different, but equivalent, methods. The first one, based on the computation of suitable moments, leads to the definition of the so-called generalised LEs. The second one is instead based on large-deviation theory. Both are described in Chapter 5, where we also illustrate a powerful numerical technique for the detection of rare fluctuations (see Section 5.4.2). Some generalised LEs are theoretically more manageable than the usual LEs; they can be used to derive approximate analytic expressions (some examples are presented in Chapter 8). In Chapter 6, we discuss the implication of LE fluctuations on the definition of the family of (generalised) fractal dimensions and entropies. The size of the fluctuations proves particularly useful to identify those borderline cases, where positive and negative finite-time LEs coexist. Finally, fluctuations arise also in spatially extended systems, where it is important to understand their scaling behaviour with the system size (this is discussed in Chapter 11).

If finite, instead of infinitesimal, perturbations are considered, the so-called finite-amplitude LEs can be defined. The usefulness and the limits of this generalisation are discussed in Chapter 7. Of particular relevance is the case of collective chaos, where such exponents allow identification of the presence of macroscopic instabilities.

LEs quantify the growth rates of generic (infinitesimal) perturbations. Further information is contained in the direction of the perturbations, i.e., in the orientation of the corresponding Lyapunov vectors. This issue is extensively discussed in Chapter 4, where different definitions and algorithms for their reconstruction are presented. The structure of Lyapunov vectors is discussed also in Chapter 11, where an analogy with rough interfaces (in spatially extended systems) allows for a fairly complete characterisation. An important property that is worth testing in high-dimensional contexts is the extended vs. localised nature of the Lyapunov vectors. The presence of localised vectors (especially the first one) helps to develop approximate expressions for the LE; some examples are discussed in Chapters 8 and 10. Furthermore, the presence of extended vectors is conjectured to be a signature of the presence of collective chaos (see Chapter 11).

Exact analytical expressions for the LEs are not typically available in generic nonlinear/disordered systems. Some results are obtained with the help of suitable perturbative techniques when the fluctuations are quite small. Other results can be derived under the assumption of relatively simple statistical properties (e.g. lack of temporal and/or spatial correlations). The corresponding theories are reviewed in Chapter 8. This chapter is particularly important for those interested in applications of LEs to intrinsically random situations, but it also provides a basis to obtain approximate results for chaotic dynamical systems.

Chapters 9, 10 and 11 deal with LEs in complex systems. In Chapter 9, we start from the relatively simple setup of two coupled systems, used as a testbed for understanding the effect of coupling on nearly identical units (e.g. coupling sensitivity and synchronisation). Chapter 10 is devoted to a general discussion of high-dimensional systems: various classes of setups are considered, but the emphasis is mostly put on systems characterised by short-range interactions in a one-dimensional physical space. One of the key properties is the extensivity of the chaotic dynamics (i.e. the proportionality between the number of active degrees of freedom and the dimension of the phase space), which manifests itself as a well-defined Lyapunov density spectrum (in the so-called thermodynamic limit). Finally, Lyapunov vectors in spatially extended systems are analysed in Chapter 11, mostly through an insightful analogy with rough interfaces. Furthermore, models characterised by a chaotic collective behaviour are used to elucidate differences (and possible connections) between the stability analysis at a microscopic and a macroscopic level.

In Chapter 12 we review several physical problems, where the application of Lyapunov exponents plays an essential role. The four appendices contain some technical details. In Appendix A, almost all models that are used as a reference across the book are properly defined. Appendix B contains the most basic pseudocodes to be used for the computation of Lyapunov exponents and vectors. In Appendix C, we derive some general formulas used in Chapter 8 for the computation of LEs in products of random matrices. Finally, rudiments of symbolic encoding are presented in Appendix D: they prove useful in implementing some techniques introduced in Chapter 5.

1.3 Notations

While writing a monograph, the use of homogeneous notations is a must to avoid unpleasant misunderstandings on the meaning of some symbols. We have made our best efforts to avoid overlaps, but, given the large number of quantities we had to introduce, this has not been always possible. However, we have tried to clearly identify and differentiate the general observables, such as time, space and phase-space variables. Particular care has been taken in differentiating the many different types of Lyapunov exponents that are discussed in the book. The main notations are summarised in the following tables.

	Acronyms
Abbreviation	Description
BV	Bred vectors
CLV	Covariant Lyapunov vector
FAE	Finite amplitude exponent
FLI	Fast Lyapunov indicator
FPU	Fermi-Pasta-Ulam
GALI	Generalised alignment index
GS	Gram-Schmidt orthogonalisation
KPZ	Kardar-Parisi-Zhang
LE	Lyapunov exponent
SVD	Singular value decomposition

General notation

Symbols used	Description	Page
$U, V, \mathbf{U}, \mathbf{V}$	Capital letters are used to refer to variables in phase space; bold letters denote vectors of variables.	10, 11
$u, v, \mathbf{u}, \mathbf{v}$	Lower-case letters refer to variables in tangent space obeying linearised equations; bold letters denote vectors in tangent space.	10
$F, \mathbf{F}$	Capital F usually denotes the velocity field for a differential equation or the map function; bold denotes a system of equations or a high-dimensional map.	10, 11
t	Time (either discrete or continuous)	10, 11
$\mathsf{J}, \mathsf{H}, \mathsf{M}$	Sans serif letters denote matrices	
$\mathsf{J} = \frac{\partial \mathbf{F}}{\partial \mathbf{U}}$	Instantaneous Jacobian of a map	10
K	Instantaneous Jacobian for continuous time differential equations	11
H	The product of many Js for a sequence of Jacobian matrices or the integration of K over a finite time for continuous time case	10
x	Spatial variable (discrete or continuous) for spatio-temporal dynamics, written as an index, e.g., U_x, when discrete	166, 172, 173, 247
L	System size for spatially extended systems	168

Different types of Lyapunov exponents

λ_k	kth Lyapunov exponent (typically ordered from the largest to the smallest one)	12
S_n	Sum of the first n Lyapunov exponents (or the sum of all Lyapunov exponents)	21, 22
$\Lambda(t_0, \tau)$	Finite time Lyapunov exponent determining the growth rate in the time interval $t_0 < t < t_0 + \iota$	39, 71
$\Gamma = \Lambda \tau$	Overall expansion factor over time interval τ	39, 71
$\mu = \exp[\lambda]$	Multipliers (i.e. their logarithms are the Lyapunov exponents)	12
$\mathcal{L}(q)$	Generalised Lyapunov exponents	74
$G(q) = \mathcal{L}(q)q$	Characteristic function of perturbation growth	74
ℓ	Finite amplitude exponent (averaged)	111
$\mathbb{L}$	Instantaneous finite amplitude exponent (related to ℓ, like Λ to λ)	111
$\mathscr{L}$	Convective (velocity-dependent) exponent	179

In this chapter we introduce the mathematical background that is necessary for defining the Lyapunov exponents (LEs). We then introduce the LEs by referring to the Oseledets multiplicative ergodic theorem. As everywhere else in the book, rather than presenting fully rigorous derivations, we prefer to propose physical and heuristic arguments with the help of examples of increasing complexity.

General properties of the LEs are also illustrated to help explain the role of LEs in general physical contexts.

2.1 The mathematical setup

The notion of Lyapunov exponents emerges while assessing the stability of generic trajectories of a dynamical system. Two classes are typically studied in the scientific literature, namely discrete- and continuous-time models.

A discrete-time dynamical system, usually referred to as a map, is represented by the recursive relation

$$\mathbf{U}(t+1) = \mathbf{F}(\mathbf{U}(t)). \tag{2.1}$$

Here, $\mathbf{U}$ is an N-dimensional state variable, t is an integer variable denoting time and $\mathbf{F}(\mathbf{U})$ is a (possibly non-invertible) function from $\mathbb{R}^N$ to $\mathbb{R}^N$. An initial condition $\mathbf{U}(0)$ uniquely defines the trajectory $\mathbf{U}(t)$, which may be generated by iterating the relation (2.1), and is assumed to be well defined for all $t > 0$. Its *stability* can be assessed by selecting a second close trajectory and thereby checking whether it remains close to the original one. It is customary to consider infinitesimal perturbations. This requires linearising the map (2.1) (this step assumes a sufficient smoothness of the map). The perturbation $\mathbf{u}(t)$ of the trajectory $\mathbf{U}(t)$ follows the linear transformation

$$\mathbf{u}(t+1) = \frac{\partial \mathbf{F}}{\partial \mathbf{U}}(t)\mathbf{u}(t) =: \mathsf{J}(t)\mathbf{u}(t), \tag{2.2}$$

where J is the so-called Jacobian matrix. The current value $\mathbf{u}(t)$ of the perturbation is generated by iterating Eq. (2.2),

$$\mathbf{u}(t) = \mathsf{J}(t-1)\mathbf{u}(t-1) = \mathsf{J}(t-1)\mathsf{J}(t-2)\mathbf{u}(t-2) = \dots$$
$$= \prod_{k=0}^{t-1} \mathsf{J}(k)\mathbf{u}(0) = \mathsf{H}(t)\mathbf{u}(0), \tag{2.3}$$

where the matrix $H(t) = \prod_{k=0}^{t-1} J(k)$ defines the evolution over t steps. The perturbation depends, of course, on the trajectory $U(t)$; in practice this dependence can be reduced to that on the initial state $U(0)$, as it uniquely identifies the forward trajectory.

In order to determine whether the perturbation either grows or decays, it is necessary to study the effect of a product of matrices on some initial vector, or, equivalently, the properties of $H(t)$ for large t.

In the case of continuous-time systems, the typical starting point is an ordinary differential equation (although we will also consider partial and delay differential equations). We write it as

$$\dot{U} = F(U, t), \tag{2.4}$$

where U is an N-dimensional state variable and the vector field F possibly depends on time (this is the case of non-autonomous dynamical systems).

The generator for the evolution of infinitesimal perturbation is again obtained by linearising the equations of motion,

$$\dot{u} = \frac{\partial F}{\partial U}(t)u(t) =: K(U, t)u. \tag{2.5}$$

By integrating this equation over a time t, one obtains

$$u(t) = H(U_0, t)u(0), \tag{2.6}$$

where $H(U_0, t) = \exp\left(\int_0^t dt' K(U(t'), t')\right)$ depends on the trajectory $U(t')$ at all intermediate times. In practice, the matrix $H(U_0, t)$ is obtained by solving the linear system of ordinary differential equations (2.5).

The relation (2.6) is fully equivalent to Eq. (2.3).

2.2 One-dimensional maps

It is convenient to start from the study of one-dimensional maps, i.e. $N = 1$. In this case, $F(U)$ is a scalar function; the Jacobian matrix J reduces to its derivative $F'(U)$ and the scalar perturbation u obeys

$$u(t) = \prod_{k=0}^{t-1} F'(U(k))u(0).$$

The perturbation can either grow or decay in absolute value and change sign. The latter effect is irrelevant for assessing the stability of a trajectory. As for the absolute value,

$$|u(t)| = |u(0)| \prod_{k=0}^{t-1} |F'(U(k))|. \tag{2.7}$$

Instead of considering products of numbers, it is more convenient to deal with sums. This can be accomplished by computing the logarithm of expression (2.7),

$$\ln |u(t)| = \sum_{k=0}^{t-1} \ln |F'(U(k))| + \ln |u(0)|. \qquad (2.8)$$

The properties of the sum depend on the correlations along the original trajectory $U(t)$. If we assume that $U(t)$ is a statistically stationary ergodic process, then the time average of the observable $\ln |F'(U(k))|$ can be typically represented – thanks to the Birkhoff ergodic theorem (see Walters (1982)) – as an average over the corresponding probability measure

$$\frac{1}{t} \sum_{k=0}^{t-1} \ln |F'(U(k))| \xrightarrow[t\to\infty]{} \langle \ln |F'(U)| \rangle.$$

As a result, the perturbation, typically, grows exponentially for large t,

$$|u(t)| \approx |u(0)| \exp[\langle \ln |F'(U)| \rangle t] = |u(0)| \exp[\lambda t], \qquad (2.9)$$

where the quantity $\lambda = \langle \ln |F'(U)| \rangle$ is called the Lyapunov exponent of the one-dimensional map.

2.3 Oseledets theorem

In the one-dimensional case, the sign of a perturbation can change irregularly during the evolution, while its absolute value generally follows an exponential law (2.9). Similarly, one does not expect regularities in the direction of the perturbation vector $\mathbf{u}(t)$ in Eq. (2.3). The amplitude of the perturbation is, in fact, determined by the norm of the vector

$$\|\mathbf{u}(t)\|^2 = \|\mathsf{H}_t \mathbf{u}(0)\|^2 = \mathbf{u}^\mathsf{T}(0) \mathsf{H}^\mathsf{T}(t) \mathsf{H}(t) \mathbf{u}(0), \qquad (2.10)$$

where H^T is the transpose of H and one can equivalently assume the time to be either discrete or continuous. Thus, only the properties of the real symmetric matrix

$$\mathsf{M}(t) = \mathsf{H}^\mathsf{T}(t) \mathsf{H}(t)$$

are relevant for the time evolution of the norm of the vector $\mathbf{u}$.

These properties have been established in a seminal theorem by Oseledets (1968, 2008), which states that for a statistically stationary, ergodic sequence of matrices (which happens if the underlying process $\mathbf{U}(t)$ is ergodic), the limit

$$\lim_{t\to\infty} [\mathsf{M}(t)]^{\frac{1}{2t}} = \mathsf{P} \qquad (2.11)$$

exists and is an N-dimensional matrix with positive eigenvalues $\mu_1 \geq \mu_2 \geq \cdots \geq \mu_N$. The N Lyapunov exponents are defined as

$$\lambda_k = \log \mu_k.$$

The set of all Lyapunov exponents is called the Lyapunov spectrum.

This definition of Lyapunov exponents can be equivalently reformulated in terms of the linear evolution of perturbations as follows: any initial perturbation $\mathbf{u}(0)$ evolves exponentially in time, the growth rate being one of the Lyapunov exponents,

$$\lim_{t\to\infty} \frac{1}{t} \ln \frac{\|\mathsf{H}(t)\mathbf{u}(0)\|}{\|\mathbf{u}(0)\|} = \lim_{t\to\infty} \frac{1}{t} \ln \frac{\|\mathbf{u}(t)\|}{\|\mathbf{u}(0)\|} = \lambda_k. \tag{2.12}$$

Which Lyapunov exponent characterises the evolution of a particular perturbation $\mathbf{u}(0)$ is discussed in Section 2.3.2.

2.3.1 Remarks

The Oseledets theorem is often referred to as the 'multiplicative ergodic theorem'. In the one-dimensional case, one deals with a product of random numbers, which, after introducing the logarithm of the local multipliers, can be reduced to a sum of suitable observables. Accordingly, the usual Birkhoff ergodic theorem applies (Eq. (2.8)), which states that the time average exists and is equal to the average over the probability distribution. Only the former of these statements can be extended to a product of matrices, namely the existence of long-time averages in the sense of Eq. (2.11). The representation as a straightforward average over the probability distribution is not possible, since information on the orientation of the perturbation is required.[1] Thus the name 'multiplicative ergodic theorem'.

Lyapunov exponents have been introduced here on an intuitive level, without paying attention to mathematical rigor (those interested in more details can look at Eckmann and Ruelle, 1985; Barreira and Pesin, 2007; Oseledets, 2008; Viana, 2014; and references therein). It is, nevertheless, useful to stress some relevant points. First of all, the matrices describing the evolution of the perturbations are assumed to be non-singular, and their norm admits an exponential bound $\|\mathsf{H}(t)\| \leq \exp[ct]$. Next, it is necessary to assume that the underlying dynamical system has nice statistical properties, i.e. the trajectory belongs to an invariant set characterised by an invariant probability measure and is ergodic with respect to it. Under such conditions, the Lyapunov exponents almost surely do not depend on the particular trajectory around which the perturbation is considered (i.e. there may exist a zero-measure set of orbits characterised by different exponents – see Chapter 5 for a discussion of the consequences of this point).

Additionally, the so-called Lyapunov regularity is assumed, which implies the existence of the long-time limits in Eqs. (2.11, 2.12). Whenever this is not the case, it is necessary to replace the "lim" operation with the more general lim sup operation ($\overline{\lim}$).

The Oseledets theorem has been formulated by referring to Jacobian matrices determined along a generic trajectory. In a more general formulation, one can consider generic matrices suitably defined in each state of the dynamical system. This mathematical construction is called a cocycle.

Moreover, the LEs have been introduced for two typical classes of dynamical systems: ordinary differential equations and discrete maps. The theory is, however, applicable to

---

[1] More and less sophisticated methods for the determination of Lyapunov exponents as ensemble averages may be, nevertheless, implemented – see, for instance, Chapter 8.

any dynamical system provided the solution of an initial value problem is unique and defined at all times, and ergodicity holds. Many infinite-dimensional systems belong to this class. This includes partial differential, differential-delay and integral equations. In such systems the number of Lyapunov exponents is infinite, but that of positive LEs is typically finite (at least when the spatial extension or the delay is finite). The dependence of the Lyapunov spectrum on the system size in spatially extended dynamical systems is discussed in Chapter 10.

Although the generation of the perturbation dynamics requires linearising the original equations of motion Eqs. (2.1, 2.4), the system itself need not be uniformly smooth. Indeed, it is sufficient to be able to define the linearised dynamics for typical trajectories. For example, a map (2.1) can be piece-wise differentable and even have discontinuities; this does not prevent the existence of well-defined Lyapunov exponents. Those trajectories which hit the singular points are indeed non generic.

In continuous-time dynamical systems, it may happen that trajectories are generically subject to discontinuous jumps at certain moments of time, such as the collisions exhibited by a particle moving inside a billiard, and yet the Lyapunov exponent is well defined. In fact, what is important for the definition of the Lyapunov exponents is the behaviour of the separation between nearby trajectories rather than that of the trajectories themselves. In practice, one can follow a small perturbation between consecutive collisions and add the effect of the collisions themselves (see the next chapter for a technical description). Trajectories may also exist along which linearisation is not possible (e.g. those trajectories that exactly hit a corner of a billiard table). As long as they are atypical, they do not contribute to the Lyapunov exponents.

It is also worth discussing the relationship between matrix eigenvalues and Lyapunov exponents. The eigenvalues of M are the singular values of H, and they generally differ from the (square root of the) eigenvalues of the latter matrix. Only when the matrix H is normal, its singular values correspond to the absolute values of the eigenvalues. The difference between eigenvalues and singular values is especially pronounced in the case of rotations. We illustrate this with the following example, adapted from I. Morris (unpublished). Let us assume that the linearised dynamics is defined by successive applications of two matrices

$$A = \begin{pmatrix} 0 & 1 \\ 1 & 0 \end{pmatrix}, \qquad B = \begin{pmatrix} 0 & 1/4 \\ 4 & 0 \end{pmatrix}.$$

It is easy to check that

$$H_{2l} = (AB)^l = \begin{pmatrix} 2^{2l} & 0 \\ 0 & 2^{-2l} \end{pmatrix}, \qquad M_{2l} = \begin{pmatrix} 2^{4l} & 0 \\ 0 & 2^{-4l} \end{pmatrix},$$

$$H_{2l+1} = B(AB)^l = \begin{pmatrix} 0 & 2^{-2l-1} \\ 2^{2l+1} & 0 \end{pmatrix}, \qquad M_{2l+1} = \begin{pmatrix} 2^{-4l-2} & 0 \\ 0 & 2^{4l+2} \end{pmatrix}.$$

On the one hand, the eigenvalues ν of the H matrices alternate,

$$\frac{1}{k} \ln \nu_k = \begin{cases} \pm \ln 2 & \text{for } k = 2l \\ 0 & \text{for } k = 2l+1 \end{cases},$$

with no convergence towards some limit value. On the other hand, the singular values μ (the eigenvalues of matrices M) behave nicely, yielding

$$\frac{1}{2k} \ln \mu_k = \pm \ln 2 \text{ for all } k.$$

This confirms the need of referring to the singular values for a proper definition of the LEs.

2.3.2 Oseledets splitting

Which of the Lyapunov exponents is selected in Eq. (2.12) depends on the choice of the initial perturbation vector $\mathbf{u}$. Suppose, for simplicity, that all Lyapunov exponents are different, i.e. that there are N different exponents $\lambda_1 > \lambda_2 > \cdots > \lambda_N$. Then, in the N-dimensional space of vectors $\mathbf{u}$ there exists a set of nested subspaces $\mathbb{D}^k, k = 1, \ldots, N$, of dimensions $N - k + 1$, such that $\mathbb{D}^{k+1} \subset \mathbb{D}^k$. Here $\mathbb{D}^1 = \mathbb{R}^N$ is the full space, while $\mathbb{D}^N$ has dimension one. Then, the Lyapunov exponent λ_k in Eq. (2.12) will hold for all initial vectors $\mathbf{u} \in \mathbb{D}^k \setminus \mathbb{D}^{k-1}$. This means that all vectors, except for those in the $(N-1)$-dimensional space $\mathbb{D}^2$, grow asymptotically with the largest Lyapunov exponent λ_1. All the vectors in the $(N-1)$-dimensional space $\mathbb{D}^2$, except for those in the $(N-2)$-dimensional subspace $\mathbb{D}^3$, grow with the second Lyapunov exponent λ_2, etc. Only vectors belonging to the one-dimensional space $\mathbb{D}^N$ grow with the smallest exponent λ_N. In case some of the Lyapunov exponents are equal, not all of the nested subspaces are well defined: if, e.g., $\lambda_2 = \lambda_3$, one must "jump" from the subspace $\mathbb{D}^2$ to $\mathbb{D}^4$. All the vectors in $\mathbb{D}^2 \setminus \mathbb{D}^4$ grow with an exponent $\lambda_2 = \lambda_3$.

Lyapunov exponents have been so far defined by referring to the forward iteration of the dynamical equations. This allows covering non-invertible dynamical systems such as maps of the interval that are often used as paradigmatic examples of a chaotic dynamics. In invertible systems, a perturbation can be followed both forwards and backwards in time, allowing for a sharper characterisation of the tangent space. We illustrate this idea by referring, for simplicity, to a discrete-time model with no degeneracy in the Lyapunov spectrum.

Consider a typical trajectory $\mathbf{U}(t)$ and assume that the evolution of an infinitesimal perturbation $\mathbf{u}(t)$ is described by Eq. (2.2), $\mathbf{u}(0)$ denoting the perturbation amplitude at time $t = 0$. Again we assume for simplicity that all the Lyapunov exponents are different. The tangent space covered by all perturbations $\mathbb{R}^N$ can be split into one-dimensional subspaces $\mathbb{R}^N = \mathbf{E}^1 \oplus \mathbf{E}^2 \oplus \cdots \oplus \mathbf{E}^N$ such that a generic initial vector in $\mathbf{E}^k$ grows according to the Lyapunov exponent λ_k, both forwards and backwards in time,

$$\lambda_k = \lim_{t \to \pm\infty} \frac{1}{t} \log \frac{\|\mathbf{e}(t)\|}{\|\mathbf{e}(0)\|} \qquad \text{if } \mathbf{e}(0) \in \mathbf{E}^k.$$

The *Oseledets splitting* generally depends on the point $\mathbf{U}(0)$ where it is determined, in the sense that the vectors $\mathbf{E}^k$ have different directions in different points of the phase space. The splitting is, however, covariant under the application of the transformation $\mathbf{F}$,

$$\mathbf{E}^k(\mathbf{U}(t+1)) = \mathbf{E}^k(\mathbf{F}\mathbf{U}(t)) = \mathbf{J}(\mathbf{U}(t))\mathbf{E}^k(\mathbf{U}(t)).$$

In the physical literature, the directions $\mathbf{E}^k$ are often called Lyapunov vectors; see Chapter 4 for a more detailed discussion. In non-invertible systems, there may be more than one trajectory coming from the past, which reaches a given point $\mathbf{U}(t)$; the Oseledets splitting is, strictly speaking, meaningless.

2.3.3 "Typical perturbations" and time inversion

Remarkably, the LEs provide a complete description of the behaviour of all possible perturbations. No perturbation can be generated whose exponential growth rate differs from all of the Lyapunov exponents of the system. This property allows one to determine explicitly some Lyapunov exponents, following the time evolution of particular perturbations and applying Eq. (2.12), as done in Section 2.5.6.

What happens to an "arbitrarily" selected perturbation? Assuming, for simplicity, that there are no degeneracies, the foliation $\mathbb{D}^k, k = 1, \ldots, N$ of the tangent space described in Section 2.3.2 implies that $\mathbb{D}^1$, $\mathbb{D}^2$, $\ldots$ have dimensions $N, N-1, \ldots$. Perturbations growing[2] with the largest exponent are "typical": they fill the whole volume in the tangent space except for the $(N-1)$-dimensional subspace $\mathbb{D}^2$. If one, instead, specifically selects a perturbation within $\mathbb{D}^2$, then such a perturbation will "typically" grow with the exponent λ_2, unless it belongs to the "exceptional" subset $\mathbb{D}^3$, etc. This property is used in the numerical evaluation of the Lyapunov exponents (Chapter 3): a randomly chosen perturbation will almost surely grow with the largest Lyapunov exponent; two randomly chosen perturbations will almost surely identify a two-dimensional subspace with a component in $\mathbb{D}^2$, i.e. growing with the second largest exponent, etc.

Based on this discussion, it is clear that finding the smallest (most negative) Lyapunov exponents is not an easy task, as they are highly "non-typical". They become, however, "typical" if the dynamics is followed backwards in time. In both the discrete and the continous time setups, inverting time means looking at the inverse ratios in expression (2.12), so that the Lyapunov exponents in the reversed time direction $\tilde{\lambda}$ are the opposite of the forwards ones, $\tilde{\lambda} \Rightarrow -\lambda$. If the exponents are ordered from the largest to the smallest, one has instead $\tilde{\lambda}_k = -\lambda_{N-k+1}$. Correspondingly, the backwards foliation $\tilde{\mathbb{D}}^k$ for $k = 1, \ldots, N$, differs from the forwards one. $\tilde{\mathbb{D}}^1$ is the full space, and "typical" initial perturbations (except for those in $\tilde{\mathbb{D}}^2$) grow with the Lyapunov exponent $\tilde{\lambda}_1 = -\lambda_N$, etc.

Noteworthy, in order to follow the perturbations backwards in time, it is not necessary to invert the original dynamical system (2.1) or (2.4). In fact, this may not even be possible if the map in (2.1) is non-invertible. It is enough to collect the coefficients of the linear equations (2.2) or (2.5) along a given trajectory and solve these linear equations (by inverting the Jacobian matrix in (2.2) or solving the ordinary differential equation (2.5) backwards in time). In this way one automatically selects the same trajectory of the nonlinear dynamical system.

Of course, for invertible systems, these "typical" conditions can be expressed in terms of the Oseledets splitting. A typical initial vector has components in all directions $\mathbf{E}^k$; thus

---

[2] For the sake of simplicity, the term "growing" is used even if the exponent is possibly negative.

it has components which grow with all Lyapunov exponents. From these components, the one corresponding to the largest Lyapunov exponent dominates at large positive times; thus a "typical" initial perturbation alignes along $\mathbf{E}^1$ and grows with the largest exponent. Backwards in time, the evolution of a "typical" perturbation is instead dominated by the largest exponent in the reversed time, λ_N.

2.4 Simple examples

2.4.1 Stability of fixed points and periodic orbits

The first and simplest context where linear stability is encountered is that of constant or periodic trajectories.

In the case of a fixed point $\mathbf{U}_0$ of a discrete-time dynamical system, the stability analysis reduces to determining the eigenvalues of a constant Jacobian matrix J defined at this point. Perturbations $\mathbf{u}$ indeed grow according to the (possibly complex) eigenvalues ν_k of this matrix: $\mathbf{u}(t) \propto (\nu_k)^t \mathbf{u}(0)$. Substituting this into Eq. (2.12) we obtain

$$\lambda_k = \log|\nu_k|.$$

Similarly, for a periodic orbit of period T one has to calculate the eigenvalues ν_k of the product of Jacobian matrices $\mathsf{J}_{T-1}\mathsf{J}_{T-2}\dots\mathsf{J}_0$ to obtain

$$\lambda_k = \frac{1}{T}\log|\nu_k|. \tag{2.13}$$

In the continuous-time case, the perturbations around a fixed point grow according to Eq. (2.5), as $\mathbf{u} \propto \exp[\nu_k t]$ where ν_k are (possibly complex) eigenvalues of the Jacobian matrix K, so that from Eq. (2.12) it follows that the Lyapunov exponents are just the real parts of the eigenvalues

$$\lambda_k = \mathrm{Re}(\nu_k).$$

For a periodic orbit $\mathbf{U}(t+T) = \mathbf{U}(t)$ in the continuous-time case, the linear equation (2.5) has periodic coefficients. Its general solution, according to the Floquet theory (see, e.g., Hale (1969)), can be written as $\mathbf{u}(t) = \exp[\nu_k t]\hat{\mathbf{u}}(t)$, where $\hat{\mathbf{u}}(t) = \hat{\mathbf{u}}(t+T)$ has period T, and ν_k are the Floquet exponents. The Lyapunov exponents are the real parts of the Floquet exponents (the expression (2.13) still holds).

Summarising, in the case of a regular dynamics, the characterisation of stability by means of the Lyapunov exponents corresponds to the standard eigenvalue analysis. The only difference is that the information on the rotation frequencies (encoded in the imaginary parts of the eigenvalues) is lost. One might wonder whether it is possible to extend the definition of Lyapunov exponents to suitably include the imaginary component. In the context of space-time chaos, this is a key point of the chronotopic approach discussed in Chapter 10. Otherwise, the only "frequency"-like observable that can be introduced is the so-called rotation number: a single number that can be at most generically defined in three-dimensional flows (Ruelle, 1985).

2.4.2 Stability of independent and driven systems

It is clear that for a large system composed of two (or many) independent subsystems, the Lyapunov spectrum is also comprised of the exponents of the subsystems.

A little more involved is the consideration of skew, or driven systems. In many physical applications, a *master* subsystem **U** drives a *slave* subsystem **V** but is not influenced by the latter. Next, we consider a discrete-time setup, but the following arguments (which rely on the orthogonalisation procedure explained in detail in the next chapter) can be easily repeated for analogous continuous-time models. In the tangent space, the dynamics of a master-slave system is described by the mapping

$$\mathbf{u}(t+1) = \mathsf{A}(t)\mathbf{u}(t), \qquad \mathbf{v}(t+1) = \mathsf{B}(t)\mathbf{u}(t) + \mathsf{C}(t)\mathbf{v}(t),$$

where the vector **u** is N-dimensional, **v** is M-dimensional (we will call these subspaces master and slave subspaces, respectively), and the three matrices fluctuate in time. Suppose that the transformation is applied to an orthonormal basis at time t, which is composed of N $\mathbf{u}_k$ vectors from the master subspace and M $\mathbf{v}_j$ vectors from the slave subspace. The application of the mapping leads to the following set of new vectors:

$$\mathbf{u}_k'(t+1) = \begin{pmatrix} \mathsf{A}(t)\mathbf{u}_k(t) \\ \mathsf{B}(t)\mathbf{u}_k(t) \end{pmatrix}, \qquad \mathbf{v}_j'(t+1) = \begin{pmatrix} 0 \\ \mathsf{C}(t)\mathbf{v}_j(t) \end{pmatrix}.$$

As shown in the following chapter, the Lyapunov exponents are obtained by orthogonalising this set of new vectors. Let us proceed by starting from the $\mathbf{v}_j'$ vectors. They all lie in the M-dimensional slave subspace and so does the resulting orthonormal basis $\mathbf{v}_j(t+1)$. One then proceeds by orthogonalising the space spanned by the vectors $\mathbf{u}_k'$. These vectors lie in both subspaces, the master and the slave one, but in the latter subspace the orthonormal basis $\mathbf{v}_j(t+1)$ already exists. This implies that all components in the slave subspace of the $\mathbf{u}_k'$ vectors (i.e. the components $\mathsf{B}\mathbf{u}_k(t)$) are to be discarded. As a result, one is left with the problem of orthogonalising the vectors $\mathsf{A}\mathbf{u}_k(t)$, i.e. the vectors describing solely the master system. Altogether, the Lyapunov exponents are obtained by separately processing the master and the slave systems: the coupling contributes only indirectly to the identification of the trajectories around which linearisation has to be performed. The matrix $\mathsf{B}(t)$ contributes to defining the direction of the Lyapunov vectors (see Chapter 4) only. Contrary to the case of independent subsystems, for skew systems the Lyapunov vectors span over both master and slave subspaces. These arguments can be generalised to chains with a unidirectional coupling; Lyapunov exponents can be determined in each element of such a chain separately.

2.5 General properties

2.5.1 Deterministic vs. stochastic systems

Originally, the Lyapunov exponents and the Oseledets theorem have been introduced while studying the linear stability properties of trajectories of deterministic dynamical systems. They have, however, an essentially statistical nature, and they can be equally applied to the description of the linear stability of either nonlinear stochastic systems or the asymptotic properties of purely stochastic linear models, like products of random matrices and stochastic linear differential equations (see also Chapter 8).

In the context of noisy dynamical systems, e.g. noise-driven maps

$$\mathbf{U}(t+1) = \mathbf{F}(\mathbf{U}(t), \boldsymbol{\xi}(t)), \tag{2.14}$$

where $\boldsymbol{\xi}(t)$ are random processes, the Lyapunov exponents measure the stability of trajectories with respect to variations of the initial condition under the action of the *same* realisation of the noise. Indeed, by linearising (2.14) one obtains

$$\mathbf{u}(t+1) = \mathsf{J}(\mathbf{U}, \xi(t))\mathbf{u},$$

so that the product of Jacobian matrices, and correspondingly the Lyapunov exponents, are defined for a particular sequence $\boldsymbol{\xi}(t)$ of random numbers.

In practice, the dependence on the realisation of the stochastic process is akin to the dependence of the Lyapunov exponents on the selection of the initial condition in purely deterministic setups.

2.5.2 Relationship with instabilities and chaos

As is clear from its definition, the largest Lyapunov exponent determines the long-term linear stability of a given trajectory. A positive largest exponent implies an exponential instability, while a negative largest exponent means stability. In the case of a vanishing largest Lyapunov exponent, the linear analysis does not give the final answer, and a more refined study is needed.[3]

A strictly positive maximal Lyapunov exponent is often considered as a definition of *deterministic chaos*. Indeed, chaos is generally defined as a stationary (in a statistical sense) dynamical process accompanied by a sensitive dependence on the initial conditions. This sensitivity means that if any given state is approximately reproduced in the course of time – which must happen due to the assumed statistical stationarity of the whole process – further evolution is only approximate and volatile, eventually deviating from the previous patterns, as a result of the sensitivity of these patterns to tiny deviations. The largest Lyapunov exponent provides a quantitative measure of this sensitivity. Note that this criterion should be applied to the typical trajectories, not the exceptional ones (e.g. unstable periodic orbits that are embedded in the invariant set under consideration). Following a suggestion of Rössler (1979), a regime with more than one positive exponent is often called *hyperchaos*.

The Lyapunov exponent is dimensionally equivalent to a frequency, or inverse time. The inverse of the largest Lyapunov exponent (sometimes called Lyapunov time) identifies the "predictability time". If the state of a chaotic system is perturbed by an amount δ, this

---

[3] For non-regular systems, where the Lyapunov exponents are defined in sense of limsup, the situation is more subtle: even the largest Lyapunov exponent does not determine the stability – see Leonov and Kuznetsov (2007) for further discussion and additional references.

perturbation grows exponentially in time $\propto \exp[\lambda_1 t]$, reaching a size Δ, after a time

$$T_{pr} \approx \frac{1}{\lambda_1} \log\left(\frac{\Delta}{\delta}\right). \tag{2.15}$$

The relation is only approximate because the Lyapunov exponent is defined as an asymptotic growth rate that is not necessarily valid at small times. One can, nevertheless, see that the predictability time T_{pr} depends weakly on the size of initial perturbation δ and it is essentially determined by the value of the largest Lyapunov exponent. Thus, practically only short-time predictions of individual states in chaotic systems are possible. This difficulty is quite often formulated as the *butterfly effect*: the flapping of wings of a butterfly in Brazil may cause a tornado in Texas. This statement first expressed by E. Lorenz means that a very minor perturbation somewhere in the phase space (here, the physical world) may amplify so much as to become extremely sizable somewhere else. It is important to stress that the inverse of the largest Lyapunov exponent is just of one of the time scales present in a generic dynamical system. The decorrelation of chaotic oscillators, for instance, depends on the diffusion of the phases, which is not related to any Lyapunov exponent (Farmer, 1981; Pikovsky et al., 1997c).

If the largest Lyapunov exponent is zero or negative, then the dynamics is non-chaotic. However, a degree of irregularity can be observed in such systems as well. Some examples are the so-called strange nonchaotic attractors (Feudel et al., 2006), some polygonal billiards (Gutkin, 1986; Artuso et al., 2000) and *stable chaos*, a phenomenon arising in high-dimensional systems, which manifests itself as exponentially long irregular transients (Politi and Torcini, 2010).

In most cases the existence of zero Lyapunov exponents can be attributed to the presence of continuous symmetries (see Section 2.5.6). The most typical symmetry is the invariance with respect to time shifts in autonomous continuous-time dynamics. In such systems, the presence of negative LEs accompanied by a single zero exponent is the signature of a stable limit cycle (periodic motion), while two or three zero exponents indicate a quasiperiodic motion with two or three incommensurate frequencies. In integrable Hamiltonian systems, all the Lyapunov exponents are typically zero. In a periodically forced continuous-time system, a periodic solution is stable if the largest Lyapunov exponent is negative, while a stable quasiperiodic regime with two incommensurate frequencies has a zero largest exponent.

In practical applications one should remember that the LE definition involves two limits: (i) linearisation of the equations (i.e. the limit of infinitesimal perturbations is taken) and (ii) infinite-time limit (taken to determine the asymptotic growth rate). Of course, these limits cannot be interchanged, and in particular applications where one needs to follow a finite perturbation over a finite time interval, special care is needed.

2.5.3 Invariance

The LEs are independent of both the metric used to determine the distance between perturbations and the choice of variables. This property implies that they are *dynamical invariants* and thereby provide an objective characterisation of the corresponding dynamics.

The independence of the metric follows from the equivalence of all norms in finite-dimensional spaces: given any two norms α and β, one can show that $A\|\mathbf{u}\|_\alpha \leq \|\mathbf{u}\|_\beta \leq B\|\mathbf{u}\|_\alpha$ for some value of the constants A and B and any vector $\mathbf{u}$. Thus, according to the definitions (2.12), the Lyapunov exponents do not depend on the norm. This is no longer true in infinite-dimensional space: this "ambiguity" lies at the heart of different classifications of supposedly chaotic regimes in infinite-dimensional systems (see the discussion on stable chaos in Chapter 10).

Similarly, one can prove the invariance of the Lyapunov exponents under a smooth change of coordinates. Suppose a dynamical system is defined with reference to the variable $\mathbf{U}$ and that the new variable $\mathbf{V} = \mathbf{G}(\mathbf{U})$ is introduced. Then the trajectory $\mathbf{U}(t)$ can be expressed, in the new variable, as $\mathbf{V}(t)$, while a small perturbation accordingly becomes $\mathbf{v}(t) = \frac{\partial \mathbf{G}}{\partial \mathbf{U}}(t)\mathbf{u}(t)$. If we assume that the transformation of variables is everywhere non-singular, i.e. $c \leq \|\frac{\partial \mathbf{G}}{\partial \mathbf{U}}\| \leq C$ for some positive constants c and C, then the norms $\|\mathbf{u}\|$ and $\|\mathbf{v}\|$ differ by no more than a finite factor. In the limit of long times, this correction provides a negligible contribution in the expressions (2.12), so that the Lyapunov exponents are the same in both variables.

2.5.4 Volume contraction

The sum $\mathcal{S}_N = \sum_{k=1}^N \lambda_k$ of all Lyapunov exponents of an N-dimensional system measures the contraction rate of volumes in the phase space on the invariant set. In the discrete case, this follows from the fact that the determinant of the Jacobian matrix $\mathbf{J}$ determines the local growth rate (close to the trajectory $\mathbf{U}(t)$) of the phase volume: $V_N(t+1) = |\det \mathbf{J}|V_N(t)$. Here the index N means that we consider the full N-dimensional volume. Since the determinant of the product of matrices is equal to the product of the determinants, the standard Birkhoff ergodic theorem holds, like in one-dimensional case (Section 2.2), yielding

$$\lim_{n\to\infty} \frac{1}{t} \ln \frac{V_N(t)}{V_N(0)} = \langle \ln|\det \mathbf{J}| \rangle .$$

On the other hand, for any t,

$$\det \mathbf{H}_t^\mathsf{T}\mathbf{H}_t = \prod_0^{t-1} |\det \mathbf{J}(\mathbf{U})|^2,$$

so that the determinant of the limiting matrix $\mathbf{P}$ in Eq. (2.11) is equal to $\exp[\langle \ln|\det \mathbf{J}|\rangle]$. This determinant is equal to the product of the eigenvalues, $\prod_k \nu_k = \exp[\langle \ln|\det \mathbf{J}|\rangle]$. By taking the logartithm, we establish the equivalence between the sum of the Lyapunov exponents and the average growth rate of volumes in phase space,

$$\mathcal{S}_N = \sum_{k=1}^N \lambda_k = \langle \ln|\det \mathbf{J}| \rangle . \tag{2.16}$$

A relation similar to (2.16) holds in the continuous-time case. Here, for large t

$$\ln \det \mathsf{P} = \lim_{t \to \infty} \ln \left[[\det \mathsf{H}(t)]^{1/t} \right] = \lim_{t \to \infty} \frac{1}{t} \ln \det \exp \left(\int_0^t dt' \mathsf{K}(\mathbf{U}(t'), t') \right).$$

Taking into account the linear-algebra relation $\det \exp \mathsf{A} = \exp \operatorname{tr} \mathsf{A}$, we obtain

$$\mathcal{S}_N = \sum_{k=1}^N \lambda_k = \ln \det \mathsf{P} = \lim_{t \to \infty} \frac{1}{t} \int_0^t dt' \operatorname{tr} \mathsf{J}(\mathbf{U}(t'), t') = \langle \operatorname{tr} \mathsf{K} \rangle.$$

In the so-called dissipative systems, where $\det \mathsf{J} < 1$ (in the discrete case) and $\operatorname{tr} \mathsf{K} < 0$ (in the continuous case), the sum $\mathcal{S}_N$ of the Lyapunov exponents is negative, meaning that volumes around generic trajectories shrink exponentially to zero. In volume-preserving (e.g. Hamiltonian) systems, since $\det \mathsf{J} = 1$ for discrete time and $\operatorname{tr} \mathsf{K} = 0$ for continuous time, the sum of the exponents vanishes, $\mathcal{S}_N = 0$. This property can be used to test the numerical precision in the calculation of Lyapunov exponents.

These relations can be extended to partial volumes as well. Let us assume, for the sake of simplicity, that the Oseledets splitting (Section 2.3.2) is valid and consider two generic initial vectors $\mathbf{u}(0)$ and $\mathbf{v}(0)$. The area spanned by their iterates is given by the vector product $\mathbf{u}(t) \times \mathbf{v}(t)$. The two initial vectors typically have components along all directions $\mathbf{E}^k(0)$. As a result of the evolution over a long time interval t, each component will be roughly multiplied by a factor $\exp(t\lambda_k)$. Accordingly, in the vector product, as the term $\mathbf{E}^1(t) \times \mathbf{E}^1(t)$, involving the largest component, vanishes, the leading contribution is provided by the cross terms $\mathbf{E}^1(t) \times \mathbf{E}^2(t)$, which is proportional to $\exp[t(\lambda_1 + \lambda_2)]$. Therefore, the area V_2 spanned by two generic initial vectors grows as the sum of the two largest Lyapunov exponents:

$$\lambda_1 + \lambda_2 = \lim_{n \to \infty} \frac{1}{t} \log \frac{V_2(t)}{V_2(0)}. \tag{2.17}$$

Similarly, for a typical M-dimensional volume V_M spanned by M generic different initial vectors,

$$\mathcal{S}_M = \sum_{k=1}^M \lambda_k = \lim_{n \to \infty} \frac{1}{t} \log \frac{V_M(t)}{V_M(0)}. \tag{2.18}$$

This relation, which holds also for infinite-dimensional systems, lies at the heart of the numerical methods for determining Lyapunov exponents discussed in Chapter 3.

2.5.5 Time parametrisation

In the Lyapunov analysis, time plays a special role. One can often change variables to find the most appropriate ones (this "game" will be often played throughout the book) for either numerical or analytical computations. There exists, however, only one *time* and usually one can at most think of rescaling it according to some characteristic scale of the problem. In relativistic dynamics, this is not so, as there is no absolute time axis: the Lyapunov exponents depend on the observer (Francisco and Matsas, 1988) and the use of them as chaos indicators has even been challenged. Occasionally, it may be useful to perform a non-trivial change of the time variable also in classical contexts.

Let us start by introducing a generic scalar variable:

$$\phi = \phi(\mathbf{U}).$$

This is a proper time-like (phase-like) variable if its time derivative

$$\frac{d\phi}{dt} = \frac{\partial \phi}{\partial \mathbf{U}} \cdot \mathbf{F}(\mathbf{U}) \tag{2.19}$$

is strictly positive everywhere (if not in the entire phase space, at least on the invariant set). If this is true, one can use ϕ to order all of the events along the true time axis. Furthermore, under this assumption, one can switch the independent variable from t to ϕ, arriving at the evolution equation

$$\frac{d\mathbf{U}}{d\phi} = \frac{\mathbf{F}(\mathbf{U})}{\mathbf{F}(\mathbf{U}) \cdot \partial\phi/\partial\mathbf{U}}. \tag{2.20}$$

Formally, one of the variables is redundant because of the link between ϕ and the position in phase space, but one can nevertheless disregard this property, select a generic initial condition $\mathbf{U}$ and let it evolve. In practice, the system (2.20) is a reformulation of the initial problem, where the time variable has been adjusted. In order to establish a full link with the original equations, it is necessary to complete the mathematical model with the equation that allows the determination of the true time as a function of ϕ. This is nothing but the inverse of Eq. (2.19):

$$\frac{dt}{d\phi} = \frac{1}{\mathbf{F}(\mathbf{U}) \cdot \partial\phi/\partial\mathbf{U}}. \tag{2.21}$$

In practice, it is easy to convince oneself that the Lyapunov exponents λ_i^ϕ of the system (2.20) are equal to those of the original model up to a multiplicative constant

$$\lambda_i = \lambda_i^\phi/\tau, \tag{2.22}$$

where

$$\tau = \lim_{\phi \to \infty} \frac{t}{\phi}. \tag{2.23}$$

An example where a meanigful variable can be introduced is the Rössler attractor (A.8) for the standard parameter values, setting $\phi = \arctan(y/x)$ as the "phase" of the oscillations.

Within relativistic dynamics ϕ is not a time-like variable, but the true time measured by some other observer (Motter, 2003). Remarkably Eq. (2.22) still holds when a Lorentz tranformation turns a bounded trajectory into an unbounded one (or vice versa) – see Motter and Saa (2009) for a more detailed discussion. As a result, the sign of the Lyapunov exponent is preserved, and this assures that a positive value is a valid criterion to identify chaos across space-time transformations.

This change of variables proves rather useful whenever one has to determine a Poincaré surface of a section. Imagine that the section is defined by the condition $\phi = C$. It is much easier to integrate Eq. (2.20) until the "time" variable ϕ reaches the preassigned value C than to incorporate this condition into the original system. In general, one cannot expect that the variable definining the surface of section is a global time-like variable. One can nevertheless introduce the new variable ϕ locally, sufficiently close to the Poincaré surface, such that one can be sure it behaves monotonically. This is basically the trick suggested long ago by Hénon (1982).

This construction also allows one to establish a relationship between the continuous- and discrete-time representations of a given dynamical system. There are two ways to reduce a continuous-time system to a map: (i) by monitoring the continuous-time system stroboscopically, with a prescribed time interval T and (ii) by constructing a Poincaré map, sampling a trajectory only when it crosses some $(N-1)$-dimensional surface of section in the N-dimensional phase space. In the case of the stroboscopic map, the dimension of the phase space is unchanged, and the N Lyapunov exponents are related by the equation

$$T\lambda_k^{\text{cont}} = \lambda_k^{\text{discr}}.$$

When a Poincaré map construction is used (typically in autonomous continuous systems), the number of Lyapunov exponents reduces to $N-1$. The zero exponent, which corresponds to the invariance of the original trajectories under time shift (see Section 2.5.6), "disappears" because this symmetry is lifted in discrete time. In fact, as discussed, the Poincaré map can be interpreted as a stroboscopic map for a time-like variable ϕ with constant $C = 1$. In this latter case, Eq. (2.23) yields $\tau = \langle T_n \rangle$ where T_n is the Poincaré return time, so that all other exponents are related by

$$\langle T_n \rangle \lambda_k^{\text{cont}} = \lambda_k^{\text{Poin}}.$$

The zero exponent may be eventually recovered by including the evolution equation (2.21) for the true time.

2.5.6 Symmetries and zero Lyapunov exponents

Symmetries and conservation laws play an important role in determining the spectrum of Lyapunov exponents. Continuous symmetries, for instance, typically yield zero Lyapunov exponents. We start by showing that autonomous continuous-time dynamical systems have a zero Lyapunov exponent, provided the trajectory does not converge to a steady state. To see this, let us consider a trajectory $\mathbf{U}(t)$ of Eq. (2.4) and select the perturbation $\mathbf{u} = \mathbf{F}$. Since in autonomous system $\mathbf{F}$ does not explicitly depend on time, time differentiaton yields

$$\frac{d\mathbf{F}}{dt} = \frac{\partial \mathbf{F}}{\partial \mathbf{U}}\frac{d\mathbf{U}}{dt} = \frac{\partial \mathbf{F}}{\partial \mathbf{U}}\mathbf{F} = \mathsf{J}(\mathbf{U})\mathbf{F},$$

which tells us that $\mathbf{u} = \mathbf{F}(\mathbf{U})$ satisfies Eq. (2.5). Thus, one can use the norm of the "velocity vector" to determine the corresponding Lyapunov exponent from Eq. (2.12). As $\|\mathbf{F}\|$ remains bounded from zero for a trajectory that does not converge to a fixed point, we conclude that the corresponding Lyapunov exponent vanishes. This fact has a simple physical interpretation: the zero Lyapunov exponent is measured by perturbing a phase point along its trajectory. Because the system is autonomous, such a perturbation on average neither grows nor decays; it just fluctuates, depending on the local velocity.

Invariance of an autonomous continuous-time dynamical system with respect to time shifts is an example of a continuous symmetry. In the presence of a continuous symmetry, one can map a trajectory onto an equivalent one. In particular, the transformation can be infinitesimal and thereby interpreted as a perturbation that neither grows nor decays and is thus associated to a zero Lyapunov exponent. An example of a discrete-time system with a continuous symmetry is the complex map $z(n + 1) = f(|z|)z$, which is symmetric with respect to a rotation of the argument $z \rightarrow e^{i\alpha}z$. In the context of spatio-temporal chaos, translational invariance of a partial differential equation (with periodic boundary conditions) yields a zero Lyapunov exponent. In the case of a quasiperiodic motion with k incommensurate frequencies, the dynamics can be properly reduced to that of k phases $\dot{\varphi}_j = \omega_j$. The equations are invariant with respect to a shift of each phase; therefore the model possesses k zero Lyapunov exponents.

Zero exponents appear also as a consequence of conservation laws. Let an N-dimensional dynamical system possess an integral of motion $C(\mathbf{u}) = $ const. Then, the dynamics live on an $(N - 1)$-dimensional manifold characterised by a specific (constant) C-value and has $(N - 1)$ Lyapunov exponents. In the original phase space, this $(N - 1)$-dimensional dynamics should be augmented by the equation $\dot{C} = 0$, which does not contribute to the contraction rate of the phase volume. Thus, the sum of all N Lyapunov exponents is the same as the sum of $(N - 1)$ essential exponents, so that the integral of motion is associated with a zero Lyapunov exponent. As an example, let us consider a chain of oscillators with nearest neighbour coupling, characterised by the Hamiltonian $H = \sum P_i^2/2 + \sum_i V(Q_i - Q_j)$. This model has four zero exponents: one is associated with invariance under time translation; one arises from the invariance under the spatial translation $(Q \rightarrow Q + \delta Q)$ and the last two zeros arise from momentum and energy conservation, respectively.

Finally, zero exponents may also (non generically) occur at bifurcation points, where some direction is (linearly) marginally stable. In such cases, it is necessary to go beyond the linear approach to determine the stability.

Symmetries have additional implications: they may induce degeneracies for positive and negative Lyapunov exponents. Let $S(\mathbf{U})$ denote a generic symmetry transformation. A dynamical regime is said to be *instantaneously symmetric* if each configuration of the invariant measure is S-symmetric, i.e. if $S(\mathbf{U}) = \mathbf{U}$. As an example, consider a set of three coupled oscillators U_1, U_2 and U_3 and the transformation $S(U_1, U_2, U_3) = (U_3, U_2, U_1)$. The system is instantaneously S-symmetric if $U_1(t) = U_3(t)$. In the presence of instantaneous symmetries, the evolution in tangent space can be decomposed into that of diagonal blocks, so that the stability analysis is greatly simplified (see, for instance, the implementation of the master stability function in Chapter 9). Whenever some of the blocks are "equivalent", a multiplicity of Lyapunov exponents emerges (Aston and Dellnitz, 1995). The appearance of degeneracies may be more subtle and related to the presence of an *average symmetry*, i.e. not the symmetry of the single phase-points but that of the entire set of points in the invariant measure. Although this latter property by itself does not typically have any implication on the structure of the Lyapunov spectrum, it may have an implication, when suitably combined with the instantaneous symmetry (Ashwin and Breakspear, 2001; Aston and Melbourne, 2006). Consider, for instance, the complex Ginzburg-Landau equation (A.18) with periodic boundary conditions in the interval $[0, 2\pi]$: it has translational and reflection symmetry, besides being invariant under rotations of the complex variable U. For $R = 4$, $\mu = -4$, in the interval $\nu \in [1.9, 2.3]$, the dynamics converge towards a state of spatial period π, thus with an instantaneous spatial translational symmetry of π, while the attractor itself exhibits a spatial translational symmetry of $\pi/2$ (and a reflection symmetry). The combination of the two ensures a strong degeneracy in the spectrum: two subsets of the Lyapunov exponents are equal to one another (Aston and Laing, 2000). Symmetries have been found to yield (multiple) degeneracies in various contexts, including a two-dimensional gas of hard disks (Eckmann et al., 2005) and among the negative exponents in coupled oscillators (see the discussion in Section 2.5.7).

2.5.7 Symplectic systems

Symplectic dynamics is defined in terms of pairs of conjugated (canonical) variables $\mathbf{q}, \mathbf{p} \in \mathbb{R}^N$, so that it is customary to refer to a symplectic system with N degrees of freedom as a dynamical system characterised by $2N$ variables and $2N$ LEs. They possess a special symmetry property: Lyapunov exponents come in pairs $\lambda_k = -\lambda_{2N-k+1}$.

The most prominent such example is Hamiltonian dynamics, described by the equations of motion

$$\dot{\mathbf{q}} = \frac{\partial H}{\partial \mathbf{p}}, \qquad \dot{\mathbf{p}} = -\frac{\partial H}{\partial \mathbf{q}},$$

where $H(\mathbf{q}, \mathbf{p})$ is the energy (Hamiltonian function) of the system.

In discrete time, one speaks of symplectic maps, which can be viewed as a canonical transformation $(\mathbf{q}(0), \mathbf{p}(0)) \rightarrow (\mathbf{q}(t), \mathbf{p}(t))$ generated by the evolution of a Hamiltonian model. They are defined so as to preserve the canonical structure of the model; i.e. they must satisfy the Poisson brackets

$$\{q_i(t), q_j(t)\} = \{p_i(t), p_j(t)\} = 0, \quad \{q_i(t), p_j(t)\} = \delta_{ij}.$$

Such conditions can be rewritten in terms of the Jacobian matrix J of the transformation as

$$\mathsf{J}\Omega\mathsf{J}^\mathsf{T} = \Omega, \quad \Omega = \begin{bmatrix} 0 & \mathsf{I}_N \\ -\mathsf{I}_N & 0 \end{bmatrix}, \tag{2.24}$$

where I_N is a unit $N \times N$ matrix and Ω is skew-symmetric. Matrices satisfying condition (2.24) are called symplectic. Since $\Omega^2 = -\mathsf{I}$, it can be easily proven that the relation

$$\mathsf{J}^\mathsf{T}\Omega\mathsf{J} = \Omega \tag{2.25}$$

holds as well. The Lyapunov exponents are defined in terms of the eigenvalues of the matrix $\mathsf{H}^\mathsf{T}\mathsf{H}$ (see Eqs. (2.10, 2.11)). Now, we show that if μ is an eigenvalue of $\mathsf{H}^\mathsf{T}\mathsf{H}$, where H is symplectic, then $(\mu)^{-1}$ is also an eigenvalue. Indeed, if $\mathbf{y}$ is the eigenvector associated to the eigenvalue μ, then

$$\mathsf{H}^\mathsf{T}\mathsf{H}\mathbf{y} = \mu\mathbf{y}.$$

By multiplying this equation on the left by $H^TH\Omega$, one obtains

$$H^TH\Omega H^THy = \mu H^TH\Omega y.$$

Since H is the product of symplectic maps, it is still symplectic. By recursively using the conditions (2.24, 2.25), it is found that

$$(\Omega y) = \mu H^TH(\Omega y).$$

This proves proves that Ωy is the eigenvector of the matrix H^TH with eigenvalue μ^{-1}. Thus, the eigenvalues of a product H^TH of symplectic matrices appear in symmetric pairs. This implies that, taking the infinite-time limit in Eq. (2.11), the Lyapunov exponents also appear in symmetric pairs, $\lambda_k = -\lambda_{2N-k+1}$.

Various generalisations of this symmetry to a wider class of dynamical systems have been discussed in the literature. Remarkably, this includes the symmetry with respect to a negative value $-\gamma$ ($\lambda_k + \lambda_{2N-k+1} = -\gamma$) that appears in some models of dissipative or suitably thermostatted particles (Dressler, 1988; Gupalo et al., 1994).

Numerical methods

By definition, the computation of Lyapunov exponents requires knowledge of the evolution equations in tangent space and, in particular, of the local Jacobians. This is, however, not always possible. The reason may be the complexity of the model itself, as for the most detailed models used for weather predictions, or the absence altogether of a quantitative model, as in many experiments. In such cases one can substitute the lack of knowledge of the Jacobian by following trajectories that are close enough to the reference trajectory to fall in the linear regime (and frequently rescaling their distance to avoid it becoming either smaller than the computational accuracy or the noise in the system, or so large as to be affected by nonlinearities).

In practice, it is necessary to reconstruct (either implicitly or explicitly) the Jacobian. In fact, this chapter is almost entirely devoted to discussing various methods to determine LEs, starting from the known local Jacobians. In Section 3.7, we nevertheless discuss some approaches that can be applied to time series in the absence of a known quantitative model.

3.1 The largest Lyapunov exponent

The numerical calculation of the largest Lyapunov exponent is a relatively simple problem. It is sufficient to implement the relation (2.12). As discussed in Section 2.3.3, a generic initial perturbation $\mathbf{u}(0)$ has a non-zero component along the most expanding direction, which will dominate the evolution over long times. One can thus compute the norm of the perturbation after a suitable transient time (which depends on the difference between the largest and the second exponent) and thereby determine λ_1 from Eq. (2.12). Practically, in order to avoid large or small numbers (for a positive or negative largest exponent, respectively), it is convenient to periodically renormalise the perturbation. As an example, we describe the procedure for the case of discrete time. Given an initial unit vector $\mathbf{u}(0)$, one first computes $\tilde{\mathbf{u}}(1) = \mathsf{J}(0)\mathbf{u}(0)$ and then renormalises it to unit norm,

$$\mathbf{u}(1) = \frac{\tilde{\mathbf{u}}(1)}{\|\tilde{\mathbf{u}}(1)\|}. \tag{3.1}$$

Next, the Jacobian matrix $\mathsf{J}(1)$ is applied to the unit vector $\mathbf{u}(1)$, etc. The largest Lyapunov exponent is determined by averaging the normalisation factors obtained at each step,

$$\lambda_1 = \lim_{t \to \infty} \frac{1}{t} \sum_{k=1}^{t} \ln\|\tilde{\mathbf{u}}(k)\|. \tag{3.2}$$

In practice, the average is performed over a finite time (possibly dropping an initial transient). This method cannot be straightforwardly extended to the computation of the second Lyapunov exponent, as one needs to choose a non-typical initial vector in the $(N-1)$-dimensional subspace L^2, which is a priori unknown.

Turning, instead, our attention to the definition (2.11), one might wish to solve the problem by determining the eigenvalues of the symmetric real matrix $M(t) = H^T(t)H(t)$, where the Jacobian matrix $H(t)$ is obtained by multiplying the single-step Jacobian matrices for a time interval from 0 to $t-1$. Although it is formally sufficient to implement a suitable linear-algebra routine, one can see that this procedure is eventually rather inaccurate.

Let us, indeed, look at the action of the $N \times N$ matrix $H(t)$ on a given set of initial vectors. Singular value decomposition (SVD) allows decomposing H_t as

$$H(t) = ODQ^T,$$

where O and Q are two orthogonal matrices, while D is diagonal. From the equality $H^T H = QD^2Q^T$, one can conclude that the eigenvalues of $M(t)$ are $\mu_i^2 = D_{ii}^2$. In practice, SVD reveals that a generic transformation H_t can be decomposed into a rotation, Q^T, followed by an expansion/contraction along various orthogonal directions (D) and by a second rotation O. Therefore, when $H(t)$ is applied to a unit N-dimensional sphere $\mathbb{S}_N$, the first rotation is irrelevant, while the application of D transforms $\mathbb{S}_N$ into an ellipsoid that is then suitably rotated. Altogether, this means that the sizes of the axes of the ellipsoid directly measure the multipliers μ_i. Unfortunately, one cannot use this method to compute the Lyapunov exponents. If the matrix $H(t)$ is computed for too long of a time (this is needed to ensure that the limit $t \to \infty$ is taken), the singular values D_{ii} cover too wide a range[1] to be estimated with enough accuracy with any method. On the other hand, one cannot break the computation into many consecutive subintervals since the outcome of the first SVD would be an ellipsoid rather than a sphere. Thus, the aforementioned arguments cannot apply, as they work only when the initial set is a unit sphere. The methods described next overcome this difficulty by breaking the overall computation into many distinct, short time-steps.

The main idea is to use the relationship between the Lyapunov exponents and the evolution of the phase volume, discussed in Section 2.5.4. Indeed, the evolution of a phase volume, Eq. (2.18), is the evolution of a scalar quantity, similar to the norm of the vector, and its growth rate can be computed in short steps, analogously to what was done in Eqs. (3.1, 3.2) for the largest Lyapunov exponent.

3.2 Full spectrum: QR decomposition

A solution to this problem is obtained by estimating the growth rate $\mathcal{S}_m$ of the volume V_m of a generic m-dimensional parallelepiped,

---

[1] In fact, a range that grows exponentially in time.

$$S_m := \lim_{t \to \infty} \frac{\ln V_m}{t} = \sum_{i=1}^{m} \lambda_i, \tag{3.3}$$

where the last equivalence relation follows from the tendency of the parallelepiped to almost surely align along the most expanding m-dimensional subspace (see Section 2.5.4).

Let us now denote with $\mathbf{Q}_0$ an $N \times m$ orthogonal matrix (i.e. a set of m orthogonal vectors in the N-dimensional tangent space). Within a time t, $\mathbf{Q}_0$ evolves into a parallelepiped identified by $\mathbf{P} = \mathbf{H}(t)\mathbf{Q}_0$. The matrix $\mathbf{P}$ admits a unique decomposition of the type

$$\mathbf{P}(t) = \mathbf{QR},$$

where $\mathbf{Q}$ is an orthogonal $N \times m$ matrix, while $\mathbf{R}$ is an upper triangular $m \times m$ matrix with positive diagonal elements.

The determinant of $\mathbf{R}$ is equal to the previously mentioned volume V_m (since $\mathbf{Q}$ involves only rotations and reflections that do not affect the volume). Therefore,

$$V_m = \prod_{i=1}^{m} R_{ii}.$$

By substituting this expression into Eq. (3.3) and applying it successively to $m = 1, m = 2$, etc., one finds that

$$\lambda_j := \lim_{t \to \infty} \frac{\ln R_{jj}}{t}.$$

The main advantage with respect to the SVD is that now one can decompose the computation of $\mathbf{R}$ into the product of many terms, each of which can be accurately determined.

In fact, let us break the time interval t into L steps of size τ and denote with $\mathbf{H}_k$ the operator resulting from the integration between time $(k-1)\tau$ and $k\tau$ $(k = 1, L)$,

$$\mathbf{P} = \prod_{k=1}^{L} \mathbf{H}_k \mathbf{Q}_0.$$

If we now define

$$\mathbf{P}_k := \mathbf{H}_k \mathbf{Q}_{k-1}$$

and perform the QR decomposition

$$\mathbf{P}_k = \mathbf{Q}_k \mathbf{R}_k,$$

a recursive iteration of the procedure leads to

$$\mathbf{P} = \mathbf{Q}_k \prod_{k=1}^{L} \mathbf{R}_k.$$

Given that the product of the $\mathbf{R}$ matrices is still upper triangular and that the decomposition is unique, one can conclude that the computation of $\mathbf{R}$ can be split into that of many single steps of finite length, and the jth Lyapunov exponent is obtained by summing the single $\ln R_{jj}$ contributions. For the computation of the Lyapunov exponents, only the diagonal terms of R are necessary. Notice also that it is not necessary to determine separately the matrix H_k and multiply it by Q_k, as it suffices to integrate the equations in tangent space, starting from the initial conditions Q_k. Altogether one can summarise the approach by stating that given a set of orthogonal vectors, they are re-orthogonalised after some time. In the process, the expansion of volumes of different dimensions is computed, while the orthogonalisation prevents the different directions to mutually align, thus avoiding problems of numerical accuracy. This approach was first proposed in the late 1970s (Shimada and Nagashima, 1979; Benettin et al., 1980a, b); its implementation allowed determining Lyapunov spectra in many relevant dynamical systems.

3.2.1 Gram-Schmidt orthogonalisation

The most popular algorithm to perform the QR decomposition is the Gram-Schmidt (GS) procedure, which is defined as follows. Let P be an $N \times m$ matrix and set $R_{11} = \|P_1\|$, where $\| \|$ denotes the norm of the vector, while P_k denotes the kth column vector of matrix P. In the first step one calculates $Q_1 = P_1/R_{11}$. The following steps are recursively defined as

$$\tilde{Q}_i = P_i - \sum_{j=1}^{i-1}(P_i \cdot Q_j)Q_j, \qquad R_{ii} = \|\tilde{Q}_i\|, \qquad Q_i = \tilde{Q}_i/R_{ii}.$$

These computations can be carried out in two different ways with different degrees of numerical accuracy. The classical approach amounts to literally following this definition. The modified GS method amounts to subtracting the components along the Q_j vectors as soon each new vector is generated. The different order of the operations guarantees a better degree of orthogonality in the resulting matrix. A pseudocode of the latter approach is sketched in Appendix B.

As for the computational complexity of this method, one can easily check that the leading term requires N^3 multiplications (in the case $m = N$).

3.2.2 Householder reflections

Householder reflections provide a more accurate tool to compute Lyapunov exponents. This was first recognised in a 1984 preprint of Johnson et al. (1987) and then implemented by Eckmann and Ruelle (1985).

A Householder transformation is a reflection around a given plane $\mathcal{A}$. In matrix form, this orthogonal operation is described by

$$O = I - 2ww^T,$$

where I is the identity matrix, while w is a unit vector orthogonal to the plane $\mathcal{A}$.

Given a matrix $P^{(0)}$, its QR decomposition is obtained by repeatedly applying a sequence of Householder transformations O_i,

$$P^{(i)} = O_i P^{(i-1)}.$$

Notice that the superscript index in parentheses is used here and below to distinguish the intermediate steps during a given QR decomposition. This is to discriminate from the subscript k used to distinguish the P matrices generated along a trajectory. These transformations are identified by the unit vectors $\mathbf{w}_i = \tilde{\mathbf{w}}_i / |\tilde{\mathbf{w}}_i|$, where

$$\tilde{\mathbf{w}}_i = \lceil \mathbf{P}_i^{(i-1)} \rceil_{i-1} + \text{sign}\left[P_{ii}^{(i-1)} \right] \mathbf{e}_i,$$

$\lceil \cdot \rceil_i$ denotes the projection operator which zeroes the first i components and $\mathbf{e}_i$ is the ith Euclidean unit vector. It is clear that the first $i-1$ components of $\mathbf{w}_i$ are equal to zero. This expresses the fact that the ith Householder transformation does not affect the first $i-1$ variables. It turns out that $\mathsf{P}^{(i)}$ has the following structure

$$\mathsf{P}^{(i)} = \begin{pmatrix} \mathsf{R}^{(i)} & \mathsf{S}^{(i)} \\ \mathbf{0} & \tilde{\mathsf{S}}^{(i)} \end{pmatrix},$$

where $\mathsf{R}^{(i)}$ is an $i \times i$ upper triangular matrix and $\mathsf{S}^{(i)}$ and $\tilde{\mathsf{S}}^{(i)}$ are two $i \times (m-1)$ and $(N-i) \times (m-i)$ matrices, respectively. Finally, $R_{ii} = \|[\mathbf{P}_i]_i\|$; here we should note that in order to guarantee that the diagonal elements of R are positive, suitable additional reflections should be considered. Such operations are, however, totally irrelevant for the computation of the Lyapunov exponents, which depend only on the absolute values.

As a result of the recursive procedure,

$$\mathsf{P}^{(m)} = \mathsf{O}_m \mathsf{O}_{m-1} \ldots \mathsf{O}_1 \mathsf{P}^{(0)} = \begin{pmatrix} \mathsf{R} \\ \mathbf{0} \end{pmatrix}$$

and therefore $\mathsf{Q}^\mathsf{T} = \tilde{\mathsf{P}}^{(m)}$, where the tilde means that only the first m rows of the matrix are considered.

An explicit expression of Q can be obtained by repeatedly applying the matrices O_i,

$$\mathbf{y}^{(i)} = \mathbf{y}^{(i-1)} - 2(\mathbf{w}_i \cdot \mathbf{y}^{(i-1)})\mathbf{w}_i.$$

By implementing the whole procedure to the column matrix $\mathsf{P}^{(0)} \equiv \mathbf{P}_k$ and setting $\mathbf{y}^{(0)} = \mathbf{e}_k$, one eventually finds that $\mathsf{Q}_k^\mathsf{T} = \mathbf{y}^{(m)}$. A pseudocode of this algorithm is reported in Appendix B.

In the case $m = N$, the Householder procedure requires $(2/3)N^3$ multiplications for the factorisation and N^3 for the reconstruction of the Q matrix, i.e. $(5/3)N^3$ altogether. The scaling behaviour is the same as for GS, while the prefactor is slightly worse in this latter case. Notice, however, that it is possible to further optimise the method, bringing the number of operations down to $(2/3)N^3$ (von Bremen et al., 1997).

So far, we have not included the computational complexity of the generation of the new matrix $\mathbf{P}_k$ from $\mathbf{Q}_{k-1}$. This step depends on the structure of the equations of motion. Assuming that N_I represents the average number of interactions between a generic variable and all the others, this task requires a number of operations that is proportional to nN_IN^2, where n is the number of integration time steps in between two QR decompositions. In many physical models, N_I is independent of N and possibly a small number (in the case of nearest-neighbour coupled oscillators, it may be as small as 3). In such cases, the computational bottleneck depends on n/N but is likely to be the QR decomposition for large systems. In systems with global coupling, the bottleneck is typically the integration of the vectors, which would require a time nN^3 (if all exponents have to be determined).

An exception is the class of mean-field models where all variables are driven by the same field, and, thereby, the cost of the integration is much reduced.

3.3 Continuous methods

The methods described in Sections 3.1 and 3.2 are equally applicable to discrete-time and continuous-time systems. The only difference is that in the continuous case, between consecutive QR decompositions, it is necessary to integrate a nonlinear differential equation together with the differential equation for the perturbations, using appropriate (possibly standard) numerical methods.

If the time variable is continuous, one can, however, determine the Lyapunov exponents by integrating directly suitable differential equations without linear-algebra manipulations. This idea was first proposed by Goldhirsch et al. (1987). The starting point is the fundamental equation

$$\dot{Y} = KY,$$

where Y is a matrix made of N independent vectors and K is the Jacobian of the differential equation (or, more in general, just a time-dependent matrix). In order to compute the Lyapunov spectrum, it is necessary to decompose Y:

$$Y(t) = Q(t)R(t),$$

where Q is orthogonal and R is upper triangular. For the sake of simplicity we assume that Q is a square matrix, i.e. all exponents are going to be computed, but the approach works generally, with some crucial differences that will be properly emphasised. By combining these two equations, one obtains

$$\dot{Q}R + Q\dot{R} = KQR,$$

which leads, after multiplying from the left with Q^{T} and from the right with R^{-1}, to

$$\tilde{K} = Q^{\mathsf{T}}KQ - Q^{\mathsf{T}}\dot{Q} = \dot{R}R^{-1}, \qquad (3.4)$$

where the last term in the r.h.s. is upper triangular, being the product of two upper triangular matrices. Accordingly, the antisymmetric matrix $L(t, Q) = Q^{\mathsf{T}}\dot{Q}$ (this property follows from the orthogonality of Q) satisfies

$$L_{i,j} = \begin{cases} [Q^{\mathsf{T}}KQ]_{ij} & i > j, \\ 0 & i = j, \\ -[Q^{\mathsf{T}}KQ]_{ji} & i < j. \end{cases}$$

In other words, Q evolves according to the differential equation

$$\dot{Q} = QL(t, Q), \qquad (3.5)$$

while the information on the volume expansion is contained in the matrix R, which satisfies the equation

$$\dot{\mathsf{R}} = \tilde{\mathsf{K}}\mathsf{R}$$

(see Eq. (3.4)). As the Lyapunov exponents depend only on the diagonal terms of R (the diagonal terms of L are equal to zero), this equation reduces to a set of independent equations, namely

$$\dot{R}_{ii} = [\mathsf{Q}^\mathsf{T}\mathsf{K}\mathsf{Q}]_{ii}R_{ii},$$

so that

$$\lambda_i = \lim_{t \to \infty} \frac{1}{t} \ln R_{ii} = \lim_{t \to \infty} \frac{1}{t} \int_0^t [\mathsf{Q}^T(s)\mathsf{K}(s)\mathsf{Q}(s)]_{ii} ds,$$

which can be integrated by quadratures. By construction, the evolution of Q is constrained to the (Stiefel) manifold of orthogonal matrices. If one is interested in the computation of all eigenvectors, the Stiefel manifold is neutrally stable (Dieci and Van Vleck, 1995), and it is harmless to integrate the equation with unitary algorithms (e.g. Gauss Runge-Kutta). Their implementation, however, is not straightforward (Dieci ct al., 1997), especially in high-dimensional spaces. On the other hand, if one is interested in only a few exponents, the Stiefel manifold is transversally unstable (as noted by Dieci and Van Vleck (1995)), and different approaches become necessary. A first solution was proposed by Christiansen and Rugh (1997), who added a suitable dissipation which vanishes on the Stiefel manifold but stabilises transversal perturbations. Unfortunately, the method is only conditionally stable (above a model-dependent parameter value) and, in some cases, just fails (Ramasubramanian and Sriram, 2000). A more sophisticated approach has been developed by Bridges and Reich (2001), where the authors have been able to stabilise unconditionally the Stiefel manifold by making use of concepts of differential geometry. In practice, this method is based on a transformation of the matrix L so as to make it strictly antisymmetric.[2]

Yet another class of methods is based on mapping Q onto a series of angles, which uniquely identify the transformation Q, and thereby writing suitable differential equations. This method was first proposed by Rangarajan et al. (1998) and later extended by Ramasubramanian and Sriram (2000). However, the number of expressions that one has to manage grow rapidly with the dimension and are unmanageable above $N = 4$. Finally, in some special flows, such as symplectic dynamics, different decompositions have been proposed (Habib and Ryne, 1995), which again suffer from the problem of scalability. Altogether the implementation of automatic unitary integrators is not straightforward, and this represents a major obstacle to the study of high-dimensional systems.

In the end, even though much less elegant, the most versatile algorithms are those based on the so-called *projected schemes*, where a standard integration algorithm is combined with an orthogonalisation routine, to remove systematically transversal components (Dieci et al., 1997).

---

[2] In fact, as noted by Dieci and Van Vleck (1995), the Stiefel manifold is unstable as long as L is only *weakly* antisymmetric, i.e. if $\mathsf{Q}^\mathsf{T}[\mathsf{H} + \mathsf{H}^\mathsf{T}]\mathsf{Q} = 0$, without being $\mathsf{H}^\mathsf{T} = -\mathsf{H}$. This is possible only for rectangular orthogonal matrices.

From a computational point of view, one should notice that even disregarding the CPU time used to orthogonalise Q (required only by the last approach), the integration of Eq. (3.5) is much more CPU-time consuming than the integration of P, since L is (except for its symmetry) a full matrix. This implies that the number of operations per time steps is of the order of N^3. This means that, unless one can use a significantly longer integration time step, such methods are necessarily more CPU-time consuming than the discrete-time ones.

3.4 Ensemble averages

The standard definition of Lyapunov exponents requires averaging a local expansion rate along a formally infinite trajectory. It is natural to ask whether LEs can also be determined by performing an ensemble, or invariant-measure, average.

In the case of one-dimensional maps (cf. Section 2.2), the local expansion rate is given by $\ln F'(U)$, and the Lyapunov exponent can be simply defined as

$$\lambda = \int \ln |F'(U)| \rho(U) dU,$$

where $\rho(U)$ is the invariant probability density (invariant measure). In other words, the LE can be expressed as the average of a local observable in phase space. In higher-dimensional systems, the problem is more difficult, since the expansion rate depends on the orientation of the perturbation. As will be illustrated in Chapter 4, Ershov and Potapov (1998) proved that each point in phase space is equipped with a unique set of *covariant* Lyapunov vectors that are the proper directions to be used. Therefore, one is entitled to define

$$\lambda_i = \int \log \|\mathsf{F_U V}_i(\mathbf{U})\| \rho(\mathbf{U}) d\mathbf{U}, \tag{3.6}$$

where the unit vector $\mathbf{V}_i(\mathbf{U})$ is aligned along the ith covariant Lyapunov vector in $\mathbf{U}$.

A direct implementation of Eq. (3.6) is generally problematic, since the covariant vectors are unknown. The only way to determine them is by setting up the same algorithm which allows determining the LEs via the standard approach (see next chapter)! This idea should not, however, be completely discarded. One can indeed follow a hybrid approach, by combining an average over the phase space with some time evolution. For instance, one can formally define

$$\lambda_i = \lim_{t \to \infty} \int \log |\mathsf{R}_{ii}| d\mu, \tag{3.7}$$

where the matrix R is determined via a standard QR decomposition along a trajectory of time length t that starts in $\mathbf{U}$ at time 0 with a random initialisation of the vectors in tangent space, while μ is the given invariant measure.

This definition looks awkward, as it involves a time limit as well as an average over the phase space. Here, however, the convergence in time is much faster than for the usual approach, since it is limited only by the relaxation time towards the proper Lyapunov vectors in tangent space. Slow processes in phase space do not affect the computation, as one assumes to deal directly with the invariant measure. A practical implementation of this idea was proposed by Aston and Dellnitz (1999) for the largest LE and later extended by Beyn and Lust (2009) to the whole spectrum.

The convergence in Eq. (3.7) can be further accelerated if, before starting the evolution in the tangent space with a set of random vectors, one first iterates the initial point $\mathbf{U}$ for some time t_b backwards in phase space. This way, the orthogonal subspaces are properly oriented at time $t = 0$. This idea was implemented by Eckhardt and Yao (1993) to determine the maximum Lyapunov exponent in the Lorenz attractor and the Chirikov-Taylor standard map.

Altogether, the advantage of this method is based on the ability to avoid statistical fluctuations by performing an integral in phase space. Therefore, its validity depends crucially on the accuracy of the known invariant measure. In Hamiltonian systems, the invariant measure is known a priori, but, more in general, it has to be determined one way or another. Typically, the phase space can be partitioned into relatively small boxes $\mathcal{B}_i$ (Dellnitz and Hohmann, 1997), to thereby build a discretized Perron-Frobenius operator P, which governs the evolution of the measure

$$\mathsf{Q}_{ij} = \frac{m(\mathcal{B}_j \bigcap \mathbf{F}^{-1}(\mathcal{B}_i))}{m(\mathcal{B}_j)},$$

where $m(\mathcal{B}_j)$ denotes the uniformly distributed mass within the box $\mathcal{B}_j$ (cf. Section 5.5). The (normalized) leading eigenvector of P provides an estimate for the suitably coarse-grained invariant measure. The bottleneck of this approach is the phase-space dimension; it can be hardly applied to high-dimensional dynamics.

The ensemble-average approach may be also implemented analytically. In the case of a weakly forced chaotic oscillator (a problem connected with the onset of phase synchronisation), it allows deriving a perturbative expression for the second Lyapunov exponent (see Section 9.1).

Finally, notice that, in the case of random processes, the integral over the phase space in Eq. (3.6) is to be replaced by two integrals: one over the possible orientations of the given covariant Lyapunov vectors and the other over the realisations of the stochastic process itself. This is a rather complex task, but the distribution of the random variables is known a priori and is typically rather smooth (at variance with the usual wild dependence of the covariant vectors on the position in phase space). In fact, analytic results for the maximum LE can be obtained for some problems involving products of random matrices (see Chapter 8).

3.5 Numerical errors

The first, obvious source of errors in the computation of Lyapunov exponents is the integration of the evolution equations in the tangent space. This problem is analogous to that of guaranteeing a good accuracy in the real-space evolution; one must, however, keep in mind that time scales in tangent space may be different. This is particularly true for the computation of very negative Lyapunov exponents that might require a time step smaller than for the nonlinear equation. Since all of these problems are not specific to Lyapunov exponents, the interested reader can look at the proper literature on this subject. Methods for the integration of general ODEs are, e.g., found in Butcher (2008), while more specific algorithms to deal with systems with symmetries (such as symplectic dynamics) are discussed in Leimkuhler and Reich (2004) and Hairer et al. (2010). Finally, a comprehensive presentation of methods for partial differential equations can be found in Ames (1992), Quarteroni and Valli (1994), and Morton and Mayers (2005).

A comparison of the various numerical methods for the computation of Lyapunov exponents can be found in Geist et al. (1990) and Ramasubramanian and Sriram (2000). Although they are somewhat outdated, one can nevertheless still find some useful information. In the following sections we discuss two major sources of errors: (i) the finite orthogonalisation time and (ii) statistical fluctuations. The last section is devoted to a specific example, where small errors may give rise to sizable effects, because of a quasi-degeneracy.

For the sake of clarity, some of our considerations are supported and exemplified by numerical simulations.

3.5.1 Orthogonalisation

The errors on the Lyapunov exponents are a direct consequence of the errors affecting the triangular matrix R. In discrete methods, the error on R depends in the errors made in the computation of the transition matrices P. The first analysis was presented by Dieci et al. (1997) and later refined by Dieci and Van Vleck (2005). There, it was found that there are two sources of errors. One is related to the "nonnormality" of the matrix R, i.e. to the size of its nondiagonal components, which may obstruct the estimate of the diagonal terms (Dieci and Van Vleck, 2005). This contribution arises from the fact that some directions may not be well separated from each other; it grows with the orthogonalisation time interval τ. Except for possibly pathological models, it should not create any problem provided that a small enough value is selected for τ. The second contribution is due to the actual value of the expansion/contraction rates. A bound was given by Dieci et al. (1997),

$$\delta\lambda_i = C\frac{\left\langle R_{ii}^{-1}(\tau)\right\rangle}{\tau}\varepsilon, \tag{3.8}$$

where C is typically an unknown prefactor, $R_{ii}(\tau)$ is the expansion factor over the orthogonalisation time τ and ε represents the numerical accuracy. The accuracy may, therefore, degrade significantly for the negative multipliers. Here we illustrate the problem by studying a chain of Rössler oscillators (see Eq. (A.17)) for parameter values where some LEs are approximately equal to -10 (see Fig. 3.1). In that case, assuming that $\langle R_{ii}^{-1}(\tau)\rangle \approx \exp(\lambda_i\tau)$, we find that for $\tau = 1$ the error is amplified by a factor 10^6, i.e. six digits are lost in the computation. As long as the single computations are carried out with an accuracy of 15–16 digits, this degradation does not create practical problems, and the full dots in Fig. 3.1 indeed provide a fairly accurate description of the whole Lyapunov

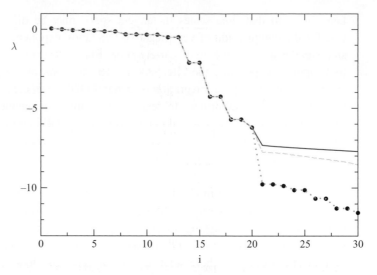

Fig. 3.1 Lyapunov spectra for a chain of 10 Rössler oscillators (A.17) for $a = b = 0.1, c = 10$ and $\varepsilon = 2$. The various symbols correspond to different protocols and values of the orthogonalisation time τ. Full dots and the solid curve have been obtained using the GS procedure with $\tau = 1$ and $\tau = 2$, respectively. The dashed line corresponds to a double GS implementation (see the text). The dotted line has been obtained by implementing Householder reflections with $\tau = 2$.

spectrum. For $\tau = 1$, the differences among the discrete methods cannot be appreciated. However, as soon as $\tau = 2$, Eq. (3.8) implies that the amplification factor becomes 10^{12} (apart from the unknown factor C). As a result, the last 10 exponents are badly estimated, if determined with the GS approach (here and almost everywhere in the book, the modified GS method is implemented) — see the solid line. This can be partially understood by noticing that there is a large gap in the spectrum between the 20th and the 21st exponent. This gap cannot be properly handled by the GS method, which grossly overestimates the spectrum. The reason is that in some limit cases, the GS approach is not able to produce a truly orthogonal basis, and some improvements can be obtained by orthogonalising the basis, i.e. applying the GS orthogonalisation twice (see the dashed line). The dotted line, obtained by using Householder reflections, proves that this method is superior to the GS orthogonalisation; one can exploit the higher accuracy by using longer orthogonalisation times.

In continuous-time methods, the error in R depends on the integration of the matrix Q, whose evolution is less unstable (Dieci and Van Vleck, 2005). Accordingly, such methods are superior in this respect. We should recall, however, that typically the major sources of errors are the statistical fluctuations discussed in Section 3.5.2.

3.5.2 Statistical error

The different degree of stability of the various regions in phase space represents an unavoidable source of fluctuations in chaotic dynamics. Equivalently, fluctuations may

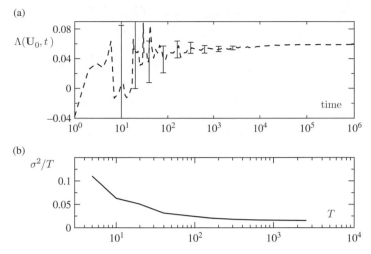

Fig. 3.2 (a) The finite-time Lyapunov exponent $\Lambda(\mathbf{U}_0, t)$ in the Rössler oscillator (A.8) for the standard parameter values, obtained from a single trajectory. The error bars are obtained by following the procedure described in the text. (b) Effective diffusion coefficient σ^2/T versus T, from the same data set.

arise in stochastic contexts, when different matrices are singled out in different time steps. In order to determine how the presence of such fluctuations affects the computation of an LE, it is necessary to sharpen the notations, introducing appropriate observables. Let $\Lambda(\mathbf{U}_0, t)$ denote the growth rate measured over a time t, along a trajectory starting from $\mathbf{U}_0$.[3] We then introduce the logarithm of the expansion factor over a time t, such as $\Gamma(t) := \Lambda(\mathbf{U}_0, t)t$. The time evolution of $\Gamma(t)$ can be viewed as a diffusive process with a drift. The average increment $\langle\Gamma\rangle(T)$ of Γ over a time T is nothing but λT. Its variance $\sigma^2(T)$ is expected to increase linearly with time, $\sigma^2(T) \approx DT$ for T large enough (unless we are in the presence of an anomalous diffusion). Accordingly, the statistical error in the Lyapunov exponent is $\sigma_\lambda = D/\sqrt{T}$. In practice, the diffusion coefficient D can be estimated by splitting the overall computation time t into a relatively large number of intervals of length T. If T is longer than the correlation time, one can determine the diffusion coefficient as $D = \sigma^2(T)/T$ and thereby extrapolate the error for longer times, including the length t of the entire simulation, without the need of repeating the simulation.

This procedure is illustrated in Fig. 3.2, which refers to the maximum exponent in a single Rössler oscillator (A.8) with the standard parameter values. The solid line in Fig. 3.2a corresponds to $\Lambda(\mathbf{U}_0, t)$ for some specific initial condition, versus time. The error bars are obtained by splitting the entire trajectory into samples of length $t_0 = 3000$ (much smaller than the total length) and thereby estimating the statistical fluctuations. Fig. 3.2b contains the effective diffusion coefficient σ^2/T. Its decrease reveals the presence of slowly decaying correlations that survive more than 1000 time units. On the basis of the D value

---

[3] In Chapter 5, Λ will be identified with the finite-time Lyapunov exponent. For the sake of simplicity, we drop the index here that identifies the LE itself.

obtained for $T = 1280$, the estimate of the Lyapunov exponent at time $t = 10^6$ is found to be affected by a statistical error of 1.2×10^{-4}.

The diffusion coefficient D, introduced here as a tool to estimate the statistical error on the Lyapunov exponent, is itself a dynamical invariant that contributes to characterising a chaotic dynamics. This concept is extensively discussed in Section 5.3. Here, we limit ourselves to recall that zero LEs are characterised by a vanishing diffusion coefficient whenever they are due to symmetries or conservation laws. The very small LEs arising in weakly chaotic systems (such as nearly integrable Hamiltonian models) is a qualitatively different case. There, the convergence of the Lyapunov exponents is affected by the long-lasting temporal correlations which characterise also the convergence of the invariant measure to its asymptotic shape. As a consequence, it is not possible to define generally valid protocols, apart from comparing the convergence of the LE, while starting from significantly different initial conditions.

3.5.3 Near degeneracies

We now address a problem that may occur in the case of nearly degenerate Lyapunov exponents. A sufficiently long simulation (10^7 time units) of a chain of Rössler oscillators (A.17) yields the set of the exponents reported in Fig. 3.3, where one can recognise that there is just one positive exponent (see the inset) and that many LEs come in pairs (the latter ones are joined by segments to make the pairs easily identifiable). As we will further clarify, the presence of pairs is to be attributed to the (discrete) translational symmetry, which, in the case of spatially periodic solutions, manifests itself as an equivalent stability against sine and cosine perturbations. Here, we focus on the four almost identical exponents indicated by the arrow in the inset (i.e. on two nearly degenerate consecutive pairs). Long simulations (up to 10^8 time units) reveal that there is a small difference between

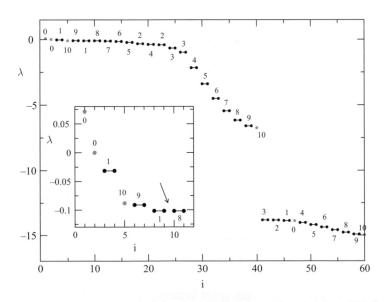

Fig. 3.3 Lyapunov spectrum for a chain of $N = 20$ Rössler oscillators (A.17) with $a = b = 0.1, c = 14$ and $\varepsilon = 2$.

the two pairs (which correspond to $-0.1009(0)$ and $-0.1025(5)$). We now show that this deviation is substantially a spurious effect, due to a subtle problem of numerical accuracy. In fact, it is easy to verify that the attractor is (for the parameter values as in Fig. 3.3) a perfectly homogenous (synchronous) state, $U_\ell(t) = U(t)$. Accordingly, the Ansatz $u_\ell(t) = \tilde{u}^{(k)}(t)e^{2\pi ik\ell/N}$ (where $k = 0, \ldots, N/2$) diaganolises the evolution equation in tangent space (see Section 9.2.3 for a general treatment); i.e. the eigendirections of the linear problem are the Fourier modes.

As a result, the dynamics of the kth mode is determined by three coupled linear equations for $\tilde{u}^{(k)} := (\tilde{x}^{(k)}, \tilde{y}^{(k)}, \tilde{z}^{(k)})$:

$$\dot{\tilde{x}}^{(k)} = -\tilde{y}^{(k)} - \tilde{z}^{(k)},$$
$$\dot{\tilde{y}}^{(k)} = \tilde{x}^{(k)} + a\tilde{z}^{(k)} + 2\varepsilon\left(\cos(2\pi k/N) - 1\right)\tilde{y}^{(k)}, \tag{3.9}$$
$$\dot{\tilde{z}}^{(k)} = \tilde{z}^{(k)}(X - c) + Z\tilde{x}^{(k)}.$$

As these equations have real coefficients, sine and cosine components are characterised by the same evolution, and this explains the existence of degenerate pairs in the spectrum. The only modes that are not characterised by this degeneracy are the 0th mode and the $(N/2)$th one. So, leaving aside the degeneracy, each Fourier mode is characterised by a triple of Lyapunov exponents. The numbers shown inside Fig. 3.3 close to the Lyapunov spectrum identify the corresponding Fourier modes. Long simulations reveal that the results obtained with the Fourier approach coincide with those obtained with the original method, except for the aforementioned pairs, where the Fourier approach leads to slightly different values, namely $-0.1016(8)$ and $-0.1018(9)$.

The analysis of the Lyapunov vectors (see Chapter 4 for a detailed discussion of these vectors) helps to explain the origin of the differences. In Fig. 3.4 we plot the instantaneous

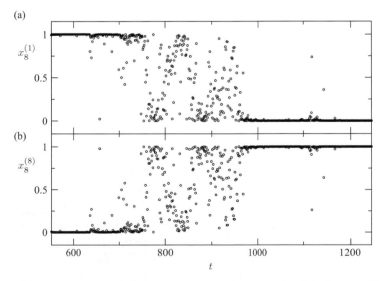

Fig. 3.4 Scalar product of the x component of the 8th Lyapunov vector (rescaled to norm 1) with the first (panel (a)) and eighth (panel (b)) Fourier mode versus time. The sum of the two terms is always equal to 1.

value of the scalar product of the x-component of the 8th Lyapunov vector (as obtained from the standard Gram-Schmidt orthogonalisation) with the 1st ($x_8^{(1)}$) and the 8th ($x_8^{(8)}$) Fourier mode: see the circles in the upper and lower panels, respectively.

Ideally, the two scalar products should be equal to 1 and to 0, respectively, because these two modes correspond to different Lyapunov exponents. However, the intermittent behaviour reveals that the vector occasionally aligns along the "wrong" direction (as the sum of the two scalar products is always equal to one, no other direction is involved in the process). The reason can be attributed to the fluctuations of the finite-time Lyapunov exponent. If, for some time, there is an inversion between the Lyapunov exponent of the first pair and that of the second pair (i.e. the vector corresponding to the smaller average LE grows temporarily faster than the one corresponding to the larger exponent), the 8th Lyapunov vector, due to numerical inaccuracies, is attracted towards the currently more expanding (but least expanding on average) direction. This results in an erroneously larger value of the Lyapunov exponent. In fact, the Lyapunov value obtained from the GS analysis, equal to $-0.01009(0)$, is larger than the true value $-0.01016(8)$. The difference is small but reveals a source of inaccuracy.

In order to further clarify this point, let us study a simple model: a product of two-by-two matrices, constructed in such a way that the eigendirections are constant, but the expansion rates fluctuate. As a result, the eigendirections temporally exchange their order. More precisely, we integrate Eq. (2.5), where the Jacobian K is given by the matrix

$$\mathsf{K} = \begin{pmatrix} s_+(t)p_2 + s_-(t)p_1 & -\sqrt{p_1 p_2}(s_+(t) - s_-(t)) \\ -\sqrt{p_1 p_2}(s_+(t) - s_-(t)) & s_+(t)p_1 + s_-(t)p_2 \end{pmatrix}, \tag{3.10}$$

where p_1 and p_2 are two positive numbers such that $p_1 + p_2 = 1$, while $s_1(t)$ and $s_2(t)$ are two time-dependent functions. One can easily check that the eigenvectors of K have the same direction independent of the value of s_1 and s_2, and, therefore, the Lyapunov exponents are trivially $\lambda_1 = \langle s_+ \rangle$ and $\lambda_2 = \langle s_- \rangle$. The presence of a noise acting on the evolution of the first Lyapunov vector can mimic a controllable accuracy of the numerical computations. We have chosen to add a bounded noise with a flat distribution between $-\delta$ and δ.

If we assume that $s_\pm = a_\pm \mp A \sin(t/T)$ with $a_+ > a_-$ and $A > (a_+ - a_-)/2$, it turns out that $\lambda_1 > \lambda_2$, but s_- becomes periodically (in some time interval) larger than s_+. Although such oscillations should not affect the value of the LE, we can see in Fig. 3.5a that the numerically computed LE varies with δ and may dramatically overestimate the expected value if $\delta > 10^{-7}$. The reason is precisely that the corresponding Lyapunov vector is not always aligned along the correct direction.

This can be seen in Fig. 3.5b, where the angle identifying the vector is plotted versus time for $\delta = 10^{-8}$. The orientation tends to switch between a low value ($-\tan^{-1}\sqrt{p_1/p_2}$), which corresponds to the expected direction of the largest Lyapunov vector, and a large value ($\tan^{-1}\sqrt{p_2/p_1}$), which corresponds to the direction of the second eigenvector. It is remarkable to see that even though the final value of the Lyapunov exponent is quite correct for this value of δ, the alignment is often wrong.

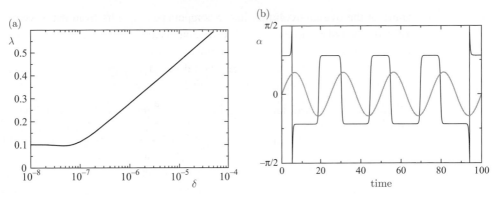

Fig. 3.5 (a) Lyapunov exponent obtained from Eq. (3.10) versus the accuracy δ for $p_1 = 0.4$, $a_\pm = \pm 0.1$, $A = 2$, $T = 25$. (b) Orientation of the first Lyapunov vector.

3.6 Systems with discontinuities

In some physical contexts, a continuous dynamics may be occasionally interrupted by sudden jumps, due to discontinuities in the basic equations, kicks which are modelled by terms containing δ-function, etc. This is typical, e.g. for billiards, where the main ingredients of the dynamics are collisions of a particle with either the boundary or another particle. Another setup where this phenomenon arises is that of neuron models, where the discontinuity is due to the idealisation of a short pulse (spike) as a δ-peak, whose absorption modifies the membrane potential instantly. Another example is that of stick-slip dynamics, where the discontinuity is induced by dry friction; many other examples can be found in engineering problems of machine dynamics (di Bernardo et al., 2008). In all such cases, the formalism for the computation of the Lyapunov exponents must be properly modified to account for the effect of the discontinuities. One might even doubt that the Lyapunov exponents are well defined in the presence of discontinuities. Fortunately, this is not the case; the reader interested in the details can consult Kunze (2000).

Here we start by analysing the simplest context where this problem arises, namely that of a kicked dynamical system, as this helps also to introduce the proper notations. Let us consider a generic dynamical system,

$$\dot{\mathbf{U}} = \mathbf{F}(\mathbf{U}, t), \tag{3.11}$$

and assume that its internal state undergoes an abrupt transition at time t_n,

$$\mathbf{U}^+(t_n) = \mathbf{G}(\mathbf{U}^-(t_n)), \tag{3.12}$$

where the superscripts "−" and "+" mean that we refer to a time just before and just after the transition. In this context, the computation of the Lyapunov exponents can be easily generalised by treating the continuous and discontinuous contributions separately. It turns out that the overall evolution in the tangent space results from the alternating application of two types of matrices,

$$P = \prod_{n=1}^{L} D(U^-(t_n)) H_n Q_0,$$

where H_n is the Jacobian arising from the integration of the differential equation (3.11) from time t_{n-1} to time t_n, and matrix $D = \frac{\partial G}{\partial U}$, appearing through linearisation of (3.12), describes the change of the perturbation at the abrupt transition.

Accordingly, the only required care is to apply the QR decomposition to the properly assembled sequence of linear transformations. Notice, also, that the kicks do not need to be evenly spaced in time for this approach to work.

When the amplitude and the direction of the kicks do not depend on the system configuration, $D \equiv I$ and no care at all is required, since the discontinuity plays no role in the tangent space. Another simple situation is when the evolution between consecutive kicks can be explicitly solved. This happens, e.g., when the only nonlinearity is contained in the kicks, while the continuous-time evolution is linear. The kicked rotor is a popular model that belongs to this class. It is described by the Hamiltonian

$$H = \frac{P^2}{2} + K \cos Q \sum_{n=-\infty}^{+\infty} \delta(t-n),$$

where the components of the state vector $U \equiv (Q, P)$ represent the position (Q) and the momentum (P), while the time scale is chosen so as to fix the time separation between kicks equal to 1. This model is sufficiently simple to allow for an exact integration of the equations of motion between consecutive kicks, since P stays constant and while Q increases linearly. If we denote with $Q(n)$ and $P(n)$ the values of the variables after the nth kick, then just prior to the next kick these quantities take values $P^-(n+1) = P(n)$ and $Q^-(n+1) = Q(n) + P(n)$. After the kick, Q is unchanged, while $P^+(n+1) = P^-(n+1) + K \sin Q^-(n+1)$. Altogether, one obtains the so-called Chirikov-Taylor standard map (A.6),

$$Q(n+1) = Q(n) + P(n) \tag{3.13}$$
$$P(n+1) = P(n) + K \sin(Q(n) + P(n)),$$

which allows the use of standard methods of LE analysis for discrete-time systems.

Another class of models, requiring more delicate handling, is the one where the occurrence of the discontinuity is not imposed externally (as in the previous case) but is self-determined, such as the collision of two disks moving in a two-dimensional plane. In full generality, the condition for the occurrence of a discontinuity can be expressed mathematically as a scalar equality $h(U) = 0$, which defines a codimension-one surface, the crossing of which triggers the discontinuity. In the collision of two circular disks, the condition is obtained by imposing that the distance between the centres of the two disks is equal to the sum of their radii.

In this more general class of models, the idea of breaking the time evolution into continuous and discontinuous components is still meaningful, but one must include the

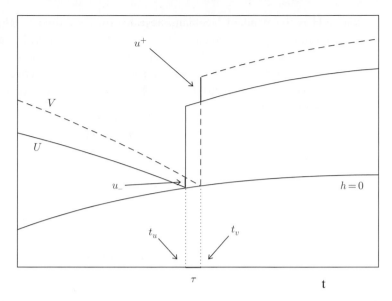

Fig. 3.6   Sketch of the evolution of two nearby trajectories (U and V) at a discontinuity represented by a collision with the line $h = 0$ (t_u and t_v denote their collision times). Here u_- and u^+ represent the separation just before and after the collision.

dependence of the collision time on the trajectory. A method to cope with this case was independently developed by Müller (1995) and Dellago and Posch (1995). Here we follow the former approach, as it is more appropriate for a general treatment. Let the scalar constraint

$$h(\mathbf{U}, t_u) = 0$$

define the condition for a discontinuity to occur at time t_u. For the sake of generality, we assume that the condition may explicitly depend on time. Moreover, let $\mathbf{F}$ and $\hat{\mathbf{F}}$ denote the velocity field before and after the collision, respectively (again, for the sake of generality, we assume that the two fields may differ, as in structurally variable systems (Müller, 1995)). Next, we consider a perturbed trajectory $\mathbf{V}(t) = \mathbf{U}(t) + \mathbf{u}(t)$, where $\mathbf{u}$ is an infinitesimal quantity. In general, it will hit the surface $h(\mathbf{U}, t_v) = 0$ at an infinitesimally different time $t_v = t_u + \tau$. Without loss of generality, we assume that $\tau > 0$, as pictorially depicted in Fig. 3.6 (if this is not the case, it is sufficient to exchange the two trajectories). The task consists in expressing the "final" value $\mathbf{u}^+$,[4] when both trajectories underwent the discontinuity, as a function of the initial condition $\mathbf{u}(0)$. Until the unperturbed trajectory reaches the surface $h(\mathbf{U}, t_u) = 0$, the perturbation evolves in a smooth way, and one can write

$$\mathbf{u}_- = \mathsf{H}(t_u)\mathbf{u}(0),$$

---

[4] In order to keep the notation as simple as possible, whenever there is no ambiguity, a subscript "–" ("+") implies that the variable is computed just before (after) the discontinuity of the unperturbed trajectory at time t_u. Superscripts refer in the same way to the second discontinuity – that of the perturbed trajectory.

where H is the standard Jacobian integrated over the time t_u. The next step consists in determining the perturbation amplitude at the "final" time t_v, by separately following the evolution of the reference and of the perturbed trajectory, respectively. On the one hand, by definition of the model, one can write, $\mathbf{U}_+ = \mathbf{G}(\mathbf{U}_-, t_u)$. Moreover, since τ is infinitesimal (and the evolution is smooth), $\mathbf{U}^+$ can be determined by expanding its evolution rule up to the first order only,

$$\mathbf{U}^+ = \mathbf{U}_+ + \tau \hat{\mathbf{F}}_+. \tag{3.14}$$

On the other hand, the perturbed trajectory evolves smoothly at first and then undergoes the discontinuity. As for the first part,

$$\mathbf{V}^- = \mathbf{V}_- + \tau \mathbf{F}(\mathbf{V}_-) = \mathbf{U}_- + \mathbf{u}_- + \tau \mathbf{F}_-, \tag{3.15}$$

where we have replaced $\mathbf{F}(\mathbf{V}_-)$ with $\mathbf{F}_- \equiv \mathbf{F}(\mathbf{U}_-)$ in the last term, since their difference would give rise only to higher (second) order corrections. After the discontinuity has occurred,

$$\mathbf{V}^+ = \mathbf{G}(\mathbf{V}^-, t + \tau) \approx \mathbf{G}\left(\mathbf{U}_- + \mathbf{u}_- + \tau \mathbf{F}_-, t + \tau\right)$$
$$\approx \mathbf{U}_+ + \mathsf{D}_-\left(\mathbf{u}_- + \tau \mathbf{F}_-\right) + \frac{\partial \mathbf{G}}{\partial t}\tau, \tag{3.16}$$

where we have again used that $\mathbf{U}_+ = \mathbf{G}(\mathbf{U}_-)$ and we assume, for the sake of generality, that $\mathbf{G}$ may have an explicit dependence on time. Finally, subtracting Eq. (3.14) from Eq. (3.16), we obtain

$$\mathbf{u}^+ = \mathsf{D}_-\mathbf{u}_- + \tau\left[\mathsf{D}_-\mathbf{F}_- - \hat{\mathbf{F}}_+ + \mathbf{G}_t\right]. \tag{3.17}$$

The first term in the r.h.s. is the same as in the previous non-autonomous regime, when the sequence of discontinuities was externally imposed at certain times rather than being self-determined during the evolution. The presence of an additional term, proportional to τ, reveals that the time shift indeed contributes to the tangent-space dynamics and must be properly taken into account. In order to complete the calculation, it is necessary to determine τ. This can be done by imposing the condition

$$h(\mathbf{V}^-, t_u + \tau) = 0. \tag{3.18}$$

By inserting Eq. (3.15) into Eq. (3.18) and noticing that $h(\mathbf{U}_-, t_u) = 0$ (this identifies the first discontinuity), one obtains

$$\left(\frac{\partial h}{\partial \mathbf{U}}\right)_- \cdot [\mathbf{u}_- + \tau \mathbf{F}_-] + \left(\frac{\partial h}{\partial t}\right)_- \tau = 0,$$

where the dot denotes a scalar product. As a result, the time separation between the discontinuities is given by the linear transformation

$$\tau = -\frac{\left(\frac{\partial h}{\partial \mathbf{U}}\right)_- \cdot \mathbf{u}_-}{\left(\frac{\partial h}{\partial \mathbf{U}}\right)_- \cdot \mathbf{F}_- + \left(\frac{\partial h}{\partial t}\right)_-}. \tag{3.19}$$

Eqs. (3.17, 3.19) provide a complete solution to the problem. One can formally combine them into a single compact relation,

$$\mathbf{u}_+ = \mathsf{S}\mathbf{u}_-,\tag{3.20}$$

where the matrix S is defined as

$$\mathsf{S} = \mathsf{D}_- - \frac{\left[\mathsf{D}_-\mathbf{F}_- - \hat{\mathbf{F}}_+ + \frac{\partial \mathbf{G}}{\partial t}\right]\left(\frac{\partial h}{\partial \mathbf{U}}\right)_-^\mathsf{T}}{\mathbf{F}_- \cdot \left(\frac{\partial h}{\partial \mathbf{U}}\right)_- + \left(\frac{\partial h}{\partial t}\right)_-}$$

and the superscript $()^\mathsf{T}$ means that the corresponding object is to be considered as a row vector, so that the numerator is altogether a matrix. The Lyapunov exponents can be computed by alternating the application of H and S matrices. Notice also that "−"s and "+"s appear only as subscripts in equation (3.20), indicating that the required variables have to be computed just before and after the jump experienced by the reference trajectory, as should be expected. From now on, since there is no ambiguity, the up/down position of $-/+$ symbol will be selected only to keep notations compact and simple.

Finally, we offer a different justification of Eq. (3.17), and for the sake of simplicity, we now drop the explicit time dependence. Given a generic trajectory and a perturbed one, they are typically characterised by a different h value at the same time. Since the discontinuity arises when the same zero h value is attained, it is necessary to "synchronise" the two trajectories; this can be done by shifting the perturbed trajectory along its own orbit,

$$\hat{\mathbf{u}} = \mathbf{u} + \tau \mathbf{F},\tag{3.21}$$

where τ is given by Eq. (3.19) and the derivatives are computed when the reference trajectory is characterised by a given but generic h value. If the condition $h = 0$ holds, one is entitled to apply the map $\mathbf{G}$, which, in tangent space, amounts to the linear transformation

$$\hat{\hat{\mathbf{u}}} = \mathsf{D}\hat{\mathbf{u}}.$$

As a last step, the trajectory must be "unfolded", bringing it back to the same time as that of the reference one. This amounts to repeating the step (3.21) with an opposite sign for the time τ and the proper velocity field in the new point in phase space (after the discontinuity has occurred). By combining the three steps together, one obtains Eq. (3.17). The presence of the first and the last step represent the difference with respect to the non-autonomous case, when the discontinuity occurs at the same time for all trajectories. In practice, the first (last) step can be performed at any time before (after) the discontinuity, provided that across the discontinuity the equations are integrated using h as the independent variable (see Chapter 2); in fact, this approach would allow maintaining the same h value, once a trajectory has been synchronised. Even more, if h turned out to be a proper global phase, one could even avoid the first and last steps, thus reducing the treatment of the discontinuity to that of the forced case.

An alternative approach consists in reducing the continuous-time evolution to a discrete-time one in between two consecutive Poincaré sections. In this case, the initial point is assumed to lie on the surface $h = 0$, which means that we deal with an $(N - 1)$-dimensional phase space. Like in the previous case, the perturbation u is smoothly evolved until time t_u, synchronised with the reference trajectory and then evolved according to the transformation D. The advantage of this method is that one does not need the third unfolding step, since the trajectory must, by definition, lie on the surface $h = 0$. This approach has been implemented in neural network dynamics (Zillmer et al., 2006). Notice also that in large systems, where each non-smooth event involves just a few variables, the advantage is no longer such, since, in the absence of the D transformation, the third step compensates the first one; i.e. the non-involved variables need not be touched in the continuous-time scheme.

3.6.1 Pulse-coupled oscillators

A large class of models where discontinuities may arise is that of pulse-coupled oscillators, where each oscillator is described by a phase-like variable which, upon reaching a threshold, is reset and, at the same time, a δ-pulse is sent to the connected oscillators. As an example, we now discuss two leaky integrate-and-fire neurons

$$\dot{U}_1 = a - U_1 - g \sum_j \delta(t - t_j^{(2)}) \tag{3.22}$$

$$\dot{U}_2 = a - U_2 - g \sum_j \delta(t - t_j^{(1)}). \tag{3.23}$$

When the variable U_1 (U_2) reaches the value 1 (at times $t_j^{1,2}$), it is reset to zero and, simultaneously, sends a spike to the other oscillator. The discontinuous transformation (we discuss the case when the first neuron reaches the threshold – the other being symmetric) is $\mathbf{G}(U_1, U_2) = (0, U_2 - g)$, so that the Jacobian is

$$\mathsf{D} = \begin{pmatrix} 0 & 0 \\ 0 & 1 \end{pmatrix}.$$

As a result,

$$\mathsf{D}\mathbf{F}_- - \mathbf{F}_+ = \begin{pmatrix} -a \\ g \end{pmatrix}.$$

The discontinuity is identified by the condition $h \equiv U_1 = 0$, so that $\frac{\partial h}{\partial \mathbf{U}} = (1, 0)$ and the time shift is

$$\tau = \frac{u_1^-}{a - 1},$$

while the final transformation reads

$$u_1^+ = -\frac{a}{1 - a} u_1^-,$$

$$u_2^+ = u_2^- + \frac{g}{a - 1} u_1^-. \tag{3.24}$$

In the case of large ensembles of neurons, at each spike emission, it is sufficient to update the perturbations of the involved variables (the one that is reset to zero and those of the neurons which receive the spike). This approach has been, e.g., implemented by Monteforte and Wolf (2010) to study chaotic properties of various neural networks.

Finally, notice that additional complications may arise if the neurons are assumed to undergo a refractory period after the reset, during which their potential stays constant to the reset value (Zhou et al., 2010).

3.6.2 Colliding pendula

In this section we consider a system of two pendula of equal mass in a gravitational field (see Fig. 3.7a). If they are sufficiently long, they can be viewed as moving horizontally in a harmonic potential. Additionally, the left pendulum is subject to a sinusoidal forcing and to a viscous damping. Altogether, in between collisions, the equations of motion are (here, $\mathbf{U} = (x_1, x_2, v_1, v_2)$)

$$
\begin{aligned}
\dot{x}_1 &= v_1 & \dot{x}_2 &= v_2 \\
\dot{v}_1 &= -\alpha x_1 - \gamma v_1 + A \cos \omega t & \dot{v}_2 &= -\alpha (x_2 - \delta),
\end{aligned}
\tag{3.25}
$$

where γ is the strength of the viscous drag, α is the amplitude of the restoring force and δ the spatial separation between the two pendula.

Whenever $x_1 = x_2$, the two pendula undergo an elastic collision, which induces a perfect exchange of their momenta, while the coordinates remain unchanged. The discontinuous transformation $\mathbf{G}$ is linear and characterised by the following 4×4 matrix

$$
\mathsf{D} = \begin{pmatrix} \mathsf{I} & 0 \\ 0 & \mathsf{E} \end{pmatrix},
$$

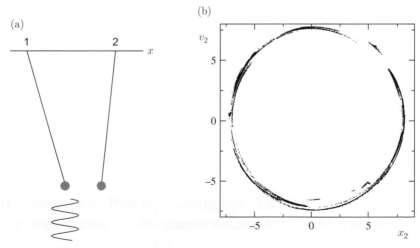

Fig. 3.7 (a) Two pendula: pendulum 1 is modulated and forces pendulum 2. (b) The chaotic dynamics exhibited by the second pendulum (position and momentum of the second particle are reported) for $\alpha = A = \delta = 1$, $\omega = 1.05$ and $\gamma = 0.13$ (see Eq. (3.25)).

where I and 0 are the identity and null 2×2 matrices, while

$$\mathbf{E} = \begin{pmatrix} 0 & 1 \\ 1 & 0 \end{pmatrix}.$$

Simple algebra shows that

$$\mathbf{DF}_- - \mathbf{F}_+ = \begin{pmatrix} \Delta v \\ -\Delta v \\ C + \gamma v_2^- \\ -C - \gamma v_1^- \end{pmatrix}, \tag{3.26}$$

where $\Delta v = v_1^- - v_2^-$ and $C = \alpha - A \cos \omega t_u$. The discontinuity is identified by the condition $h(\mathbf{U}) = x_1 - x_2 = 0$, so that $h_{\mathbf{U}} = (1, -1, 0, 0)$ and the infinitesimal time shift is

$$\tau = -\frac{u_1^- - u_2^-}{\Delta v}. \tag{3.27}$$

By inserting Eqs. (3.26, 3.27) into Eq. (3.17), one finally obtains the linear transformation induced by the discontinuity:

$$
\begin{aligned}
u_1^+ &= u_2^-, \\
u_2^+ &= u_1^-, \\
u_3^+ &= u_4^- - (C + \gamma v_2^-)(u_1^- - u_2^-)/\Delta v, \\
u_4^+ &= u_3^- + (C + \gamma v_1^-)(u_1^- - u_2^-)/\Delta v.
\end{aligned} \tag{3.28}
$$

When the oscillations of both pendula are small enough, there are no collisions and the dynamics is ordered: the first pendulum oscillates with the period of the forcing term, while the second one according to its natural frequency. However, as soon as collisions come into play, a chaotic dynamics may be induced by the nonlinear character of the kicks, as shown in Fig. 3.7b.

The implementation of this described algorithm shows that the first LE is positive, $\lambda_1 \approx 0.0084$. It is instructive to notice that the correct implementation is crucial even from a qualitative point of view: if one disregards the collision, the maximum Lyapunov exponent would vanish, while a treatment analogous to the non-autonomous case (i.e., without synchronisation and unfolding) would even yield a negative result ($\lambda_1 \approx -0.022$).

Finally, notice that in the case of many particles, it is sufficient to transform only the variables of the particles involved in the given collision. A chain of elastically colliding harmonic oscillators has been investigated by Sano and Kitahara (2001), while a Lorentz gas has been studied by Dellago and Posch (1995).

3.7 Lyapunov exponents from time series

The methods discussed in the previous sections for the computation of the Lyapunov exponents assume that the evolution rule is known (either as a differential equation or as a recursive map). In typical experimental contexts, however, the equations of motion are, at best, known only approximately. It is therefore desirable to develop methods for the determination of the LEs directly from the time series of suitable observables.

LEs are dynamical invariant; i.e. they are independent of the variables chosen to describe a given physical system. It is, however, necessary to access sufficiently many variables to reconstruct the underlying dynamics. Takens (1981) proposed an "embedding" technique, which allows proceeding even when a single scalar observable is available. Given a sequence $\{U(i\tau)\}$ of measurements of a D-dimensional attractor, one can construct the m-tuple

$$\mathbf{U}_i^{(m)} = (U(i\tau)), U((i+1)\tau), \ldots, U((i+m-1)\tau),$$

where, for the sake of simplicity, we have assumed that the time separation between consecutive elements coincides with the sampling time τ. Accordingly, the original scalar time series is embedded into $\mathbb{R}^m$. If $m > 2D$, the attractor itself is faithfully unfolded; i.e. distinct points are parametrised by different m-tuples.

As a result, the corresponding evolution rule can be expressed as a mapping from time $k\tau$ to $(k+1)\tau$,

$$\mathbf{U}_{k+1}^{(m)} = \mathbf{F}^{(m)}[\mathbf{U}_k^{(m)}]$$

where $\mathbf{F}^{(m)}$ denotes the unknown transformation of $\mathbb{R}^m$. The corresponding evolution in tangent space is ruled by the recursive equation

$$\mathbf{u}_{k+1}^{(m)} = \frac{\partial \mathbf{F}^{(m)}}{\partial \mathbf{U}^{(m)}} \mathbf{u}_k^{(m)} \equiv \mathsf{J}_k^{(m)} \mathbf{u}_k^{(m)},$$

where the Jacobian has the following simple structure

$$\mathsf{J}_k^{(m)} = \begin{pmatrix} 0 & 1 & 0 & \cdots & 0 \\ 0 & 0 & 1 & \cdots & 0 \\ \vdots & \vdots & \vdots & \ddots & \vdots \\ 0 & 0 & 0 & \cdots & 1 \\ a_1 & a_2 & a_3 & \cdots & a_m \end{pmatrix}.$$

In practice, the only nontrivial components are those of the last row, where we have dropped the dependence on the index k for the sake of simplicity.

Let $\mathbf{U}_k^{(m)}$ and $\mathbf{U}_n^{(m)}$ denote two generic points on the attractor. If they are close enough, one can expand the evolution rule, writing

$$\mathbf{U}_{n+1}^{(m)} - \mathbf{U}_{k+1}^{(m)} = \mathsf{J}_k^{(m)}(\mathbf{U}_n^{(m)} - \mathbf{U}_k^{(m)}) + \text{h.o.t.}$$

To the extent that the higher order terms (h.o.t.) can be neglected, this equation tells us that the Jacobian can be reconstructed from the evolution of nearby points. In principle, m nontrivial components of the Jacobian can be determined by identifying and following in time m neighbours of the reference point $\mathbf{U}^{(m)}(k)$. In practice, it is more reliable to identify a larger number of neighbours and thereby proceed with a least-square fit by minimising

$$S_k = \sum_n \left[U((n+m)\tau) - U((k+m)\tau) - \mathbf{a} \cdot (\mathbf{U}_n^{(m)} - \mathbf{U}_k^{(m)}) \right]^2,$$

where the sum is restricted to all points $n \neq k$ that are close enough to the reference point k.

In some cases the computation of the LEs is affected by relatively large errors. This is particularly true for the negative exponents, as they refer to directions that are poorly explored by the dynamics. Some improvements can be obtained by adopting a series of precautions such as the selection of a not-too-small τ-value (to avoid Jacobian eigenvalues too close to 1, which degrade the accuracy) or the discard of temporally close points (which may bias the neighbour statistics). The reader interested in such details can refer to Kantz and Schreiber (2004).

A more fundamental difficulty arises from the fact that m may be larger than the dimension of the space actually spanned by the invariant measure. As a result, the reconstructed dynamics is characterised by a certain number of additional "spurious" exponents associated with directions that are everywhere transversal to the attractors. They have no physical meaning, as they do not correspond to any degree of freedom but just follow from the selection of variables adopted to represent the underlying attractor. A theoretical study performed within the limits of infinitesimal perturbations suggests that the spurious exponents are suitable combinations of the true ones (Sauer et al., 1998). This property, however, breaks for finite perturbations, even small ones.

More robust information can be extracted from the local direction of the corresponding covariant Lyapunov vectors (see Chapter 4 for their definition). In fact, so long as a given Lyapunov vector is transversal to the invariant measure, the corresponding LE is spurious, and it has thereby to be discarded. This idea, first proposed by Brown et al. (1991), has been revisited by Kantz et al. (2013). Here, we illustrate the approach with reference to the simple Hénon map (A.4) for the standard parameter values $a = 1.4$ and $b = 0.3$.

A time series containing $N = 6 \times 10^5$ points is used to embed the attractor in a space of dimension $m = 5$, while the local Jacobian is reconstructed by using all neighbours within a ball of radius $r = 0.001$. The Jacobians are thereby multiplied in a standard way to determine five Lyapunov exponents (after discarding a suitable transient). The five values correspond to the full circles in Fig. 3.8 (see the x coordinate); the dashed lines mark the true LEs, while the dotted-dashed lines correspond to the linear combinations that are

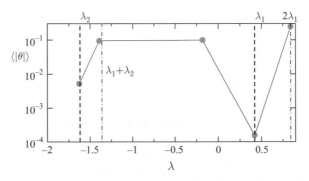

Fig. 3.8   Average angles between covariant vectors and the invariant measure – obtained from a five-dimensional reconstruction of the Hénon map – versus the actual value of the Lyapunov exponents. (Data: courtesy of H. Kantz, G. Radons and H. Yang.)

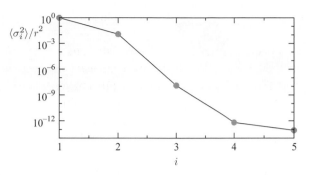

Fig. 3.9 Rescaled singular values for the Hénon map (A.4) in boxes of radius $r = 0.001$. (Data: courtesy of H. Kantz, G. Radons and H. Yang.)

predicted by the theory (Sauer et al., 1998). One can see that there exists a slightly negative LE (≈ -0.1) that is not predicted by the theory.

As a next step, the covariant vector corresponding to each LE is estimated (see Chapter 4 for a description of the method). Simultaneously, the angle θ between the vector itself and the manifold locally spanned by the experimental points is calculated. The underlying idea, originally proposed by Broomhead et al. (1987), is to locally approximate the manifold covered by the attractor with a linear subspace to thereby identify the directions actually covered by the attractor. In practice, given the points which fall within each prescribed ball, singular value decomposition is applied to determine the variance σ^2 along the main axes.

The singular values are thereby rescaled and averaged over the entire attractor. The results are plotted in Fig. 3.9. There, we see that the last three are really negligible (they are non-zero because of the nonlinearities, which induce a bending of the linear subspace; moreover in real applications we can expect the presence of observational noise, which induces fluctuations along any direction, whether spurious or not). In this case, it is clear that the physically relevant invariant measure is restricted to the first two principal axes. One can therefore locally determine the angle θ between the ith covariant vector and the subspace identified by the first two principal axes. The average absolute value $\langle|\theta|\rangle$ is reported in Fig. 3.8 (see the vertical coordinate). There we see that only two of them are close to zero and therefore correspond to physically relevant directions. This is confirmed by the actual values of the LEs that coincide with the known exponents of the Hénon map. The other three exponents are spurious: these include those two values that could be identified as linear combinations of the true values and the intermediate negative exponent that could not be spotted otherwise.

In full generality, it should be stressed that this criterion also suffers some problems when applied to real world data. In particular, we refer to the identification of the directions that are truly spanned by the invariant measure. Already in the simple setup of the Hénon map, in the absence of observational noise, the average angle between the negative exponent and the attractor manifold is not as small as one might hope for. The determination of whether a direction is covered or not by a given attractor is certainly more tricky in truly experimental data.

4 Lyapunov vectors

The linear stability analysis of fixed points involves the computation not only of eigenvalues but also of eigenvectors. Equivalently, a complete characterisation of chaotic or stochastic dynamical systems requires going beyond the knowledge of the LEs, including the identification of the (local) orientation of stable and unstable manifolds.

An eigenvector of a given linear transformation can be identified as a direction that is mapped onto itself. This definition cannot be straightforwardly extended to contexts where different transformations are applied at different times, as no invariant direction is expected to exist. It is, however, possible to rephrase the definition in such a way that a generalisation becomes possible. Eigenvectors can, in fact, be viewed as the only directions which, if iterated forwards and backwards in time, are accompanied by an expansion rate which coincides with the eigenvalues of the given matrix (here, for the sake of simplicity, we assume that no complex eigenvalues exist). In this definition, the very fact that the direction itself is invariant becomes a secondary property. Accordingly, it can be extended to any sequence of matrices, requiring that the observed average expansion rate has to coincide with one of the LEs of the given system. Such directions, often referred to as *covariant Lyapunov vectors* in the physics literature, are nothing but the vectors $\mathbf{E}^k$ introduced in Section 2.3.2, while referring to the Oseledets splitting. They had been introduced already by Oseledets (1968) and later formalised as tangent directions of invariant manifolds (Ruelle, 1979) but for many years escaped the attention of researchers, probably due to the lack of effective algorithms to determine them. The first computation of covariant vectors was performed in the context of time-series analysis (see Section 3.7) (Bryant et al., 1990; Brown et al., 1991). Since then, covariant vectors have been occasionally used as a tool to determine Lyapunov exponents via a transfer matrix approach (Politi et al., 1998) or to characterise spatio-temporal chaos (Kockelkoren, 2002). Only after the development of two effective computational methods (Ginelli et al., 2007; Wolfe and Samelson, 2007), the usefulness of covariant vectors was eventually recognised.

Here, we provide a heuristic discussion of the subject, while a more formal introduction is given in Section 4.1. For simplicity, we assume that all of the LEs of an N-dimensional system are different. We start by considering a generic initial perturbation. One expects it to have components along all of the $\mathbf{E}^k$ directions; therefore its forwards evolution is dominated by the largest Lyapunov exponent; i.e. after some transient, the perturbation aligns along the most unstable direction $\mathbf{E}^1$. Accordingly, the determination of $\mathbf{E}^1$ is a trivial task as it "attracts" generic initial conditions. Analogously, a generic perturbation, when iterated backwards in time, aligns along the most "expanding" direction in negative times, i.e., along $\mathbf{E}^N$, which corresponds to the smallest Lyapunov exponent of the original system. Therefore, if $N = 2$, this completes the task of finding the covariant Lyapunov vectors. Notice that if one is interested in knowing the two covariant vectors in a given point of phase space, it is necessary to know a long trajectory, which includes the point under consideration and extends into the far past as well as into the far future.

In higher-dimensional systems, new approaches are necessary. A possible strategy is as follows. As discussed in Chapter 3, the second Lyapunov exponent λ_2 can be determined by following two linearly independent vectors and orthogonalising them to avoid the alignment of both of them along $\mathbf{E}^1$. As a result, one obtains two vectors that span the plane $(\mathbf{E}^1, \mathbf{E}^2)$, where the second vector is not aligned along $\mathbf{E}^2$ but is orthogonal to $\mathbf{E}^1$. This information can, however, be used to determine $\mathbf{E}^2$: it is sufficient to consider an arbitrary vector in the plane $(\mathbf{E}^1, \mathbf{E}^2)$ and iterate it backwards. In fact, because $\mathbf{E}^2$ is the least expanding direction in this restricted space, it "attracts" the evolution of a generic initial condition when iterated backwards in time, where it becomes the most expanding direction. Thus, the direction of the second covariant vector can be determined by combining forward and backward iterations of suitable perturbation vectors. This is the core of the dynamical algorithm that is extensively illustrated later in this chapter (together with other approaches).

4.1 Forward and backward Oseledets vectors

We will start by revisiting the definition of Lyapunov exponents given in Chapter 2. As, here, we consider backward and forward transformations, it is necessary to slightly change the notations.

So, let us denote with $\mathsf{H}(t', t'')$ the Jacobian matrix ruling the evolution in tangent space from time t' to time t'', along the trajectory $\mathbf{U}(t)$. From Chapter 2 and, in particular, from Eq. (2.12), the Lyapunov exponents λ_j can be identified from the eigenvalues of the matrix $\mathsf{M}^+(t', t'') = \mathsf{H}^\mathsf{T}(t', t'')\mathsf{H}(t', t'')$ (in the limit $(t'' - t') \to +\infty$),

$$\mathsf{M}^+(t', t'')\boldsymbol{\xi}^{(j)}(t', t'') = \mathrm{e}^{2\lambda_j(t''-t')}\boldsymbol{\xi}^{(j)}(t', t''), \tag{4.1}$$

where the eigenvectors $\boldsymbol{\xi}^{(j)}(t', t'')$ belong to the tangent space in $\mathbf{U}(t')$. In fact, the action of $\mathsf{M}^+(t', t'')$ corresponds to the application of $\mathsf{H}(t', t'')$, which maps the initial vector from time t' to time t'', followed by the application of $\mathsf{H}^\mathsf{T}(t', t'')$, which maps it back to time t' (the superscript "+" in M^+ helps to remind us that the matrix operates in the future). The eigenvectors $\boldsymbol{\xi}^{(j)}(t', t'')$ are nothing else as than the *right singular vectors* of the matrix $\mathsf{H}(t', t'')$.

Similarly, one can define the *left singular vectors* $\boldsymbol{\psi}^{(j)}$ of $\mathsf{H}(t', t'')$ from the eigenvalue equation

$$\mathsf{M}^-(t', t'')\boldsymbol{\psi}^{(j)}(t', t'') = \mathsf{H}(t', t'')\mathsf{H}^\mathsf{T}(t', t'')\boldsymbol{\psi}^{(j)}(t', t'') = \mathrm{e}^{2\lambda_j(t''-t')}\boldsymbol{\psi}^{(j)}(t', t''). \tag{4.2}$$

The vectors $\boldsymbol{\psi}^{(j)}$ are associated to the tangent space in $\mathbf{U}(t'')$, as also confirmed by the following standard equations

$$\mathsf{H}(t',t'')\boldsymbol{\xi}^{(j)}(t',t'') = e^{\lambda_j(t''-t')}\boldsymbol{\psi}^{(j)}(t',t'')$$
$$\mathsf{H}^{\mathsf{T}}(t',t'')\boldsymbol{\psi}^{(j)}(t',t'') = e^{\lambda_j(t''-t')}\boldsymbol{\xi}^{(j)}(t',t''),$$

(4.3)

which show how the two vectors are related to one another.

Since the matrix $\mathsf{M}^+(t',t'')$ is symmetric, its eigenvectors are mutually orthogonal, but they depend on the coordinates chosen to represent the dynamical evolution. This indeterminacy disappears in the limit $t'' \to \infty$, when the right singular vectors converge to the so-called *forward Oseledets* vectors $\mathbf{v}_+^{(j)}(t') = \boldsymbol{\xi}^{(j)}(t',+\infty)$. This is telling us that the the Oseledets vectors are well-defined objects attached to the initial point $\mathbf{U}(t')$ (see Ershov and Potapov (1998) for a rigorous analysis).

Similarly, with reference to the matrix $\mathsf{M}^-(t',t'')$, if one takes the limit $t' \to -\infty$, the left singular vectors converge to the *backward Oseledets* vectors $\mathbf{v}_-^{(j)}(t'') = \boldsymbol{\psi}^{(j)}(-\infty,t'')$.

Rigorously speaking, the Oseledets vectors can be uniquely identified only in the absence of degeneracies, i.e. when no two (or more) consecutive LEs are equal to one another. In such cases, one should, more properly, refer to suitable subspaces of dimension equal to the level of degeneracy. Since taking into account the presence of degeneracies would involve the introduction of heavier notations, here, for the sake of simplicity, we assume they are absent (the reader interested in a more complete treatment can consult Ginelli et al. (2013)).

As a result, in any point $\mathbf{U}(t)$ of the phase space, one can define two sets of Oseledets vectors: by considering the time interval (t',t) $[(t,t'')]$ and taking the limit $t' \to -\infty$ $[t'' \to +\infty]$ one obtains $\mathbf{v}_-^{(j)}(t)$ $[\mathbf{v}_+^{(j)}(t)]$. Moreover, from the relation between the left and right singular vectors (4.3), there exists a connection between them,

$$\mathbf{v}_-^{(j)}(t'') = \mathsf{H}(t',t'')\mathbf{v}_+^{(j)}(t'),$$

up to a scaling factor, if $t'' - t'$ is large enough.

In practice, as already noted in the previous chapter, it is not convenient to determine Lyapunov exponents by diagonalising the matrices in Eqs. (4.1, 4.2). It is preferable to implement some form of QR decomposition, which also generates orthogonal bases. The bases obtained by following the evolution forwards in time (with the help of either the Gram-Schmidt orthogonalisation or Householder reflections) converge to the backward Oseledets vectors (in the same phase points). This seemingly odd relationship is due to the fact that the "backward" vectors are termed as such because they depend on the past, so they are naturally obtained by reaching a given point from previous states. Analogously, if one generates a long trajectory in the far future and then moves backwards in tangent space, implementing again a QR decomposition up until the point of interest, the forward Oseledets vectors are obtained (although their identification requires moving backwards!).

The most serious criticism that can be made against both sets of vectors is that, with the exception of the first vector, they are not covariant, i.e. $\mathbf{v}_\pm^{(j)}(t'') \neq \mathsf{H}(t',t'')\mathbf{v}_\pm^{(j)}(t')$.

This can be appreciated in Fig. 4.1, where one can compare the evolution of the first two backward vectors. There, we see that $\mathbf{v}_-^{(1)}(n+1)$ is, by construction, nothing but a rescaled version of $\mathsf{H}(t_n,t_{n+1})\mathbf{v}_-^{(1)}(n)$; so it is true that it is mapped onto itself (except for a scaling factor). The vector $\mathbf{v}_-^{(2)}(n+1)$ is instead obtained after subtracting from $\mathsf{H}(t_n,t_{n+1})\mathbf{v}_-^{(2)}(n+1)$ its component parallel to $\mathbf{v}_-^{(1)}(n+1)$.

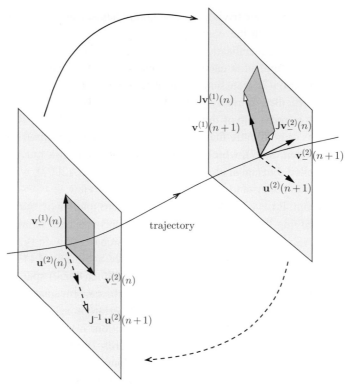

Fig. 4.1 Schematic representation of the evolution of the backward Oseledets vectors $\mathbf{v}_-^{(j)}(n)$ and of the dynamical algorithms for the determination of the covariant Lyapunov vectors (because here we depict one-step evolution, we refer to the local Jacobian J instead of the product H).

4.2 Covariant Lyapunov vectors and the dynamical algorithm

A truly covariant Lyapunov vector (CLV) of an N-dimensional system can be defined by intersecting the subspaces identified by the $\mathbf{v}_\pm$ vectors. Let $(\Xi^{(i)}(t))^-$ be the subspace spanned by the first i eigenvectors of M^-. This subspace at time t contains all perturbations that grow with the Lyapunov exponents $\lambda_1, \ldots, \lambda_i$; i.e. it coincides with the subspace that, in terms of the Oseledets splitting is $E^1 \oplus \cdots \oplus E^i$. Then, let $(\Xi^{(i)}(t))^+$ be the space spanned by the last $N+1-i$ eigenvectors of M^+; in terms of the Oseledets splitting it corresponds to $E^i \oplus \cdots \oplus E^N$. The (one-dimensional) intersection of the two subspaces yields the covariant or characteristic Lyapunov vector as it has been defined by Ruelle (1979):

$$\mathbf{u}^{(i)}(t) = (\Xi^{(i)}(t))^+ \cap (\Xi^{(i)}(t))^-. \tag{4.4}$$

Invariance of the vector $\mathbf{u}^{(i)}(t)$ under time evolution follows from the invariance of the spaces $(\Xi^{(i)})^\pm$.

When the same matrix $\mathsf{J}(t) = \mathsf{A}$ is applied at all times, it is easy to convince oneself that $\mathbf{u}^{(i)}(t)$ reduce to the standard eigenvectors (provided that no degenerate eigenvalues are present): they are trivially mapped onto themselves (apart from an irrelevant scaling factor associated with the corresponding eigenvalue). In general, the vectors $\mathbf{u}^{(i)}(t)$ represent the proper generalisation of the concept of eigenvectors to a context where a different matrix is applied at each time step. In order to transform this definition into an effective algorithm, it is necessary to generate forward and backward Oseledets vectors. Because of this difficulty, for many years CLVs have been determined only sporadically, as mentioned in the introduction of this chapter.

It is only recently that two effective algorithms appeared (Ginelli et al., 2007; Wolfe and Samelson, 2007), which make the computational task much easier. Here, we describe the so-called dynamical algorithm from Ginelli et al. (2007), which is convincingly stable and easier to understand from a physical point of view.

From Eq. (4.4), we see that the determination of the second CLV $\mathbf{u}^{(2)}(t)$ requires identifying the covariant (one-dimensional) subspace obtained by intersecting $(\Xi^{(2)}(t))^-$ with $(\Xi^{(2)}(t))^+$. The former two-dimensional subspace is easily obtained by iterating the first two backward Oseledets vectors (see again Fig. 4.1). As for $(\Xi^{(2)}(t))^+$, there is no need to build the entire (typically, high-dimensional) space, but it is sufficient to restrict the study to $(\Xi^{(2)}(t))^-$ when following the tangent space evolution backwards in time. In fact, for the same reason that a generic perturbation, when iterated forwards in time, tends to align along the most expanding direction, it tends to align along the least expanding (most contracting) when iterated backwards. If the evolution is constrained to the subspace $(\Xi^{(2)}(t))^-$, this means that it aligns along the second most expanding directions (since $-\lambda_2 > -\lambda_1$), i.e. along the second covariant direction.

Now, we transform this way of reasoning into a general algorithm. Let $\mathbf{w}^{(i)}(t)$ denote a generic vector embedded in the subspace spanned by the first i backward Oseledets vectors $\mathbf{v}_-^{(j)}(t)$. As the latter vectors represent an orthonormal basis, one can write

$$\mathbf{w}^{(i)}(t) = \sum_{j=1}^{i} c^{(j,i)}(t)\mathbf{v}_-^{(j)}(t), \tag{4.5}$$

where $c^{(j,i)}(t) = \langle \mathbf{w}^{(i)}(t)|\mathbf{v}_-^{(j)}(t)\rangle$. By denoting with $\mathbf{c}^{(i)}(t)$ the set of coordinates of $\mathbf{w}^{(i)}(t)$, one can express the iteration from time t to time $t+1$ as

$$\mathbf{c}^{(i)}(t+1) = \mathsf{R}^{(i)}(t)\mathbf{c}^{(i)}(t),$$

where the $i \times i$ upper triangular matrix $\mathsf{R}^{(i)}(t)$ corresponds to the upper-left block of the matrix $\mathsf{R}(t)$, determined by means of the QR decomposition.

The ith covariant vector is obtained by iterating backwards this equation,

$$\mathbf{c}^{(i)}(t) = (\mathsf{R}^{(i)})^{-1}\mathbf{c}^{(i)}(t+1), \tag{4.6}$$

for a sufficient number of times to allow for the transient to die out. $(\mathsf{R}^{(i)}(t))^{-1}$, too, is an upper triangular matrix, whose diagonal elements are the inverse of those of $\mathsf{R}^{(i)}(t)$; they, in fact, contain the information on the volume-contraction rates. Its off-diagonal elements are crucial to identify the orientation of the CLV. In practice, it is not necessary to perform an explicit matrix inversion; backward iteration can be efficiently performed with back-substitution algorithms, starting from the last ith component of $\mathbf{c}^{(i)}(t)$.

4.3 Dynamical algorithm: numerical implementation

The various steps of the dynamical algorithm can be summarised as follows:

Step 1 – The first step is fully equivalent to the one needed for the computation of the Lyapunov exponents. Here, we briefly summarise the procedure for later convenience. A given initial condition in phase space and a set of $m \leq N$ orthogonal tangent-space vectors $\mathbf{v}^{(i)}(0)$, $i = 1, \ldots, m$, are allowed to evolve in phase- and tangent-space dynamics (by applying the QR decomposition described in Chapter 3), respectively. This evolution should last a time interval t' long enough to allow the trajectory and the tangent vectors to converge to the attractor (in dissipative models) and to the backward Oseledets vectors, respectively.

Step 2 – Both the trajectory and the backward Oseledets vectors are evolved forwards, from time t' until time t'''. During this evolution the components of the backward Oseledets vectors $\mathbf{v}_-(t)$, as well as the elements of the triangular matrices $\mathsf{R}(t)$, are recorded at each step. The stored information is necessary for the execution of the following two steps.

Step 3 – A random set of vectors $\mathbf{c}^{(i)}(t''')$ ($i = 1, m$) is selected (notice that $\mathbf{c}^{(i)}(t''')$ has i components) and evolved backwards via Eq. (4.6) until time t'', which lies between t' and t'''. The backward transient time $t''' - t''$ should be long enough to allow each vector to converge to the corresponding covariant vector. At variance with the forward evolution, here each vector is iterated independently and is only rescaled (from time to time) to avoid too large numbers.

Step 4 – The backward evolution continues, yielding the proper covariant vectors for $t' \leq t \leq t''$; notice that the $\mathbf{c}^{(i)}(t)$ ($i = 1, m$) coordinates are automatically expressed with reference to the local backward Oseledets basis.

A nice property of this algorithm (shared also by the methods described next) is that the computation of the first m CLV requires only the iteration (forwards and backwards) of m vectors; there is no need to explore other directions. If the dynamics is invertible, the last m vectors can be equivalently obtained without evolving the first $N - m$ vectors. It is necessary first to evolve the real space dynamics to generate a faithful trajectory and thereby follow this scheme by exchanging forward with backward evolution.

If the Lyapunov spectrum is degenerate, some covariant subspaces have a dimension larger than one and the individual vectors have no physical meaning. In this case, this algorithm will simply return some arbitrary set of independent vectors which depend on the selection of tangent vectors made at time t''. A pseudocode describing the entire procedure is illustrated in Appendix B.

Memory

The algorithm requires the storage of the $\mathsf{R}(t)$ matrices and the $\mathbf{v}_-^{(i)}(t)$ vectors to be used during the backward evolution from t''' to t'. The $\mathsf{R}(t)$ matrices needed to run the backward dynamics involve $m(m+1)/2$ floating-point numbers, while the $\mathbf{v}_{-}^{(i)}(t)$ vectors require mN floating-point numbers. Thus, taking into account that $h = |t''' - t'|$ sampling points are to be considered, the total number M_T of floating-point numbers is

$$M_T = h\,m\left[N + \frac{m+1}{2}\right].$$

This burden can be sensibly reduced if one is not interested in the structure of the CLV in the original basis, but just in the mutual angles, since the $\mathbf{v}_{-}^{(i)}(t)$ vectors stored in the second step are not needed for the backward iteration. In this case, the hmN term in the r.h.s. of this equation can be disregarded. A large amount of memory may, nevertheless, be needed, if the phase-space dimension is large and/or many vectors are required, exhausting the capacity of the fast-access memory.

In such a case, in order to avoid frequent calls to a slow-access memory (e.g. disk storage), it is convenient to split the computation into separate blocks. Once the maximum amount of data M_b that can be stored in the fast-access memory is determined, divide the total number h of forward steps into n_b blocks of length h/h_b such that $M_T/n_b \leq M_b$. Then, step 2 of the algorithm is replaced by a forward run, where the $\mathsf{R}(t)$ matrices and the $\mathbf{v}_{-}^{(i)}(t)$ vectors are not stored; the current phase-space configuration and the backward Oseledets vectors are instead saved every h_b time step, at the beginning of each block. This data can be typically stored on a disk, as it will not be accessed too frequently. Once this step has been completed, perform a series of block-by-block forward and backward iterations, starting from the last block. At the end of each pair of steps, the phase-space position and the corresponding Oseledets vectors are recovered and used to generate (a second time) the $\mathsf{R}(t)$ matrices and the $\mathbf{v}_{-}^{(i)}(t)$, which are now stored in the fast memory. At the end of a forward step, a backward iteration begins from the $\mathbf{c}^{(i)}(t)$ vectors obtained at the end of the previous block (with the exception of the very first step, where random initial conditions are selected, instead). This way, the CLVs are kept properly oriented. The overall price paid to study large systems is, therefore, the need to repeat the forward iterations twice.

Computational complexity

Suppose the dynamical system has N degrees of freedom and we are interested in computing the first m CLVs. The number of operations required in the forward step is the same as for the computation of the Lyapunov exponents.

To calculate the forward dynamical evolution one must run both phase and tangent space dynamics for k_f time steps and then perform a single QR decomposition. Dynamical systems such as those with finite-range or mean-field interactions (*easy dynamics*) require $\mathcal{O}(N)$ operations for a single step of phase-space dynamics and $\mathcal{O}(mN)$ for the tangent space evolution. Dynamical systems with long-range interactions (*hard dynamics*) typically require $\mathcal{O}(N^2)$ and $\mathcal{O}(mN^2)$ operations for phase-space and tangent-space evolution, respectively. This leads to either $\mathcal{O}(m\,k_f N)$ (for easy dynamics) or $\mathcal{O}(m\,k_f N^2)$ (for hard dynamics) operations to be performed between consecutive QR decompositions, which is an $\mathcal{O}(m^2 N)$ algorithm in itself. So far the computation is fully equivalent to the complexity of LE computation.

A single step of the backward evolution of the CLV coefficients via Eq. (4.6) requires $\sim m^3/3$ operations (by resorting to a back-substitution algorithm), while the number of operations needed to express the CLV in the phase-space coordinate basis via Eq. (4.5) is $\sim m^2 N/2$. As a result, neglecting the relatively small number of iterations needed to rescale the CLVs, we find that the total number of operations is

$$T_{\text{tot}} \approx h\, m^2 \left(\frac{m}{3} + \frac{N}{2} \right).$$

Accordingly, in systems with finite-range or mean-field interactions, even when the full set of CLVs is computed, i.e. $m = N$, the backward evolution is approximately 2.4 times faster than the forward one (see Chapter 3), or 6 times if one is not interested in expressing the CLVs in the phase-space coordinate basis. The situation is slightly less favorable for systems with long-range interactions, but it is nevertheless clear that the computational time is not a more serious issue than in the computation of the Lyapunov exponents.

Transient time

The convergence in both forward and backward evolutions is typically exponential and is related to the difference between consecutive Lyapunov exponents (see also Ginelli et al. (2013)). More precisely, the convergence to the ith CLV ($i > 1$) depends on the difference $\lambda_i - \lambda_{i-1}$. Dynamical systems with many degrees of freedom are typically characterised by a piece-wise continuous limit spectrum $\lambda(i/N)$ (see Chapter 10). This implies that the difference between consecutive exponents scales to zero as $1/N$, so that it is advisable, when performing a finite-size analysis of such systems, to scale the transient time with the number of degrees of freedom.

4.4 Static algorithms

Other algorithms have been proposed which do not make use of the intrinsic stability of the backward evolution, when restricted to the Oseledets subspace, but, rather, determine the CLVs as linear combinations of either forward or backward Oseledets vectors. More precisely, at each point along a given trajectory

$$\mathbf{u}^{(i)} = \sum_{j=1}^{i} \langle \mathbf{v}_-^{(j)} | \mathbf{u}^{(i)} \rangle \mathbf{v}_-^{(j)} = \sum_{j=i}^{N} \langle \mathbf{v}_+^{(j)} | \mathbf{u}^{(i)} \rangle \mathbf{v}_+^{(j)}, \tag{4.7}$$

where $\mathbf{v}_+^{(j)}$ and $\mathbf{v}_-^{(j)}$ are, respectively, the forward and backward Oseledets vectors. For the sake of simplicity, we are dropping the time dependence. The relation (4.7) follows from the fact that the CLV $\mathbf{u}^{(i)}$ is the intersection of the space spanned by the first i forward Oseledets vectors with the ones spanned by the last $(N - i + 1)$ backward Oseledets vectors.

4.4.1 Wolfe-Samelson algorithm

Upon taking the scalar product of $\mathbf{u}^{(i)}$ with $\mathbf{v}_{\pm}^{(k)}$, one can write

$$\langle \mathbf{v}_{-}^{(k)} | \mathbf{u}^{(i)} \rangle = \sum_{j=1}^{i} \langle \mathbf{v}_{+}^{(j)} | \mathbf{u}^{(i)} \rangle \langle \mathbf{v}_{-}^{(k)} | \mathbf{v}_{+}^{(j)} \rangle, \quad k \geq i, \tag{4.8}$$

$$\langle \mathbf{v}_{+}^{(k)} | \mathbf{u}^{(i)} \rangle = \sum_{j=i}^{N} \langle \mathbf{v}_{-}^{(j)} | \mathbf{u}^{(i)} \rangle \langle \mathbf{v}_{+}^{(k)} | \mathbf{v}_{-}^{(j)} \rangle, \quad k \leq i. \tag{4.9}$$

By now substituting Eq. (4.8) into Eq. (4.9), one obtains the following set of equations for the components of the covariant vectors in the forward basis:

$$\langle \mathbf{v}_{+}^{(k)} | \mathbf{u}^{(i)} \rangle = \sum_{j=1}^{i} \left[\sum_{l=i}^{N} \langle \mathbf{v}_{+}^{(k)} | \mathbf{v}_{-}^{(l)} \rangle \langle \mathbf{v}_{-}^{(l)} | \mathbf{v}_{+}^{(j)} \rangle \right] \langle \mathbf{v}_{+}^{(j)} | \mathbf{u}^{(i)} \rangle, \quad k \leq i. \tag{4.10}$$

The solution of these equations requires the knowledge of N vectors, no matter the size of i. However, as noticed by Wolfe and Samelson (2007), one can simplify the numerics, by exploiting the relationship

$$\sum_{k=1}^{N} \langle \mathbf{v}_{+}^{(i)} | \mathbf{v}_{-}^{(k)} \rangle \langle \mathbf{v}_{-}^{(k)} | \mathbf{v}_{+}^{(j)} \rangle = \delta_{ij},$$

which follows from the fact that $\mathbf{v}_{-}^{(i)}$ and $\mathbf{v}_{+}^{(i)}$ are two complete sets of orthonormal vectors. As a result,

$$\sum_{l=i}^{N} \langle \mathbf{v}_{+}^{(k)} | \mathbf{v}_{-}^{(l)} \rangle \langle \mathbf{v}_{-}^{(l)} | \mathbf{v}_{+}^{(j)} \rangle = \delta_{kj} - \sum_{l=1}^{i-1} \langle \mathbf{v}_{+}^{(k)} | \mathbf{v}_{-}^{(l)} \rangle \langle \mathbf{v}_{-}^{(l)} | \mathbf{v}_{+}^{(j)} \rangle.$$

By replacing this expression into Eq. (4.10), one obtains the set of equations

$$\sum_{j=1}^{i} \left[\sum_{l=1}^{i-1} \langle \mathbf{v}_{+}^{(k)} | \mathbf{v}_{-}^{(l)} \rangle \langle \mathbf{v}_{-}^{(l)} | \mathbf{v}_{+}^{(j)} \rangle \right] \langle \mathbf{v}_{+}^{(j)} | \mathbf{u}^{(i)} \rangle = 0, \quad k \leq i, \tag{4.11}$$

which is much simpler, as the two vector-indices j and l run up to i only. As a result, we see that Eq. (4.11) produces the expansion coefficients of the ith CLV as the kernel of a matrix computed from the first i forward and $i-1$ backward Oseledets vectors.

4.4.2 Kuptsov-Parlitz algorithm

Kuptsov and Parlitz (2012) have developed another static approach, which makes use of LU factorisation.

In matrix notations, Eq. (4.7) can be rewritten as

$$\mathbf{V}_{+}\mathbf{C}_{+} = \mathbf{V}_{-}\mathbf{C}_{-}, \tag{4.12}$$

with the plus and minus indices referring to forward and backward dynamics, respectively, and where the upper and lower triangular matrices $\mathbf{C}_{\pm}$ provide a compact description of all components of the CLVs in the forward and backward bases, respectively. Since V_- is an orthogonal matrix, one can recast Eq. (4.12) as

$$(V_-^T V_+) C_+ = P C_+ = C_-.$$

If one is interested only in the ith CLV, that is, in the ith column of the matrix C_+, only the $(i-1) \times i$ upper left corner of matrix P is needed. Since C_- is lower triangular and the first i entries of its ith column are all zeros, we are left with the following system of $(i-1)$ linear homogeneous equations in i variables

$$\sum_{j=1}^{i} [P]_{k,j} [C_+]_{j,i} = 0, \quad k = 1, 2, \ldots, i-1,$$

which defines the ith vector up to a rescaling factor.

Analogously to the dynamical approach, in both of the previous schemes, the first i forward and $(i-1)$ backward vectors are needed if the first i CLV are required. In these static methods, however, one is forced to perform the QR decomposition twice, which, as shown before, is the most computationally demanding part of any algorithm. A second concern arises for large systems, which have to be dealt with by means of singular value decomposition (SVD) to attain a satisfactory numerical accuracy. SVD is more time consuming than back substitution by a factor of 18, as it requires $\sim 6m^3$ operations for an $m \times m$ matrix (Golub and Van Loan, 1996).

Finally, for what concerns memory requirements, only the forward Oseledets vectors need to be stored by the static algorithm. While this reduces the memory requirement to about 2/3 of what is needed by the dynamical algorithm (where one has to store both V and R matrices), this memory advantage is lost whenever one is interested only in the angles between vectors, for which the dynamical algorithm requires the storage of only the upper triangular matrices R.

4.5 Vector orientation

Covariant Lyapunov vectors carry information on the (local) geometrical structure of the attractor. For example, it is useful to know the angle between the stable and unstable manifolds, since this helps to check whether they are everywhere transversal as required for hyperbolic systems.

It is therefore important to determine the angles between the different vectors. It is noteworthy that the angles are not dynamical invariants, but do depend on the choice of variables used to describe the underlying chaotic attractor. Moreover, in general, the task is not to compare single vectors but higher-dimensional spaces. Given any two linear subspaces S_1 and S_2 of dimension, respectively, N_1 and N_2 (with $N_1 + N_2 \leq N$ and $N_1 < N_2$), there exists a number N_1 of principal angles between them which can be defined as follows (see section 12.3.4 in Golub and Van Loan, 1996). Let A_1 (A_2) be an $N \times N_1$ ($N \times N_2$) matrix, whose column vectors span S_1 (S_2). Then, the QR decomposition is applied to both matrices,

$$A_1 = Q_1 R_1, \qquad A_2 = Q_2 R_2.$$

The N_1 principal values $s^{(i)}$ of $Q_1^T Q_2$ are the cosines of the principal angles $\theta^{(i)}$,

$$s^{(i)} = \cos \theta^{(i)}.$$

The minimum angle $\bar{\theta} = \min_i \theta^{(i)}$ measures the degree of transversality between the two subspaces.

When S_1 and S_2 are the unstable and the stable manifolds, respectively, it is convenient to refer to the backward Oseledets basis. This simplifies the calculation of principal angles, since the matrix A_1 is upper triangular, and its corresponding orthogonal matrix Q_1 is just the identity matrix (Kuptsov and Kuznetsov, 2009).

4.6 Numerical examples

In this section we illustrate the concept of CLVs in a few examples, starting from the Hénon map (A.4). In this case, since the phase space is two-dimensional, there are two CLVs, which can be easily determined by iterating forwards and backwards in tangent space.

In Fig. 4.2a, seven consecutive points of a trajectory are superposed on the Hénon attractor. For each point the solid segment is oriented along the direction of the first CLV, while the dashed segment corresponds to the second CLV. In some cases the solid segment is hardly visible (see, e.g., point 1); this happens whenever the curvature of the unstable manifold is locally small. In fact, the first CLV is naturally oriented along the unstable manifold that is, in this case, one dimensional. In some other points, the dashed segment is hardly visible as it is almost parallel to the solid one (see, e.g., points 4 and 5). This is the

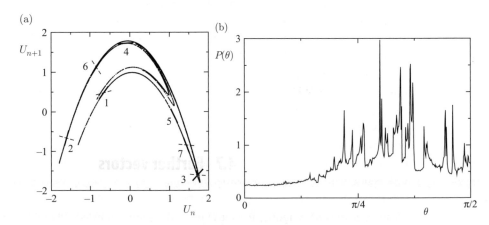

Fig. 4.2 (a) Covariant Lyapunov vectors in seven consecutive points of a trajectory of the Hénon map (with $a = 1.4$ and $b = 0.3$). (b) Distribution of the angles for the same map.

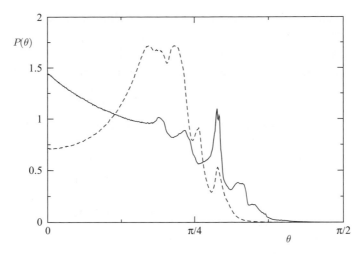

Fig. 4.3 Distribution of the minimal angle between stable and unstable manifolds in a chain of four Hénon maps (solid line, $a = 1.4, b = 0.3, \varepsilon = 0.025$) and of four symplectic maps (dashed line, $\mu = 4$).

indication that the trajectory is close to a homoclinic tangency. A compact description of how frequently such tangencies are encountered can be obtained by computing the probability density of the angle θ between two such manifolds. This is illustrated in Fig. 4.2b. There we see that a finite density of angles is observed around $\theta = 0$, which implies that the integrated probability of observing an angle smaller than θ is proportional to θ itself. This is a manifestation of deviations from strict hyperbolicity. The several peaks, instead, correspond to the regions where the unstable manifold is rather straight, and thereby the angle with the stable manifold is almost constant.

As less trivial examples, we now discuss two higher-dimensional systems: a dissipative chain of Hénon maps (see Eq. (A.11)) and a chain of symplectic maps (see Eq. (A.12)).

The results are reported in Fig. 4.3, where both models have been studied with four maps and assuming periodic boundary conditions. The angle θ is determined as the minimal angle between two four-dimensional subspaces in the former case and between two three-dimensional subspaces in the latter. In fact, in the case of the symplectic maps there are two zero Lyapunov exponents (arising from the conservation of the sum of the Z variables). We see that again the distribution extends to zero with a finite height, suggesting that deviations from hyperbolicity do not become more dangerous in higher-dimensional systems.

4.7 Further vectors

Besides covariant, singular, forward and backward vectors already discussed, other forms of Lyapunov vectors have been proposed in connection to various questions. In this section we briefly review two further classes of vectors.

4.7.1 Bred vectors

Whenever the dynamical model is highly complex, it may be inappropriate to develop algorithms for the integration of the equations in the tangent space. This is, for instance, the case of the atmospheric models typically used for weather forecasts. In such a context, it is convenient to consider finite-amplitude perturbations (see also Chapter 7).

Given a generic initial condition $\mathbf{U}$ and a randomly chosen nearby point, they are allowed to evolve for a time Δt. As a result,

$$\delta \mathbf{U}'(t + \Delta t) = \mathbf{M}\left[\mathbf{U}(t) + \delta \mathbf{U}(t)\right] - \mathbf{M}[\mathbf{U}(t)],$$

where $\delta \mathbf{U}$ is the initial perturbation with small amplitude $A = \|\delta \mathbf{U}\|$, $\mathbf{M}$ denotes the evolution operator over a time Δt.

The perturbation $\delta \mathbf{U}'$ is then rescaled to keep its norm equal to A (a free parameter one can play with),

$$\delta \mathbf{U}(t + \Delta t) = A \frac{\delta \mathbf{U}'(t + \Delta t)}{\|\delta \mathbf{U}'(t + \Delta t)\|},$$

and the procedure is repeated forwards in time. The resulting direction of the perturbation is called a *bred vector* (BV). In the small A limit, this is nothing but the method to determine the largest Lyapunov exponent. In this case, the BV converges towards the most expanding direction.

Alternatively, given the reference initial condition $\mathbf{U}$, the distance $\delta \mathbf{U}$ can be measured by comparing the evolution of two randomly perturbed trajectories; this "two-sided self-breeding" has the advantage of maintaining the linearity of the perturbation dynamics to the second order compared with the original one-sided scheme (Toth and Kalnay, 1997).

In practice, all that can be said about BVs, including the considerations reported here, applies also to the standard first covariant Lyapunov vector, except when the amplitude A is not too small (in this case, see the finite amplitude Lyapunov exponents discussed in Chapter 7).

Several numerical studies have revealed that BVs appear to be less sensitive to the selection of the norm with respect to singular vectors. This is, for instance, found when comparing potential-entrophy with the stream-function norm (see Corazza et al. (2003)). A more interesting property is the distribution of directions taken by the BVs (when computed over not-too-long times). In fact, it has been found that in global atmospheric models, as well as in other strongly nonlinear models, BVs remain distinct, rather than converging to a single leading direction; whether due to the presence of nearly degenerate spectra or of long transients, this observation has important consequences for the predictive power of the underlying model.

In order to quantify the spreading among BVs, the concept of BV-dimension has been proposed. If there are k BVs, each composed of N components (where N is the number of variables in the model), all variables can be organised into an $N \times k$ matrix $\mathbf{B}$. Afterwards, the principal component analysis, one computes the (positive) eigenvalues σ_j^2 and the eigenvectors of the $k \times k$ covariance matrix $\mathbf{C} = \mathbf{B}^{\mathsf{T}} \mathbf{B}$. So long as all BVs have converged towards the same direction, only the leading eigenvalue is significantly different from zero. On a qualitative level one can introduce the effective dimension as

$$D_{BV} = \frac{\left[\sum_{j=1}^{k} \sigma_j\right]^2}{\sum_{j=1}^{k} \sigma_j^2}.$$

If only one eigenvalue differs from zero, $D_{BV} = 1$, and if all are equal, $D_{BV} = k$. Therefore D_{BV} roughly tells us how many directions are spanned by the BVs. A large dimension is typically observed either in weakly chaotic regions, where many directions are characterised by similar (small) expansion rates, or when the amplitude A is large, allowing perturbations to be very different from one another. Interestingly, it has been found that in some cases the Earth's atmosphere is characterised by a low D_{BV} dimensionality ($D_{BV} < 2.5$) (Patil et al., 2001).

4.7.2 Dual Lyapunov vectors

The covariant Lyapunov vectors represent the generalisation of the *right* eigenvectors of a constant Jacobian J to a situation where one deals with a time-varying Jacobian. In some circumstances, it is convenient to consider the *left* eigenvectors as well, which, within linear algebra, are often referred to as *dual vectors*. In the context of this book, the ith dual Lyapunov vector is orthogonal to the space spanned by all CLVs $\mathbf{u}_j$ with $j \neq i$. In other words, dual Lyapunov vectors coincide with the CLVs only when the Jacobian is an orthogonal operator.

Dual vectors prove useful for the reconstruction of the linear response of a given dynamical system to a small external stimulus when the dynamics is periodic (and stable). Let δ denote a generic perturbation acting on a given periodic dynamics at time t_0. It can be decomposed into a transversal component that is eventually absorbed (because of the stability of the limit cycle) and a longitudinal component, which, being associated to a zero-Lyapunov exponent, is transformed into a finite phase shift (either negative or positive)

$$\delta = c_T \mathbf{u}_T + c_L \mathbf{u}_L.$$

Here $\mathbf{u}_T$ is a suitable combination of the stable CLVs, while $\mathbf{u}_L$ coincides with the zero-exponent CLV and corresponds to a shift along the trajectory. The task is to determine c_L, given δ. Upon applying (from the left, as usual) the Jacobian $H(t_0, t)$ for a time t long enough, one obtains

$$H(t_0, t)\delta = c_L H(t_0, t)\mathbf{u}_L,$$

as the transversal component dies out. By now multiplying from the left by the dual vector $\bar{\mathbf{u}}_L$ of $H(t_0, t)$, and solving for c_L, it is found that

$$c_L = \frac{\bar{\mathbf{u}}_L \cdot \delta}{\bar{\mathbf{u}}_L \cdot \mathbf{u}_L},$$

where we have made use of the fact that $\mathbf{u}_L$ is characterised by an eigenvalue equal to 1 (zero Lyapunov exponent). In practice, c_L corresponds to the expected phase shift when a perturbation δ is applied at time t_0. Notice that in the case of an orthogonal Jacobian, the denominator is equal to 1 as the CLV coincides with its dual vector. At the opposite limit of a near degeneracy, when the two vectors are almost orthogonal, the denominator can, instead, be very small, thus implying a large phase response; this happens when the zero-CLV is nearly colinear with the stable directions, so that a relatively small perturbation decomposes into two large but antilinear components.

Now, we illustrate the problem in the case of the FitzHugh-Nagumo model (A.7) for parameter values, where the evolution is asymptotically periodic. The shape of the limit cycle can be appreciated in Fig. 4.4a (see the figure caption for the parameter values); the various segments show the orientation of the second, i.e. stable, CLV. Within a linear

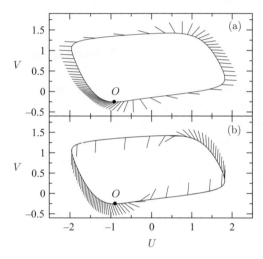

Fig. 4.4 The FitzHugh Nagumo model for $I = 0.5, \tau = 10, a = 0.7$ and $b = 0.8$ (see the solid line in both panels). The segments in panel (a) are oriented along the locally stable direction, while those in panel (b) are oriented along the covariant vector, which corresponds to the zero exponent.

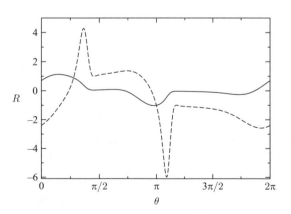

Fig. 4.5 Phase-response curve for the FitzHugh-Nagumo model: solid and dashed lines refer to a perturbation acting along the U and V variables, respectively.

approximation, phase points lying along the same segment all converge to the same trajectory.

In this 2-dimensional model, the zero dual Lyapunov vector is orthogonal to the stable CLV. The amplitude of the phase shift (the so-called phase-response curve) due to a perturbation which acts along the U (V) direction is reported in Fig. 4.5 as a function of the phase θ along the limit cycle (where θ is defined as the rescaled elapsed time, starting from the origin O).

More sophisticated applications to spatially extended systems can be found in Biktashev (2005), where the problem is approached by integrating the adjoint linearised equations backwards.

5 Fluctuations, finite-time and generalised exponents

In this chapter we revisit the concept of LE introduced in Chapter 3, starting from the definition of finite-time Lyapunov exponents. Any average performed over a finite time is not a well-defined quantity per se since it is typically affected by fluctuations. In the long-time limit, however, it turns out that the residual fluctuations, rather than representing an obstacle for an accurate estimate of the LEs, offer the opportunity to extract useful information on the degree of homogeneity of the instabilities in phase space. Such information can be properly encoded into suitable dynamical invariants, either in the form of generalised Lyapunov exponents or as a large deviation function.

Finite-time Lyapunov exponents can be used to introduce further numerical tools as well as powerful techniques for the evaluation of the LEs. On the numerical side, we briefly review some quick methods suitable for a qualitative assessment of the presence of instabilities, and then we present an advanced Monte Carlo technique to quantify large deviations. On the theoretical side, on combining the concept of generalised LEs with the ensemble-average approach introduced in Chapter 3, we show that the problem of evaluating the largest LE can be reduced to that of determining the largest eigenvalue of a suitable evolution operator. We thereby illustrate the potentiality and difficulties of the various methods with the help of some simple examples. Another theoretical approach, based on the stability of the periodic orbits embedded in a chaotic attractor, is also discussed.

The last part of this chapter is devoted to the discussion of several examples of fluctuations that may emerge in generic dynamical systems. This includes the relationship between fluctuations and deviations from hyperbolic behaviour (the Hénon map being a rather instructive reference model) and several borderline cases, where the maximal LE may be equal to zero (weak chaos) or even negative (mixed dynamics), and yet positive fluctuations are generated and induce some form of irregularity in the resulting dynamics.

5.1 Finite-time analysis

There are different ways of defining finite-time Lyapunov exponents.[1] We start with a definition that is naturally associated with the concept of covariant Lyapunov vectors.

---

[1] Sometimes finite-time exponents are called local exponents; we avoid such a terminology since locality may be considered in both time and space (see the discussion of space-time dynamics in Chapters 10 and 11).

Let $\mathbf{u}_k(t_0)$ denote a vector oriented along the direction of the kth CLV in the phase point $\mathbf{U}(t_0)$. One can, accordingly, define the finite-time Lyapunov exponent

$$\Lambda_k(t_0, \tau) = \frac{1}{\tau} \log \frac{\|\mathsf{H}(t_0, t_0 + \tau)\mathbf{u}_k(t_0)\|}{\|\mathbf{u}_k(t_0)\|} := \frac{\Gamma_k(t_0, \tau)}{\tau}, \tag{5.1}$$

where the Jacobian $\mathsf{H}(t_0, t_0+\tau)$ corresponds to the integration in tangent space from time t_0 to $t_0 + \tau$ (we use the same notations as in Section 4.1). The finite-time Lyapunov exponent is denoted with Λ_k because we want to distinguish it from the true Lyapunov exponent λ_k, defined in the limit $\tau \to \infty$. The LE λ_k can be thought of as the time-averaged finite-time Lyapunov exponent Λ_k. On the other hand, the quantity $\Gamma_k(t_0, \tau)$ is the overall exponential growth rate of the perturbation integrated over a time interval of length τ.

For finite times, $\Lambda_k(t_0, \tau)$ generally depends on the selection of coordinates/norm and on the chosen initial condition (through t_0). We start by discussing the first dependence.

If one introduces a new variable $\mathbf{V} = \mathbf{G}(\mathbf{U})$, the corresponding Jacobian can be written with help of the matrix $\mathsf{D} = \frac{\partial \mathbf{G}}{\partial \mathbf{U}}$ as

$$\mathsf{H}_V(t_0, t_0 + \tau) = \mathsf{D}(t_0 + \tau)\mathsf{H}(t_0, t_0 + \tau)\mathsf{D}^{-1}(t_0),$$

where $\mathbf{v}_k(t) = \mathsf{D}\mathbf{u}_k(t)$ so that

$$\Lambda'_k(t_0, \tau) = \Lambda_k(t_0, \tau) + \frac{1}{\tau}\left[\ln\frac{\|\mathsf{D}\mathbf{u}_k(t_0 + \tau)\|}{\|\mathbf{u}_k(t_0 + \tau)\|} + \ln\frac{\|\mathsf{D}^{-1}\mathbf{v}_k(t_0)\|}{\|\mathbf{v}_k(t_0)\|}\right].$$

The term in square brackets is a bounded correction that does not systematically grow with the integration time τ. As a result of the division by τ, the correction becomes increasingly negligible upon increasing τ. Similar corrections arise if a different norm is selected for the computation of the Lyapunov exponents.

The dependence on the initial condition can be appreciated in Fig. 5.1, where $\Lambda_1(t_0, \tau = 10)$ (dots) is plotted versus time for the Hénon map (A.4). Here, we see that even though Λ_1 is typically positive, occasional negative values are spotted as well. Strong fluctuations can be observed even for arbitrarily large τ values, although their probability decreases upon increasing τ. This is an example of the large deviations that are discussed later in this chapter. One of the sources of the fluctuations is the existence of periodic orbits characterised by different sets of Lyapunov exponents.

Finite-time Lyapunov exponents can also be defined in a simpler way, by making reference to the QR decomposition (implemented either through the Gram-Schmidt orthogonalisation or the Householder reflections – see Section 3.2)

$$\Lambda_k^{\mathrm{QR}}(t_0, \tau) = \frac{\log\|R_{kk}(t_0, t_0 + \tau)\|}{\tau}, \tag{5.2}$$

where R_{kk} are the diagonal elements of the triangular matrix R arising from the integration over a time τ.

The second Lyapunov exponent is the simplest example to illustrate the differences between the two protocols. In the QR case, the sum of the first two finite-time Lyapunov exponents $\Lambda_1^{\mathrm{QR}}(\tau) + \Lambda_2^{\mathrm{QR}}(\tau)$ corresponds to the growth rate of the most expanding area.

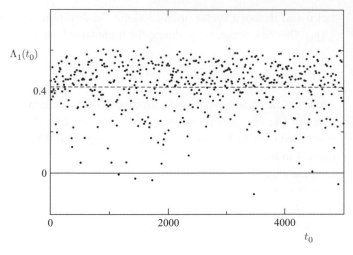

Fig. 5.1 First finite-time Lyapunov exponent for the Hénon map. Dots refer to $\Lambda_1(\tau = 10)$; each value is computed over ten consecutive points of a given trajectory, and the dashed line corresponds to the asymptotic (average) value.

In the CLV context[2], $\Lambda_1^{\mathrm{CLV}}(\tau) + \Lambda_2^{\mathrm{CLV}}(\tau)$ is characterised by additional fluctuations, since the elementary area defined by the first two (non-orthogonal) CLV varies in time as well.

The Lozi (A.5) and the Hénon (A.4) maps provide the opportunity to illustrate the difference, since in both cases, a single iterate of the map induces everywhere in phase space a volume (area) contraction by a factor b. Therefore, the QR definition implies that $\Lambda_1^{\mathrm{QR}}(\tau) + \Lambda_2^{\mathrm{QR}}(\tau) = \ln b$. It is, therefore, convenient to introduce $\Delta\Gamma^{\mathrm{CLV}}(\tau) = \Gamma_1^{\mathrm{CLV}}(\tau) + \Gamma_2^{\mathrm{CLV}}(\tau) - \tau \ln b$ to denote a fluctuation of the expansion rate, integrated over a time τ.

The outcome of numerical simulations performed with the covariant Lyapunov vectors is reported in Fig. 5.2. Panel (a) refers to the Lozi map: the distribution $P(\Delta\Gamma^{\mathrm{CLV}})$ has been obtained for $\tau = 40$. One can, first of all, appreciate that, at variance with the QR definition, $\Delta\Gamma^{\mathrm{CLV}}$ takes non-zero values. The deviations are harmless, however, as they are independent of τ (the probability distribution for $\tau = 20$ is indistinguishable from the curve reported in Fig. 5.2a). The scenario is slightly different for the Hénon map, where $P(\Delta\Gamma^{\mathrm{CLV}})$ is again independent of τ, but it now exhibits long, supposedly infinite tails (see Fig. 5.2b). This phenomenon is related to the existence of homoclinic tangencies, where the stable manifold is tangent to the unstable manifold (see Section 5.7.3 for a more detailed discussion).

The Chirikov-Taylor standard map is a widely studied model of chaos, used as a testbed for symplectic dynamics. In this case, volumes are preserved over any time interval. Nevertheless, in Fig. 5.2c we see that $\Delta\Gamma^{\mathrm{CLV}}$, now defined as $\Delta\Gamma^{\mathrm{CLV}}(\tau) = \Gamma_1^{\mathrm{CLV}}(\tau) + \Gamma_2^{\mathrm{CLV}}(\tau)$, reveals fluctuations distributed as in the Hénon map, the origin being again the non-hyperbolicity of the map.

In conclusion, we have seen that two definitions of finite-time Lyapunov exponents are available that are almost, but not exactly, equivalent. The definition (5.1), based on CLVs,

---

[2] The superscript "CLV" has been added to stress that we are referring to the definition given in (5.1).

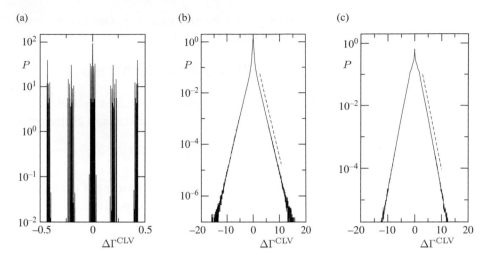

Fig. 5.2 Probability distribution $P(\Delta\Gamma^{\text{CLV}})$ of the volume contraction rate in the Lozi map with $a = 1.4$ and $b = 0.3$ (panel (a)), the Hénon map with $a = 1.4$ and $b = 0.3$ (panel (b)) and the Chirikov-Taylor symplectic map for $K = 1.5$ (panel (c)). The dashed lines in panels (b) and (c) correspond to an exponential decay with an exponent -1.

is the natural extension of the concept of eigenvalue of linear operators, but it loses the nice property of volume conservation laws that are, instead, kept by the definition (5.2), based on the QR decomposition. The differences between the two definitions, however, manifest themselves only in the tail of the distribution of the LEs, as it will become clear in the following sections.

5.2 Generalised exponents

In the previous section we saw that finite-time Lyapunov exponents do typically fluctuate. Such fluctuations are more than a hindrance for an accurate computation of the LEs. They also carry information on how the chaotic dynamics is structured, or the degree of stability of a stochastic system (which can be destabilised by rare fluctuations). There are various ways to characterise such fluctuations. Here, we proceed by introducing the so-called generalised Lyapunov exponents in an abstract manner. In fact, the formalism applies to any observable that needs to be suitably averaged over the phase space. Moreover, for the sake of simplicity, we refer to a discrete-time dynamical system, the extension to continuous time being straightforward.

So, let $\upsilon(\mathbf{U})$ be the scalar observable of interest and $\mathbf{U}_0$ a suitable initial condition. One can thereby define the time-integrated observable,

$$\Upsilon^\tau(\mathbf{U}(0)) = \sum_{k=0}^{\tau-1} \upsilon(\mathbf{F}^k(\mathbf{U}(0))),$$

so that Υ^τ/τ is the average over the time interval τ. One can accelerate the convergence by averaging also over the initial distribution

$$\langle\Upsilon^\tau\rangle = \int d^N\mathbf{U}\,\rho_0(U(0))\Upsilon^\tau(U(0)).$$

A more detailed description of the process is obtained by looking at the moments and cumulants of Υ^τ. They can be determined by introducing a characteristic function:

$$\left\langle e^{q\Upsilon^\tau}\right\rangle = \int \rho_0(\mathbf{U})e^{q\Upsilon^\tau(\mathbf{U})}\,d^N\mathbf{U}. \tag{5.3}$$

For large τ, this quantity is expected to be independent of the initial distribution and to grow exponentially in time with a rate $G(q)$ that depends on q,

$$G(q) = \lim_{\tau\to\infty}\frac{1}{\tau}\ln\left\langle e^{q\Upsilon^\tau}\right\rangle. \tag{5.4}$$

Notice that the conservation of probability implies that[3]

$$G(0) = 0.$$

The characteristic function $G(q)$ allows calculation of the cumulants of the distribution of Υ^τ as derivatives in zero. In particular,

$$\langle v\rangle = \frac{dG}{dq}\bigg|_{q=0}, \qquad \lim_{\tau\to\infty}\frac{\langle(\Upsilon^\tau - \tau\langle v\rangle)^2\rangle}{\tau} = \frac{d^2G}{dq^2}\bigg|_{q=0}, \tag{5.5}$$

where the latter quantity is the diffusion coefficient of the process Υ^τ.

The Lyapunov exponents can be treated with this formalism; it is sufficient to identify Υ^τ with $\Gamma_k(\tau)$. As a result, $\lambda = \langle v\rangle$ (for the sake of simplicity, from now on the subscript index k will be dropped). By combining Eq. (5.1) with Eq. (5.4), one obtains

$$q\mathcal{L}(q) = \lim_{\tau\to\infty}\frac{1}{\tau}\ln\left\langle\frac{\|\mathsf{H}(t_0, t_0+\tau)\mathbf{u}(t_0)\|^q}{\|\mathbf{u}(t_0)\|^q}\right\rangle, \tag{5.6}$$

where, instead of referring to the characteristic function, we have introduced the customary definition of the generalised Lyapunov exponent

$$\mathcal{L}(q) = G(q)/q. \tag{5.7}$$

For $q\to 0$, $\mathcal{L}(0)$ reduces to the standard Lyapunov exponent λ (the limit may be computed by applying the L'Hôpital's rule to the expression (5.6)). $\mathcal{L}(q)$ is a monotonously increasing function of q (see, e.g., Fig. 5.3b). Its variation with q reflects the existence of a finite range of exponential growth rates, as discussed next. Of particular relevance is $\mathcal{L}(1)$, which coincides with the topological entropy (see Chapter 6).

---

[3] In the case of chaotic repellers, the initial measure decreases exponentially, and one has to thereby modify the equation slightly.

Large deviation function

The process $\upsilon(t)$ can be equivalently characterised from the scaling behaviour of the probability $P(\Lambda, \tau)$ to observe a finite-time exponent Λ over the time interval τ (see Eq. (5.1)),

$$P(\Lambda, \tau) \underset{\tau \to \infty}{\propto} e^{-S(\Lambda)\tau}, \tag{5.8}$$

where the large deviation function $S(\Lambda)$ is positive-definite with a minimum (equal to zero) at the asymptotic value of the Lyapunov exponent λ. In the absence of long-time correlations, the minimum is quadratic, meaning that the central limit theorem holds and one can approximate the distribution of finite-time Lyapunov exponents with a Gaussian for $\Lambda \approx \lambda$ (see also the next section).

From Eqs. (5.7, 5.8), one can write

$$e^{\mathcal{L}(q)q\tau} \approx \int d\Lambda e^{[q\Lambda - S(\lambda)]\tau},$$

which implies the Legendre transformation (or, in the case of non-smoothness and/or non-concavity, the Legendre-Fenchel transform),

$$q\mathcal{L}(q) = \max\{q\Lambda - S(\Lambda)\}. \tag{5.9}$$

So long as $S(\Lambda)$ is a smooth function, the maximum is attained for the special value Λ^* where

$$S'(\Lambda^*) = q,$$

so that

$$q\mathcal{L}(q) = q\Lambda^* - S(\Lambda^*). \tag{5.10}$$

This transformation has a simple geometrical interpretation. Given a generic distribution $S(\Lambda)$ (see, for instance, Fig. 5.3a), its tangent in the point of abscissa Λ^* intersects the vertical axis in a point of ordinate equal to $-q\mathcal{L}(q)$. The resulting generalised LE $\mathcal{L}(q)$ is plotted in panel (b) of the same figure.

Finally, Eq. (5.9) can be inverted. As for generic Legendre-type transforms, given $\mathcal{L}(q)$, one can relate q with Λ through

$$\Lambda = (q\mathcal{L})',$$

while $S(\Lambda)$ is obtained by inverting Eq. (5.10):

$$S(\Lambda) = q(\Lambda - \mathcal{L}(q)).$$

This means that the fluctuations of the finite-time Lyapunov exponents, often referred to as the stability spectrum, can be equally characterised by either the large deviation function $S(\Lambda)$ or the generalised exponents $\mathcal{L}(q)$.

Notice in Fig. 5.3a that the range of possible Λ-values is finite: it is delimited by the points where $S(\Lambda) = \Lambda$. In fact, in a deterministic dynamical system, $\exp[\Lambda - S(\Lambda)t]$ corresponds to the number of trajectories of length t which are characterised by the exponent Λ (Badii and Politi, 1997): wherever $S(\Lambda) > \Lambda$, such a multiplicity vanishes,

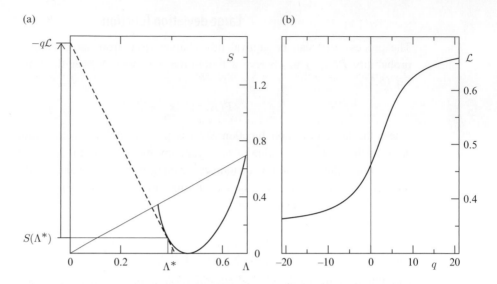

Fig. 5.3 Large deviation function (a) and generalised Lyapunov exponents (b) for the model (5.15) with $p = 1/2$. The action of the Legendre-Fenchel transform (5.9) is also schematically illustrated. The tilted solid line in panel (a) corresponds to the bisectrix $S = \lambda$.

meaning that the corresponding finite-time LE Λ cannot be observed at all. The minimum and maximum LEs are typically attained by simple (fixed point, periodic or quasiperiodic) trajectories. Indeed, if such a special, non-typical trajectory (or, more generally, an invariant subset inside chaos) possesses a certain Lyapunov exponent, then a typical trajectory can potentially come close to this non-typical set, yielding the possibility for the corresponding Lyaponov exponent to be observed for a finite time (which can be in fact arbitrarily long). Thus, the domain of the large deviation function describes the spectrum of instabilities of invariant sets inside chaos. We will see applications of this in Section 9.2.

In the presence of degeneracies (i.e. when different periodic orbits are characterised by the same exponent), the ending points of $S(\Lambda)$ may lie strictly below the bisectrix. On the other hand, in stochastic processes and in random dynamical systems, $S(\Lambda)$ can take any positive value, as $S(\Lambda)$ is determined by the properties of the given stochastic process.

In a more general context, one could consider the whole set of finite-time Lyapunov exponents $\boldsymbol{\Lambda} = \{\Lambda_1, \Lambda_2, \ldots, \Lambda_N\}$, thereby introducing the probability distribution $P(\boldsymbol{\Lambda}, \tau)$ and the corresponding large deviation function,

$$P(\boldsymbol{\Lambda}, \tau) \underset{\tau \to \infty}{\propto} e^{-S(\boldsymbol{\Lambda})\tau}.$$

$S(\boldsymbol{\Lambda})$ being still a positive-definite scalar quantity is now a function of N variables; it provides the most detailed characterisation of LE fluctuations. It can, in principle, be mapped onto a suitable characteristic function, but now it would be necessary to introduce a set of N indices q_i.

In many applications the observable of interest is a linear combination of various Lyapunov exponents (see, e.g., the Kolmogorov-Sinai entropy, which is the sum of the positive LEs, Chapter 6),

$$\upsilon(\mathbf{\Lambda}) = \sum c_i \Lambda_i(\tau).$$

Assuming knowledge of $S(\mathbf{\Lambda})$, one can easily determine the appropriate large deviation function as

$$S_\upsilon(\Upsilon) = \max\{S(\mathbf{\Lambda}, \tau)|\upsilon(\mathbf{\Lambda}) = \Upsilon\}.$$

5.3 Gaussian approximation

In general, determining S is too ambitious a task – especially for high-dimensional systems (see, however, the next section for a powerful method for detecting extremely improbable events). One can nevertheless infer relevant properties already from the curvature of S around the minimum λ, i.e. from the first terms in the expansion

$$S(\mathbf{\Lambda}) \approx \frac{1}{2}(\mathbf{\Lambda} - \lambda)\mathbf{Q}(\mathbf{\Lambda} - \lambda)^\mathsf{T}. \tag{5.11}$$

This is nothing but the Gaussian approximation, as it is readily seen in the case of a scalar variable, when $\mathbf{Q}$ reduces to the single component Q_{11},

$$P(\Lambda, \tau) \approx e^{-\frac{(\Lambda-\lambda)^2 \tau Q_{11}}{2}}.$$

From this equation it is transparent that $P(\Lambda, \tau)$ is indeed a Gaussian distribution centred around the asymptotic value of the Lyapunov exponent with a variance $D = 1/(Q_{11}\tau)$. This result is consistent with the central limit theorem, which states that the standard deviation of the average of M independent variables is on the order of $1/\sqrt{M}$ (here τ plays the role of M). In other words, as we have made no assumption on the distribution of the Λ-values, we learn that the aforementioned quadratic approximation is equivalent to assuming that correlations are sufficiently weak. This does not exclude that in some cases S may not have a quadratic minimum (see, e.g., the case of weak chaos discussed in Section 5.7.2).

In general, it is preferable to refer to the symmetric diffusion matrix $\mathbf{D} = \mathbf{Q}^{-1}$. Its elements D_{ij} can be easily determined from the (linear) growth rate of the (co)variances of $(\Gamma_i(\tau) - \lambda_i\tau)$,

$$D_{ij} = \lim_{\tau \to \infty} \left(\overline{\Gamma_i(\tau)\Gamma_j(\tau)} - \lambda_i\lambda_j\tau^2\right)/\tau. \tag{5.12}$$

Here the overline denotes time average. The information contained in $\mathbf{D}$ can be expressed in a more compact form by determining its (positive) eigenvalues μ_k ($k = 1, \dots, N$), i.e. the so-called principal components, which correspond to the fluctuation amplitudes along the principal axes.

The knowledge of the eigenvalues can indeed help to identify, if not to discover, constraints in the tangent space dynamics. For instance, if the total-volume contraction rate is constant in the phase space, and the QR definition of the finite-time LEs is used, it follows that $\sum_i \Lambda_i(\tau) = $ const, i.e. all $\{\Lambda_i\}$ n-tuples lie in the same hyperplane (see the next section for a discussion of possible differences between the two definitions of finite-time Lyapunov exponents given in Section 5.1). As a result, one eigenvalue of D must be equal to zero, as it corresponds to the (vanishing) transversal width of the hyperplane itself. Another instructive case is that of symplectic dynamics (again, the QR definition of the finite-time LEs is used). Since the LEs come in pairs whose sum is zero, the fluctuations of the negative LEs are perfectly anticorrelated with those of the positive ones. As a result D_{ij} has an additional symmetry, $D_{N+1-i,j} = D_{i,N+1-j} = -D_{ij}$, and half of the principal components are exactly equal to zero. Altogether, the possible existence of zero eigenvalues reinforces the choice of studying D rather than its ill-defined inverse Q. Moreover, since the matrices Q and D are diagonal in the same basis, and the eigenvalues of Q are the inverse of those of D, one can infer the scaling behaviour of the former ones from that of the latter. One must simply be careful and discard the *redundant* variables associated with the zero eigenvalues.

5.4 Numerical methods

In this section we briefly review some tools based on finite-time Lyapunov exponents and the so-called Lyapunov weighted dynamics, which allows the study of large deviations.

5.4.1 Quick tools

Sometimes, it may be useful to rapidly assess the chaotic character of a given trajectory. This is the case of Hamiltonian systems in the presence of a mixed phase space, when there is, e.g., the need to identify regular islands in a chaotic sea. It is also the case of celestial mechanics (see Section 12.5), where chaos is weak and the trajectories are unavoidably "relatively" short.

The overall goal is to extract as much information as possible from finite-time Lyapunov exponents. In this context, the dependence on the initial condition is not considered as a drawback but rather as an opportunity to classify different regions, according to their "short-term" stability. The most serious difficulty is the lack of knowledge of the covariant Lyapunov vectors; a randomly chosen initial direction may indeed be accidentally close to one of the less expanding directions.

In order to minimise such an effect, various approaches have been proposed. Here, we briefly summarise two of them: the Fast Lyapunov Indicator (FLI) (Froeschlé et al., 1997) and the Generalised Alignment Index (GALI) (Skokos et al., 2007). Both of them make a reference to the parallel evolution of p unit vectors $\{\mathbf{u}_1, \mathbf{u}_2, \ldots, \mathbf{u}_p\}$ in the tangent space.

In the case of the FLI, these unit vectors are initially chosen as an orthonormal basis and are thereby allowed to freely evolve for some time interval τ. The indicator is finally defined as

$$L_{FLI}(\tau) = \max_{j} \ln \|\mathbf{u}_j(\tau)\|.$$

The max operation allows the discarding of those "unfortunate" initial conditions in the tangent space, which may initially be improperly aligned.

In the case of the GALI, the tangent verctors are rescaled (but not orthogonalised!) to obtain unit vectors $\hat{\mathbf{u}}_j$ at all times. The order-p indicator is thereby defined as the volume of the parallelepiped identified by the tangent vectors,

$$V^p_{GALI}(t) = \|\hat{\mathbf{u}}_1 \wedge \hat{\mathbf{u}}_2 \wedge \ldots \hat{\mathbf{u}}_p\|.$$

In Hamiltonian systems, where a zero maximum exponent implies that all other exponents vanish as well, this indicator allows the distinguishing of chaotic from ordered dynamics. In fact, in the case of a chaotic regime (and in the absence of degeneracies), the GALI decreases exponentially in time:

$$V^p_{GALI} \approx \exp\left[(\lambda_1 - \lambda_2) + (\lambda_1 - \lambda_3) + \cdots + (\lambda_1 - \lambda_p)\right]t.$$

In the case of a regular dynamics, the order-p GALI either remains finite or, at most, decays as a power law, when p is larger than the number of frequencies that characterise the given trajectory (Skokos et al., 2007). Of course, this method with $p = 2$ cannot recognise the presence of a chaotic dynamics, if the two largest positive LEs are nearly equal.

5.4.2 Weighted dynamics

The previous discussion of finite-time LEs shows that the true exponents are the averaged ones, but there are deviations, due to the existence of phase space regions characterised by an either anomalously large or weak local instability (or even local stability). Large deviations are, by definition, improbable, and it is therefore difficult and time-consuming to collect enough statistics for a reliable estimate of their contribution. Tailleur and Kurchan (2007) have proposed an effective algorithm that is inspired by the sequential or diffusion Monte Carlo dynamics, often used in statistical mechanics.

A set of suitably selected trajectories (replicas) is simulated simultaneously to properly determine average observables even when they are dominated by large deviations. An example is the generalised Lyapunov exponent $\mathcal{L}(q)$ for q sufficiently away from 0 (for the sake of simplicity, we refer to the maximum LE, the extension to the other exponents being conceptually straightforward).

The method proceeds by alternating two steps: (i) the parallel evolution of N_c trajectories (including the position in real space and the corresponding Lyapunov vector) for a given time t to determine $\Gamma_j(t)$ (see Eq. (5.1) – notice that the subscript here labels different trajectories) and (ii) a suitable adjustment of the pool of trajectories, performed in such a way as to favour those which provide the most important contribution to $\mathcal{L}(q)$.

In practice, at the end of the first step, the expansion factor $w_j(t,q) = \mathrm{e}^{\Gamma_j q t}$ is computed for each trajectory and thereby averaged to obtain

$$\langle W(t,q) \rangle = \frac{1}{N_c} \sum_j w_j(t,q).$$

Afterwards, each replica is replaced by $M = \lfloor \eta_j - w_j(t,q)/\langle W(t,q) \rangle \rfloor$ copies, where η_j is a random number uniformly distributed between 0 and 1, while $\lfloor x \rfloor$ denotes the largest integer smaller than x. In practice, if $M = 0$, the original trajectory is removed; if $M = 1$, nothing is done, while if $M > 1$, $M - 1$ clones of the trajectory are added to the pool.

This new set of trajectories is then iterated for another time lap t (step i). In order to prohibit the daughter trajectories from evolving exactly as the parents, either a small amount of noise is added to the evolution in real space or some noise is added just at the cloning moment. In both cases the noise must be sufficiently small as not to modify substantially the invariant measure.

The generalised exponent $\mathcal{L}(q)$ can be eventually estimated as

$$\mathcal{L}(q) = \frac{1}{qt} \ln\langle W(t, q)\rangle,$$

where the average $\langle \cdot \rangle$ is performed over the total number of steps performed. As usual, t must be large enough to ensure that the correlations have died out.

Although the number of replicas is, on average, equal to N_c, it is expected to diffuse because of the randomness of the deletions/insertions. It is, therefore, wise to either randomly prune or clone some of them, to ensure a constant number.

Experience with this type of Monte Carlo algorithm suggests that the numerical estimates may be affected by metastability (a local optimum is selected which differs from the global one). In order to minimise such problems it is advisable to perform different runs and/or to increase the number of samples in the pool to check the stability of the results.

The method works because the trajectory-resampling does not modify the average value $\langle W(t, q)\rangle$ but makes the estimated value more statistically reliable by increasing the weight of the appropriate trajectories. This can be better appreciated by looking at an equivalent description of the resampling process. Given the weights w_j, let us order them from the minimum to the maximum and introduce the integral variable $C(i) = \sum_{j<i} w_j$. Given the value $C(N_c)$ of the total sum, let us now randomly extract N_c values uniformly distributed between 0 and $C(N_c)$. Depending on the number M of such values which fall in the interval $[C(i), C(i+1)]$, $C(i+1)$ may be pruned ($M = 0$), left unchanged ($M = 1$) or cloned $M - 1$ times ($M > 1$). As a result, the highly dense regions are pruned, while the sparser ones are re-populated so as to leave the total sum (the quantity we are interested in) unchanged. Much care has to be taken when the fluctuations of the Lyapunov exponent lead to very small or even negative values, as the replicas of the phase points cannot rapidly generate statistically independent trajectories.

The reader interested in the formal aspect of the Lyapunov weighted dynamics and in further details is invited to consult Tailleur and Kurchan (2007), Vanneste (2010) and Laffargue et al. (2013).

5.5 Eigenvalues of evolution operators

The generalised Lyapunov exponents correspond to the asymptotic scaling behaviour of suitable observables. The problem of their computation can be recast as the identification of the leading eigenvalue of a suitable evolution operator. We first illustrate this idea by considering piecewise linear maps of the interval in the presence of a Markov partition (see Appendix D): this simplification, besides making the analysis more understandable, helps to elucidate the difficulties encountered in more general contexts, before they manifest themselves. The existence of a Markov partition means that the unit interval can be split into disjoint subintervals I_i ($1 \le i \le B$), such that $F(I_i)$ is equal to the union of one or more such subintervals. If, moreover, we assume that the map is linear within I_i, it is easy to verify that the invariant distribution is constant within each subinterval, and one can reduce the iteration of the Perron-Frobenius operator to a matrix multiplication,

$$\rho(t+1) = Q\rho(t), \qquad (5.13)$$

where ρ is a vector of dimension B. Its components correspond to the probability densities within the interval I_i at time t, while Q_{ij} is the transition rate from interval j to interval i. In practice $Q_{ij} = 1/|F_j'|$ if $I_i \cap F(I_j) \ne \varnothing$ and $Q_{ij} = 0$ otherwise.

The generalised Lyapunov exponent $\mathcal{L}(q)$ can then be obtained from the scaling behaviour of the moments

$$\langle M(\tau)^q \rangle = \int M(\tau, U)^q \rho(U) dU,$$

where $M(\tau, U)$ is the multiplier along a trajectory of length τ, located in U at time τ. As the density is piecewise constant, one can express the moment as the sum of a finite number of components,

$$\langle M(\tau)^q \rangle = \sum_i \Delta_i \rho_i \langle M_i^q(\tau) \rangle,$$

where $\langle M_i^q(\tau) \rangle$ is an average over the trajectories ending in the interval I_i at time τ, while Δ_i is the width of the same interval. From Eq. (5.13), it is found that the vector $\langle \mathbf{M}^q(\tau) \rangle$ follows the evolution equation,

$$\langle \mathbf{M}^q(\tau+1) \rangle = \mathbf{G}(q) \langle \mathbf{M}^q(\tau) \rangle, \qquad (5.14)$$

where the entries of the matrix $\mathbf{G}(q)$ are simply given by $G_{ij}(q) = |F_j'|^{q-1}$. As a result, from Eq. (5.7)

$$\mathcal{L}(q) = \lim_{\tau \to \infty} \frac{1}{q\tau} \ln \langle M(\tau)^q \rangle = \frac{\ln \alpha_q}{q},$$

where α_q is the largest eigenvalue of $\mathbf{G}(q)$.

From the definition of $\mathbf{G}(q)$ (see Eq. (5.14)) it follows that $G_{ij}(1)$ is either equal to 1 if $I_i \cap F(I_j) \ne \varnothing$ or zero otherwise. In fact, $\mathcal{L}(1)$ identifies the topological entropy, an indicator that quantifies the growth rate of the number of different trajectories, irrespective of their probability. In particular, any linear or nonlinear map characterised by the same Markov partition is characterised by the same topological entropy $\mathcal{L}(1)$.

As an example, we now consider the map

$$U(t+1) = \begin{cases} 1 - p + \frac{p}{1-p} U(t) & \text{if} \quad U(t) < 1 - p \\ 1 - \frac{U(t) - 1 + p}{p} & \text{if} \quad U(t) > 1 - p. \end{cases} \qquad (5.15)$$

The corresponding evolution operator is the 2×2 matrix

$$Q = \begin{pmatrix} 0 & p \\ \frac{1-p}{p} & p \end{pmatrix},$$

whose eigenvalues are $\alpha = 1$ and $\alpha = -1 + p$. The first eigenvalue follows from the conservation of the probability, while the second eigenvalue accounts for the exponential convergence towards the stationary distribution. The evolution of the multiplier moments is instead controlled by

$$G(q) = \begin{pmatrix} 0 & p^{1-q} \\ \left(\frac{1-p}{p}\right)^{1-q} & p^{1-q} \end{pmatrix}.$$

Once again, the matrix is characterised by two eigenvalues. The generalised Lyapunov exponent is defined by the largest one, i.e.

$$\mathcal{L}(q) = \frac{1}{q} \log \left[\left(p^{1-q} + \sqrt{p^{2(1-q)} + 4(1-p)^{1-q}} \right) /2 \right].$$

Its shape and the corresponding large-deviation function S for $p = 0.5$ are drawn in Fig. 5.3. The minimum and the maximum exponents correspond to a period-2 and a fixed-point orbit, respectively. The separation between the maximum and the second exponent determines the convergence rate of $\mathcal{L}(q)$, if one were to determine it from Eq. (5.6) for a finite τ. This is true in general, irrespective of the dimensionality of the matrix $G(q)$.

So far we have analysed a setup where the evolution operator reduces to a finite matrix and the generalised Lyapunov exponents can be thereby exactly determined. In general, one cannot get rid of the infinite dimensionality of the Perron-Frobenius operator, and it is necessary to rely on suitable perturbative techniques. This happens already in one-dimensional maps, which, although admitting a Markov partition, have a nonlinear structure, so that the invariant measure is not piecewise constant, as in the previous example. In such a case, it is natural to approximate the original map with a piecewise linear one, by refinining the Markov partition.

This goal can be achieved by further splitting the unit interval with the help of the preimages of the borders of the initial atoms. The strategy is briefly illustrated in Fig. 5.4, where the standard generating partition, made of the two intervals $[P_0, P_2]$ and $(P_2, P_1]$, is Markov since $F(P_0) = P_2$ (while $F(P_1) = P_0$ and $F(P_2) = P_1$, by construction). A second-order refinement leads to the addition of three points, namely P_3, P_4 and P_5 (with $F(P_3) = P_0$ and $F(P_4) = F(P_5) = P_3$), and thereby to a partition made of five atoms; see $I_1, \ldots, I_5$ in Fig. 5.4, where the dotted line describes the corresponding piecewise linear approximation of the original map.

The convergence of the asymptotic Lyapunov exponents and of the large deviation function can be appreciated in Fig. 5.5, where we report the results for increasingly accurate partitions (see the figure caption).

Notice that all approximations yield the same value of $\mathcal{L}(1)$. In fact, as mentioned previously, $\mathcal{L}(1)$ corresponds to the topological entropy, a concept that is associated with a counting of orbits, irrespective of their probability.

(a)

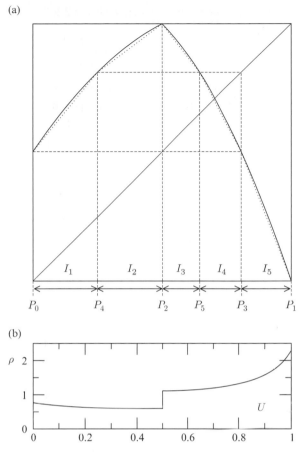

(b)

(a) Piecewise nonlinear map: $U(t+1) = a + (1-a)U(t)/a + U(t)(a - U(t))$ (for $U(t) < a$); $U(t+1) = 1 - (U(t) - a)/(1-a) + b(1 - U(t))(U(t) - a)$ (for $U(t) > a$), with $a = 1/2$, $b = 2$. The diagonal corresponds to the bisectrix. (b) The corresponding probability density.

In generic dynamical systems, however, no finite Markov partition exists at all, and even the topological entropy cannot be exactly determined. In such cases, the very first objective consists in building (either implicitly or explicitly) an approximate Markov partition. An effective, though entirely numerical, strategy consists in identifying (with the help of a generating partition) the irreducibile "forbidden" words of increasing length. A word is called forbidden if no trajectory exists which can generate such a sequence; an example is the word "00" in the map in Fig. 5.4. The word is termed *irreducibile* if it does not contain any shorter subsequence that is itself forbidden. Given a list of such words up to length n, it is possible to build a suitable Markov process which generates a sequence of words (a language) with exactly the same set of irreducible words (see Badii and Politi (1997)). By construction, the resulting topological entropy provides an upper bound to the true value since the missing information amounts to additional constraints that are not taken into account. The corresponding (approximate) Markov partition can be then refined, as in the previous case, to better approximate the nonlinearity of the map.

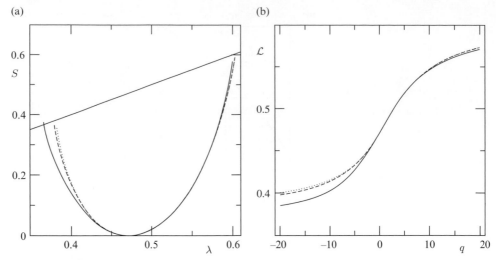

Fig. 5.5 (a) Large deviation function and (b) generalised exponents for the map described in the previous figure for different levels of approximation. Solid, dashed and dotted lines refer to partitions built with all preimages up to orders 8, 12 and 16, respectively. The tilted straight line in panel (a) corresponds to the bisectrix.

5.6   Lyapunov exponents in terms of periodic orbits

In the previous sections we have seen that Lyapunov exponents can be determined by averaging some observable or determining the (leading) eigenvalue of suitable evolution operators. As a third option, one can exploit the information contained in the stability of the (unstable) periodic orbits that are embedded in a generic chaotic set.[4]

In this section we mainly follow the book by Cvitanović et al. (2013), with a slight adjustment of the notations. We start by introducing in a more formal way the relationship between the computation of (generalised) Lyapunov exponents and the identification of the largest eigenvalue of a suitable evolution operator.

Given the general Eq. (5.3), if one adds the integration of an n-dimensional δ-function over the set of all final points $\mathbf{V} = \mathbf{F}^t(\mathbf{U})$ (omitting the argument of $\mathbf{U}(0)$ in the initial density as well as in Υ^t), it is found that

$$\left\langle e^{q\Upsilon^\tau} \right\rangle = \int d^N\mathbf{U}\,\rho(\mathbf{U}) \int d^N\mathbf{V}\,\delta^N(\mathbf{V} - \mathbf{F}^\tau(\mathbf{U}))e^{q\Upsilon^\tau(\mathbf{U})}. \tag{5.16}$$

This relation can be considered as the action of the evolution operator $\mathcal{J}^\tau$ with kernel $\mathcal{J}^\tau(\mathbf{V}, \mathbf{U})$,

$$\mathcal{J}^\tau(\mathbf{V}, \mathbf{U}) = \delta(\mathbf{V} - \mathbf{F}^\tau(\mathbf{U}))e^{q\Upsilon^\tau(\mathbf{U})}, \quad \mathcal{J}^\tau\phi(\mathbf{V}) = \int d^N\mathbf{U}\,\mathcal{J}^\tau(\mathbf{V}, \mathbf{U})\phi(\mathbf{U}).$$

---

[4] Or hidden in the different realisations of a stochastic process – see the application of the zeta function formalism to products of random matrices in Section 8.1.

Its action on the initial density and the subsequent integration over the final point leads precisely to relation (5.16), which can be rewritten as

$$e^{\tau G(q)} = \left\langle e^{q \Upsilon^\tau} \right\rangle = \int d^N \mathbf{V} \left[\mathcal{J}^\tau(\mathbf{V}, \mathbf{U}) \rho(\mathbf{U}) d^N \mathbf{U} \right]. \qquad (5.17)$$

The main advantage of working with the operator $\mathcal{J}^\tau$ is its multiplicativity along the time evolution: $\mathcal{J}^{t_1 + t_2} = \mathcal{J}^{t_2} \mathcal{J}^{t_1}$. As a result, Eq. (5.17) implies that the characteristic function $G(q)$ indeed corresponds to the largest eigenvalue of $\mathcal{J}$, as more heuristically seen in the previous section.

Now, we focus on the trace of $\mathcal{J}$. The trace, being the sum of the eigenvalues, is dominated by the largest one, which is needed to determine $G(q)$. The trace emerges naturally by setting $\mathbf{V} = \mathbf{U}$,

$$e^{\tau G(q)} \sim \text{tr} \mathcal{J}^\tau = \int d^N \mathbf{U} \, \mathcal{J}^\tau(\mathbf{U}, \mathbf{U}) = \int d^N \mathbf{U} \delta^N(\mathbf{U} - \mathbf{F}^\tau(\mathbf{U})) e^{q \Upsilon^\tau(\mathbf{U})}. \qquad (5.18)$$

Thanks to the presence of a δ-function, the integral in this equation can be easily computed. The only contributions are those from the points $\mathbf{U}_s = \mathbf{F}^\tau(\mathbf{U}_s)$, i.e. from the fixed points of the τth iteration of the map $\mathbf{F}$. These are all periodic orbits with a period τ/m with $m \geq 1$. Supposing that there are $M(\tau)$ such points, then one can write

$$e^{\tau G(q)} \approx \int d^N \mathbf{U} \, \mathcal{J}^\tau(\mathbf{U}, \mathbf{U}) = \sum_{s=1}^M \frac{e^{q \Upsilon^\tau(\mathbf{U}_s)}}{|\det(\mathsf{I} - \mathsf{J}_\tau)|}. \qquad (5.19)$$

The factor $|\det(\mathsf{I} - \mathsf{J}_\tau)|$, where $\mathsf{J}_\tau = \frac{\partial \mathbf{F}^\tau}{\partial \mathbf{U}}$ is the Jacobian matrix of the map $\mathbf{F}^\tau$ at its fixed points, appears via the integration of the δ-function. Using the identity $\det(\mathsf{I} - \mathsf{J}_\tau) = \prod_{j=1}^N (1 - \mu_j)$, where μ_j are the eigenvalues of J_τ, i.e. the multipliers of the fixed points (some of them can be complex), we can rewrite (5.19) as

$$e^{\tau G(q)} \approx \sum_{s=1}^{M(\iota)} e^{q \Upsilon^\tau(\mathbf{U}_s)} \prod_{j=1}^N |1 - \mu_j|^{-1}.$$

Under the assumption of strict hyperbolicity, this expression can be simplified for large τ. If there are N_+ unstable directions, then N_+ multipliers of $\mathbf{F}^\tau$ are very large, while $(N - N_+)$ multipliers are very small. By thereby approximating $\prod_{j=1}^N |1 - \mu_j| \approx \prod_{j=1}^{N_+} |\mu_j|$, one obtains

$$G(q) \approx \frac{1}{\tau} \ln \left[\sum_{s=1}^M e^{q \Upsilon^\tau(\mathbf{U}_s)} \prod_{j=1}^{N_+} |\mu_j|^{-1} \right].$$

With the help of (5.5), this expression of the characteristic function allows one to determine the average of a generic observable in terms of averages performed over the periodic orbits

$$\langle v \rangle \approx \frac{\sum_{s=1} \tau^{-1} \Upsilon^\tau(\mathbf{U}_s) \prod_{j=1}^{N_+} |\mu_j|^{-1}}{\sum_s \prod_{j=1}^{N_+} |\mu_j|^{-1}}. \qquad (5.20)$$

This equation means that the weight of the contribution Υ^τ/τ of an orbit of period τ (i.e. a fixed point of the map $\mathbf{F}^\tau(\mathbf{U})$) is inversely proportional to the product of the unstable multipliers along the orbit itself. Thus, the more unstable an orbit is, the smaller its contribution to the general average. In the case of the largest Lyapunov exponent, Υ^τ is the logarithm of the largest multiplier μ_1, so that

$$\lambda \approx \frac{\sum_{s=1}^{M} \tau^{-1} \ln |\mu_1(\mathbf{U}_s)| \prod_{j=1}^{N_+} |\mu_j|^{-1}}{\sum_{s=1}^{M} \prod_{j=1}^{N_+} |\mu_j|^{-1}}. \tag{5.21}$$

In one-dimensional maps there is only one multiplier, and the expression (5.21) reduces to

$$\lambda \approx \frac{\sum_{s=1}^{M} \tau^{-1} \ln ||\mu(U_s)|| \mu(U_s)|^{-1}}{\sum_{s=1}^{M} |\mu(U_s)|^{-1}}. \tag{5.22}$$

Increasingly accurate approximations are obtained by considering larger τ-values in Eqs. (5.21, 5.22).

Eq. (5.18) can be better handled with the help of the zeta-function formalism. We start by rewriting Eq. (5.19) as

$$\mathrm{tr}\mathcal{J}^\tau = \sum_{s=1}^{M} \frac{e^{q\Upsilon^\tau(\mathbf{U}_s)}}{|\det(\mathsf{I} - \mathbf{J}_\tau)|} \tag{5.23}$$

and introduce the *prime cycles* as the non-repeating ones. If a periodic orbit P has a minimal (basic) period p, then it is also periodic with period rp for any integer r. Thus, a fixed point contributes to (5.23) for any τ, while period-2 orbits for any even τ, etc. These contributions can be easily summed. Since Υ^τ is additive along a trajectory of basic period p, we have that $\Upsilon^{rp} = r\Upsilon^p$. Similarly, $\mathbf{J}_{rp} = [\mathbf{J}_p]^r$ since the Jacobian matrices multiply along a trajectory. Furthermore, as all the points along a prime cycle are equivalent, they give p equal contributions to the sum in (5.23). This allows expressing the trace as a sum over the prime cycles P only,

$$\mathrm{tr}\mathcal{J}^\tau = \sum_{P} p \sum_{r} \frac{e^{rq\Upsilon(P)}}{|\det(\mathsf{I} - \mathbf{J}_p^r)|} \delta_{\tau,rp}, \tag{5.24}$$

where $\Upsilon(P)$ is a shorthand notation to refer to a generic prime-cycle of period p. The Kronecker δ selects only those prime cycles whose period p divides τ.

It is now convenient to introduce the *dynamical* or *Ruelle zeta function*,

$$\zeta(z,q) = \exp\left(\sum_{\tau=1}^{\infty} \frac{z^\tau}{\tau} \mathrm{tr}\mathcal{J}^\tau\right). \tag{5.25}$$

In the limit of large τ, $\mathrm{tr}\mathcal{J}^\tau \approx e^{G(q)\tau}$, where $G(q)$ is the largest eigenvalue of $\mathcal{J}$. We, therefore, see that the problem of determining the characteristic function $G(q)$ can be reduced to that of finding the smallest zero $\hat{z}$ of $\zeta(z,q)$, which corresponds to the radius of convergence of the zeta function, namely $\hat{z} = e^{-G(q)}$. It is also useful to recall the additional relationship

$$\frac{1}{\zeta(z,q)} = \prod_{k}(1 - \frac{z}{z_k}) = \det(1 - z\mathcal{J}) = \exp\left[\mathrm{tr}\ln(1 - z\mathcal{J})\right],$$

where z_k are the eigenvalues of $\mathcal{J}$.

With the help of (5.24), one finds that

$$\frac{1}{\zeta(z,q)} = \sum_P \sum_r \frac{z^{rp} e^{rq\Upsilon(P)}}{r|\det(I - J_p^r)|}.$$

Notice that the sum over τ in (5.25) has allowed removing the Kronecker δ in Eq. (5.24); this is the consequence of having introduced the auxiliary variable z.

In order to get rid of the nasty "det" term, the same approximation as in Eq. (5.20) can be used, namely $\det|(I - J_p^r)| \approx \mu(P)$, where $\mu(P) = \prod_{j=1}^{N_+} |\mu_j|$ is the product of all multipliers of the cycles P that are larger than 1 in absolute value. With the following simplifying notation

$$Q_P = \mu^{-1} z^p \exp[q\Upsilon(P)],$$

one can finally write

$$\frac{1}{\zeta(z,q)} = \exp\left(-\sum_P \sum_{r=1}^{\infty} \frac{Q_P^r}{r}\right) = \prod_P (1 - Q_P), \tag{5.26}$$

where the last step follows from the equality $\sum_{r=1}^{\infty} r^{-1} Q^r = -\ln(1 - Q)$.

At first glance it looks as if the zeros of the r.h.s. of (5.26) are given by the condition $Q_P = 1$. This is not true, as one has an infinite product of the factors, which can be written as

$$\prod_P (1 - Q_P) = 1 - \sum_{\{P_1, P_2, \dots, P_k\}}' (-1)^{k+1} Q_{P_1} Q_{P_2} \dots Q_{P_k}, \tag{5.27}$$

where the prime means that the sum is over all distinct non-repeating combinations of the prime cycles. This expression is basically a polynomial expansion in powers of z (notice that Q_{P_i} is a monomer proportional to z^{p_i}).

The average of the generic observable υ can be finally obtained from Eq. (5.5), by recalling that $\hat{z}(q) = \exp[-G(q)]$ is the leading zero of $\zeta^{-1}(z,q)$ and computing the total derivative of the expression $\zeta^{-1}(\hat{z}(q), q) = 0$,

$$\langle \upsilon \rangle = \left.\frac{\partial_q \zeta^{-1}}{\partial_z \zeta^{-1}}\right|_{(1,0)} = \left.\frac{\sum' \Upsilon_\pi Q_\pi}{\sum' p_\pi Q_\pi}\right|_{(1,0)}. \tag{5.28}$$

Here the subscript in ∂ denotes a partial derivative with respect to that variable, while

$$Q_\pi = (-1)^{k+1} Q_{P_1} Q_{P_2} \dots Q_{P_k}\Big|_{(1,0)} = [\mu_{p_1} \mu_{p_2} \dots \mu_{p_k}]^{-1},$$

$$p_\pi = p_1 + p_2 + \dots p_k, \tag{5.29}$$

$$\Upsilon_\pi = \Upsilon_{p_1} + \Upsilon_{p_2} + \dots \Upsilon_{p_k}.$$

In the case of the largest Lyapunov exponent, $\Upsilon_{p_k} = \ln|\mu_1(p_k)|$, where μ_1 is the largest multiplier of the prime cycle p_k.

For the sake of simplicity, hereafter we consider the simplest nontrivial case of prime cycle organisation, namely that of a dynamical system characterised by the two symbols $(0, 1)$, where all sequences are possible (this is the case of the Bernoulli map, the tent map and the logistic map in the fully chaotic regime). The prime cycles are: two fixed points (0 and 1), one period-2 cycle (01), two period-3 cycles (100 and 110), three period-4 cycles (1000, 1100 and 1110) and so on. Writing explicitly only cycles up to length 4, one obtains a 4th-order polynomial approximation of the zeta-function

$$\frac{1}{\zeta(z(q), q)} = 1 - Q_0 - Q_1 - [Q_{10} - Q_0 Q_1]$$
$$- [Q_{100} - Q_{10} Q_0] - [Q_{110} - Q_1 Q_{10}] -$$
$$- [Q_{1000} - Q_{100} Q_0] - [Q_{1110} - Q_1 Q_{110}]$$
$$- [Q_{1100} - Q_1 Q_{100} - Q_{110} Q_0 + Q_1 Q_{10} Q_0] \quad - \cdots$$

Here we ordered the terms according to powers of z and additionally organised them into groups characterised by the same symbolic sequence, though composed of cycles with different periods. This grouping leads to a rapidly converging result, as the terms in square brackets nearly cancel each other. Substituting this expansion into (5.28) and (5.29), one can finally obtain the Lyapunov exponent as a ratio from Eq. (5.28),

$$\partial_q \zeta^{-1} = \frac{\ln \mu_0}{\mu_0} + \frac{\ln \mu_1}{\mu_1} + \left[\frac{\ln \mu_{10}}{\mu_{10}} - \frac{\ln \mu_0 + \ln \mu_1}{\mu_0 \mu_1} \right] \tag{5.30}$$
$$+ \left[\frac{\ln \mu_{100}}{\mu_{100}} - \frac{\ln \mu_{10} + \ln \mu_0}{\mu_{10} \mu_0} \right] + \left[\frac{\ln \mu_{110}}{\mu_{110}} - \frac{\ln \mu_{10} + \ln \mu_1}{\mu_{10} \mu_1} \right] + \cdots$$
$$\partial_z \zeta^{-1} = \frac{1}{\mu_0} + \frac{1}{\mu_1} + \left[\frac{2}{\mu_{10}} - \frac{2}{\mu_0 \mu_1} \right] + \left[\frac{3}{\mu_{100}} - \frac{3}{\mu_{10} \mu_0} \right] + \left[\frac{3}{\mu_{110}} - \frac{3}{\mu_{10} \mu_1} \right] + \cdots,$$

where μ_α is the absolute value of the multiplier of the prime cycle with symbolic sequence α.

If the map is one-dimensional and piecewise linear, the slope is constant in each of the two atoms corresponding to 0 and 1; then $1/\mu_0 + 1/\mu_1 = 1$, $\mu_{10} = \mu_1 \mu_0$ etc., and all the terms in square brackets cancel exactly, so that we find $\lambda = (\ln \mu_0)/\mu_0 + (\ln \mu_1)/\mu_1$, as expected.

The convergence of the expressions (5.22) and (5.30) for the skew nonlinear tent map

$$F(U) = \begin{cases} \frac{U}{a} + U(a - U) & 0 \le U \le a \\ \frac{1-U}{1-a} + (U - 1)(a - U) & a < U \le 1 \end{cases}, \tag{5.31}$$

can be appreciated in Table 5.1, where one can see that the second more sophisticated approach is not substantially more accurate.

In Chapter 8, while discussing noisy systems we will see that the convergence of the zeta function formalism can be accelerated. This is made possible by including the information about the orientation of the corresponding Lyapunov vectors.

Table 5.1 Lyapunov exponent for the skew nonlinear tent map (5.31) with $a = 0.6$ for increasing approximations. The first line refers to Eq. (5.22); the second line refers to Eqs. $(5.28, 5.30)$.

2	3	4	5	6	7	8
0.704215	0.678295	0.678407	0.680967	0.681766	0.681785	0.681724
0.718362	0.684812	0.679436	0.680542	0.681482	0.681727	0.681732

5.7 Examples

5.7.1 Deviation from hyperbolicity

The analysis of large deviations helps to clarify whether the underlying dynamics is hyperbolic. In particular, the occurrence of occasional exchanges in the order of the finite-time LEs is related to the concept of dominated Oseledets splitting (Bochi and Viana, 2005): the splitting is dominated with index i if there exists a *finite* τ_0 such that $\Lambda_i(\tau) > \Lambda_{i+1}(\tau)$ for all $\tau > \tau_0$. This property implies the absence of tangencies between the corresponding Oseledets subspaces spanned by the ith and the subsequent vector (Bochi and Viana, 2005). The probability of order-exchanges is quantified by the large deviation function

$$S_i(\delta) = \min S(\Lambda_1, \Lambda_2, \ldots, \Lambda_i, \Lambda_i - \delta, \ldots, \Lambda_N),$$

where the minimum is taken over all possible values of $(\Lambda_1, \ldots, \Lambda_i)$ and of $(\lambda_{L+2}, \ldots, \Lambda_N)$. In practice, for the Oseledets splitting to be dominated, it is necessary that the domain of $S_i(\delta)$ does not extend to negative values. In general, it is quite hard to verify whether this condition is satisfied. One can nevertheless extract some semi-quantitative information with the help of a Gaussian approximation. In practice, it is necessary to determine the diffusion coefficient K_i of $\delta_i = \Lambda_i - \Lambda_{i+1}$; it can be expressed in terms of the diffusion matrix D (5.12),

$$K_i = D_{ii} + D_{i+1,i+1} - 2D_{i,i+1}. \tag{5.32}$$

Since the probability of an exchange of finite-time LEs is equal to the probability P_i^- of observing a negative δ_i, we have, in the Gaussian approximation,

$$P_i^- \approx \frac{1}{2}\mathrm{erfc}\left\{(\Lambda_i - \Lambda_{i+1})\sqrt{\frac{\tau}{2K_i}}\right\},$$

where "erfc" is the complementary error function. In practice, if $\sqrt{K_i}$ is substantially smaller than $\Lambda_i - \Lambda_{i+1}$, one can conclude that the given dynamical system is effectively hyperbolic, as the probability of inversion is negligible.

5.7.2 Weak chaos

LEs are the appropriate tool to quantify the degree of stability of a given dynamical system, when perturbations behave exponentially in time. There is, however, a relatively wide class of systems where perturbations behave sub-exponentially (Zaslavsky, 2007). This type of behaviour is often referred to as *weak chaos* or *pseudochaos*. In the context of Hamiltonian models, it includes models with mixed phase space (Zaslavsky and Edelman, 2004) and billiards (Artuso et al., 1997), but it arises also at the transition to chaos (either via period-doubling or intermittency).

Whenever the perturbation growth is sub-exponential, the LE cannot discriminate between instability and neutral stability. Given the different type of dynamics that can hide itself behind a zero Lyapunov exponent (see, e.g., power-laws vs. stretched exponentials), one cannot build a general theory.

In some cases, however, in spite of an average sub-exponential growth, it is still possible to use the large-deviation theory to extract useful information. A rather general and relatively simple example is intermittency, described by one-dimensional maps of the Pomeau-Manneville type

$$U(t+1) = F(U(t)) = U(t) + aU^z(t) \quad \text{mod } 1 \tag{5.33}$$

with $z > 1$. The map, depicted in Fig. 5.6, is of a Bernoulli type, with the important difference that one of the two fixed points ($U = 0$) is marginally (neutrally) stable. It is important to understand to what extent the presence of a neutrally stable point affects the overall chaotic properties. The Pomeau-Manneville dynamics with integer z typically arises at bifurcation points, when a fixed point loses stability, in which case one generically expects $z = 2$ (or $z = 3$ if some symmetry is present). Here, we discuss the case of a general z value, as it helps to better clarify the underlying mechanisms.

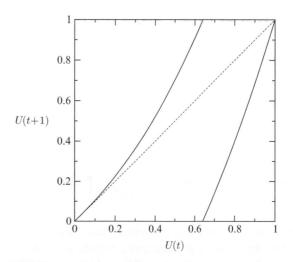

Fig. 5.6   Pomeau-Manneville map (5.33) for $a = 1$ and $z = 2.3$.

The computation of the LE in one-dimensional maps is relatively simple. As already seen in Section 3.4, λ can be determined as an ensemble average

$$\lambda = \int dU \rho(U) \log |F'(U)|,$$

where $\rho(U)$ is the invariant measure. Even under such a simplifying hypothesis, an analytic expression for $\rho(U)$ is rarely known (one of the few exceptions can be found in Pikovsky (1991)). In this case, however, it is sufficient to determine the critical behaviour of ρ in the vicinity of $U = 0$. For small U, the dynamics is well approximated by the differential equation

$$\dot{U} = aU^z.$$

One can assume that this evolution equation holds up to $U = 1$, without causing any change of the critical behaviour. Therefore, the probability distribution satisfies the equation

$$\frac{\partial \rho}{\partial t} = -\frac{\partial}{\partial U}[aU^z \rho] + S, \tag{5.34}$$

where the phenomenological uniform term S accounts for the re-injection of points from the second branch of the map. In principle S should vary in time in order to ensure that the overall probability is constantly equal to 1. Since the fluctuations of the transient length do not affect our scaling arguments, we assume S to be constant.

This equation admits the simple stationary solution

$$\rho_0(U) = \frac{C}{U^{z-1}},$$

where C is a suitable normalisation constant, to be determined. For $z > 1$, $\rho_0(U)$ exhibits a divergence in the origin. Let us now study the convergence of an initially flat distribution $\rho(U, 0) = 1$. Upon neglecting the spatial derivative of ρ in Eq. (5.34), one finds that

$$\frac{d\rho}{dt} \approx -aU^{z-1}\rho + S.$$

This equation reveals that the relaxation rate of ρ depends strongly on U (it vanishes for $U \to 0$). As a consequence, at any finite time t, the probability density has relaxed to its aymptotic value only for $U > \delta U$, where $\delta U(t)$ can be estimated as the point where the relaxation time is equal to the elapsed time t, obtaining $\delta U(t) \approx t^{-1/(z-1)}$. For $U < \delta U$, the relaxation is so slow that one can safely approximate $\rho(U, t)$ with $\rho_0(\delta U(t))$. Since $\int_0^{\delta U} dU \rho(U, t) \approx 1/t$ (independently of z), one can approximate $\rho(U, t)$ with ρ_0 for $\delta U < U < 1$. Accordingly, the normalisation constant is

$$\frac{1}{C(t)} = \frac{1}{2 - z}[1 - (\delta U)^{2-z}] = \frac{1}{2 - z}[1 - t^{-(2-z)/(z-1)}].$$

For $1 < z < 2$, $C(t)$ converges to a finite value; this is no surprise, as the singularity in zero is integrable. For $z > 2$, instead,

$$C(t) \approx t^{(2-z)/(z-1)}.$$

We can now go back to the problem of computing the LE. For small U, the logarithm of the local multiplier vanishes as U^{z-1}, thus matching the divergence of the probability density.

Therefore, for $z < 2$ (when the normalisation constant has a finite value) the integral in Eq. (5.33) is not singular, and a finite (positive) Lyapunov exponent is expected. For $z > 2$, the presence of a time-dependent normalisation constant implies that

$$\Lambda(t) \approx t^{(2-z)/(z-1)},$$

or, equivalently, that the overall exponential growth is

$$\Gamma(t) = t\Lambda \approx t^{1/(z-1)} = t^{\alpha}. \tag{5.35}$$

This equation implies that Γ does not increase linearly with time, as usual in exponentially unstable regimes, but slower, with a power $\alpha < 1$; in other words we are facing a stretched exponential behaviour.

Here, we show that in spite of the anomalous scaling of the perturbation, some useful information can be nevertheless extracted by means of the large-deviation approach. We start by considering the case $z = 1.5$. The corresponding generalised LE $\mathcal{L}(q)$ is plotted in Fig. 5.7a.

There, we see that for $q \geq 0$, $\mathcal{L}(q)$ behaves normally, while it is identically equal to zero for negative q-values, with a discontinuity in $q = 0$. The corresponding large-deviation function $S(\Lambda)$ is plotted in panel (b). Below Λ_0 (see the vertical dashed line), $S(\Lambda) = 0$; above, $S(\Lambda) > 0$ grows quadratically and eventually reaches the diagonal in a point which corresponds to the most unstable fixed point of the map ($U = 1$), whose

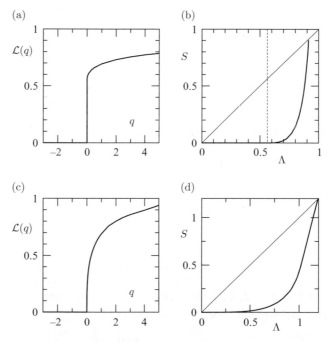

Fig. 5.7 The generalised Lyapunov exponent $\mathcal{L}(q)$ and the large deviation function $S(\Lambda)$ for the Pomeau-Manneville map (5.33) with $a = 1$. Panels (a) and (b) refer to $z = 1.5$, while panels (c) and (d) correspond to $z = 2.3$. The dashed line in panel (b) signals the value Λ_0, where S starts being strictly larger than 0.

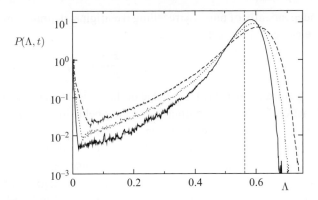

Fig. 5.8 Probability density of the Lyapunov exponents for the Pomeau-Manneville map (5.33) for $a = 1$ and $z = 1.5$. Dashed, dotted and solid lines correspond to $t = 250, 500$ and 1000, respectively. The vertical dashed line corresponds to the true LE.

Lyapunov exponent is $\Lambda = \ln 2.5$. This means that the probability density $P(\Lambda, t)$ decreases slower than exponentially in the interval $0 < \Lambda < \Lambda_0$, so that one cannot conclude where the LE is located along the interval itself. A more refined analysis is needed. In Fig. 5.8 we plot $P(\Lambda, t)$ for three different time values. There we see that the distribution becomes increasingly peaked and the peak position converges towards Λ_0 (see the vertical dashed line). Moreover, one can recognise that the low-density region does not decrease exponentially with time but rather as a power law (approximately as $1/t$).

For $z > 2$, the presence of a marginally stable fixed point $U = 0$ has deeper consequences on the overall stability of the dynamical system. In Fig. 5.7c, $\mathcal{L}(q)$ is plotted for $z = 2.5$. The most substantial difference with the previous case is the disappearance of the discontinuity for $q = 0$. This implies that $S(\Lambda)$ is now strictly larger than zero for any $\lambda > 0$, signalling that the LE is equal to zero, as previously anticipated. Numerical studies performed for different z values reveal that $\mathcal{L}(q) \approx q^\beta$ for small q values, while $S(\Lambda) = \Lambda^\gamma$ (for $\Lambda \ll 1$). From the Legendre relationship linking the two observables, it follows that

$$\gamma = 1 + \frac{1}{\beta}.$$

By then recalling the definition of $S(\Lambda)$ in terms of $P(\Lambda, t)$,

$$\mathcal{P}(\Gamma, t) \approx \exp[-b\Gamma^\gamma t^{1-\gamma}],$$

where $\Gamma = \Lambda t$, while b denotes some unknown constant. This equation implies that the growth prefactor (sometimes called generalised exponent)

$$\xi = \frac{\Gamma}{t^\alpha}, \tag{5.36}$$

where $\alpha = 1 - 1/\gamma$, is distributed as

$$Q(\xi) \approx \exp[-b\xi^\gamma],$$

independently of time. By recalling the original definition of α in terms of z (see Eq. (5.35)) we can conclude that

$$\gamma = \frac{z-1}{z-2}, \quad \beta = z - 2.$$

As a consequence, the knowledge of the critical behaviour exhibited by the large-deviation function S allows us to infer the scaling behaviour of the perturbation even in a case like this, where the LE is equal to zero.

Finally, notice that the growth prefactor ξ defined by Eq. (5.36) has no specific limit value for $t \to \infty$; it keeps fluctuating, no matter how long the computation time is. The functional dependence outlined in the previous equation is valid in the limit of large ξ values. A globally valid expression can be found in Korabel and Barkai (2010), where a more refined treatment has been implemented. A rigorous analysis of the large deviation function can be found in Pires et al. (2011) and in Venegeroles (2012).

Finally, let us recall that different instances of weak chaos are typically characterised by a different scaling behaviour and, thereby, require different treatments. For instance, in the transition to chaos via period-doubling bifurcations, a power-law behaviour is found rather than the stretched-exponential behaviour discussed in this section.

5.7.3 Hénon map

The Hénon map provides a simple, but nevertheless general, setup for a fruitful test of the methods discussed in this chapter. We start by determining the large deviation function $S(\lambda)$ for the largest exponent with the help of the Lyapunov weighted dynamics. In Fig. 5.9a, the characteristic function $G(q) = q\mathcal{L}(q)$ is plotted versus q. Above $q = -1$ the method converges quite rapidly, even sufficiently away from the canonical value $q = 0$. Fluctuations become rather nasty below $q = -1$. This is due not to their low probability but to the fact that, as pointed out by Grassberger et al. (1988), $q = -1$ corresponds to a transition point from the hyperbolic phase to the phase dominated by homoclinic tangencies with strong fluctuations.

The Legendre transform of $G(q)$ is plotted in panel (b): the function S is quite smooth and well converged above a certain value of the Lyapunov exponent, while, below, it is affected by a large uncertainty. In practice, it appears that the weighted dynamics approach is well suited to describe the hyperbolic phase, while difficulties arise in the characterisation of the non-hyperbolic one.

There are no ambiguities in the definition of the maximum finite-time LE, as the two approaches discussed at the beginning of this chapter (through CLV or the QR decomposition) coincide. The same is not true for the second exponent. In fact, in Fig. 5.2b, we have seen that $\Delta\Gamma = (\Lambda_1 + \Lambda_2 - \ln b)\tau$ is not exactly equal to zero in the CLV approach. This quantity obeys an exponential distribution

$$P(\Delta\Gamma) \approx e^{-|\Delta\Gamma|}$$

for $\Delta\Gamma$ sufficiently different from 0 (the decay rate visible in Fig. 5.2b is indeed equal to 1 – see the dashed line). One can transform such fluctuations into those of the intensive

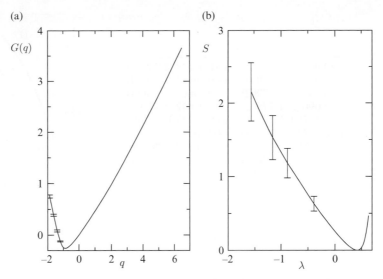

(a)

$G(q)$

(b)

S

q

λ

Fig. 5.9 (a) The generalised Lyapunov exponent $G(q) = q\mathcal{L}$ versus q for the Hénon map and (b) the corresponding large-deviation function S for the standard parameter values. The size of the error bar is only approximate. In the region where there are no error bars, the accuracy is better than the line size. Both curves have been obtained by implementing the weighted dynamics for $\tau = 20$.

variable $\delta\lambda = \Delta\Gamma/\tau$, obtaining

$$P(\delta\lambda) \approx \mathrm{e}^{-|\delta\lambda|\tau}.$$

This implies that the corresponding large deviation function is $S(\delta\lambda) = |\delta\lambda|$; i.e. differences between the two definitions of this compound observable are to be expected when $|q| > 1$. Such differences are a signature of the presence of homoclinic tangencies.

Finally, we consider the generalised Lyapunov exponent $\mathcal{L}(1)$. We have already commented – with reference to the one-dimensional map (5.15) – that $q = 1$ is a special value, where the LE does not depend on the probability distribution but only on the geometric structure of the attractor. In fact, as further explained in Chapter 6, it corresponds to the so-called topological entropy, a quantity that measures the growth rate of the number of "different" trajectories of a given length. The evaluation of $\mathcal{L}(1)$ greatly simplifies when it is possible to faithfully represent trajectories in phase space as sequences of symbols, through the implementation of a generating partition. In the case of the Hénon map, this can be done by following the approach sketched in Appendix D. In practice, one can split the phase space into two different symbolic regions, by constructing a suitable dividing line. The partition drawn in Fig. 5.10 is able to correctly identify all irreducible forbidden sequences up to length 20.

As mentioned previously, given a list of irreducible forbidden words, one can construct a suitable Markov process and thereby determine the topological entropy as the maximum eigenvalue. By using a more refined partition than that reported in the table, and

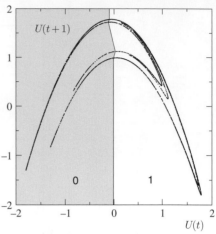

(−0.09648, 2.)	(−0.09648, 1.78027)
(−0.09645, 1.78000)	(−0.09641, 1.77979)
(−0.09673, 1.77902)	(−0.09669, 1.77883)
(−0.09672, 1.77802)	(−0.09672, 1.77799)
(−0.11750, 1.72371)	(−0.11743 1.72290)
(−0.11744, 1.72283)	(0.02682, 1.12013)
(0.02670, 1.11999)	(−0.00895 0.98676)
(−0.00893, 0.98653)	(−0.01004, 0.98484)
(−0.01004, 0.98450)	(−0.01004 −2.)

Fig. 5.10 Generating partition of the Hénon map, for $a = 1.4$ and $b = 0.3$. The eighteen points reported in the table correspond to the edges of the dividing line.

considering irreducible forbidden words of progressively longer length, it is possible to extrapolate that $\mathcal{L}(1) = 0.464936 \pm 0.000003$.[5]

5.7.4 Mixed dynamics

In the previous section we encountered a large deviation function whose domain of definition covers positive as well as negative Λ-values, indicating that the dynamics of the Hénon map is a mixture of stable and unstable trajectories. Non-hyperbolic chaos is one of the many examples that can be encountered in nonlinear dynamics. Another example is the bubbling transition, associated with the loss of transversal stability of a chaotic invariant manifold (Pikovsky and Grassberger, 1991; Ashwin et al., 1994) (see also the synchronisation transition discussed in Section 9.2). The reader interested in further details can consult Politi (2014b).

Next, we analyse two amazing cases, where rare positive fluctuations can induce an irregular dynamics even in the presence of a negative Lyapunov exponent.

Strange nonchaotic attractors

Strange nonchaotic attractors (Romeiras et al., 1987) are non-autonomous systems driven by quasiperiodic forces, which possess a non-smooth structure in spite of being characterised by a negative maximal LE. We refer, here, to the Lyapunov exponent of the driven, nonautonomous subsystem – the quasiperiodic driver which is obviously characterised by two zero Lyapunov exponents. According to the usual definition of chaos, such systems are nonchaotic, but the domain of definition of the corresponding large deviation function

---

[5] This estimate is implicitly contained in D'Alessandro et al. (1990). The use of all irreducibile forbidden words up to length 31 yields also the upper bound 0.4649418.

$S(\Lambda)$ extends to positive as well as to negative Λ-values. This is, indeed, the reason why they are characterised by a fractal structure (Pikovsky and Feudel, 1995).

The connection can be analytically unravelled by studying a simple one-dimensional, quasiperiodically driven map (in our presentation we follow Feudel et al. (2006))

$$U(t+1) = f(U(t), \theta(t)),$$
$$\theta(t+1) = \theta(t) + \omega \quad (\text{mod } 1), \tag{5.37}$$

where θ is a phase-like variable which describes a quasiperiodic motion (provided the frequency ω is irrational), while the function f is periodic in θ. If U is, on the attractor of the map (5.37), a smooth function of θ, then the attractor itself is a standard torus. Smoothness can be assessed from the derivative $\partial U/\partial\theta$. From Eq. (5.37), one obtains the recursive relation

$$\frac{\partial U}{\partial \theta}(t+1) = \frac{\partial f}{\partial x}(t)\frac{\partial U}{\partial \theta}(t) + \frac{\partial f}{\partial \theta}(t),$$

whose iteration leads to

$$\frac{\partial U}{\partial \theta}(t) = \sum_{k=0}^{t-1} \frac{\partial f}{\partial \theta}(k)X(k+1,t), \qquad X(k+1,t) = \prod_{j=k+1}^{t-1} \frac{\partial f}{\partial x}(j) = \frac{\partial U(t)}{\partial U(k+1)}. \tag{5.38}$$

The last expression involves the derivative of the variable U at time t with respect to its value at a previous time $k+1$; this quantity is related to the finite-time Lyapunov exponent,

$$\frac{\partial U(t)}{\partial U(k+1)} = \pm \exp[(n-k-1)\Lambda(k+1,t)]. \tag{5.39}$$

According to Eq. (5.8), the probability density of Λ is $\sim e^{-S(\Lambda)(t-k-1)}$, where its domain of definition is $[\Lambda_{min}, \Lambda_{max}]$, and its minimum (equal to zero) is in $\Lambda = \lambda < 0$. If $\Lambda_{max} < 0$, then all of the factors X are bounded exponentially $|X(k+1,t)| < \exp[(t-k-1)\Lambda_{max}]$, and the sum in (5.38) converges unconditionally as $t \to \infty$. This means that the derivative $\partial U/\partial\theta(t)$ is bounded and the function $U(\theta)$ is a smooth torus.

If $\Lambda_{max} > 0$, arbitrarily large values of X are possible (although they can be rare events). Such large events dominate the sum in (5.38). In order to estimate the largest factor X, we have to compare the probability of rare events with the observation time t. Within a long but finite time interval t, an exponent Λ (defined over a time τ) can be observed, if its probability is at least of order $1/t$, i.e. if $\exp[-S(\Lambda)\tau] \approx 1/t$. This yields $\tau \approx (\log t)/S(\Lambda)$. Substituting this into the expression (5.39), we obtain $|X| \approx t^{\frac{\Lambda}{S(\Lambda)}}$, and one has now to take the maximum over all possible Λ values,

$$t^\mu \sim \max_{k<t} |X(k+1,t)| \qquad \mu = \max\left(\frac{\Lambda}{S(\Lambda)}\right).$$

As $S(\Lambda)$ extends to positive Λs, $\mu > 0$ and arbitrarily large contributions to the sum (5.38) can be present. This indicates that $U(\theta)$ cannot be a smooth, but rather, a fractal curve. Therefore it is called a strange nonchaotic attractor. In Fig. 5.11 we present some numerical estimates of $S(\Lambda)$ for the system (5.37) with $f(U,\theta) = 3\tan U \cos(2\pi\theta)$ and $\omega = (\sqrt{5}-1)/2$ (the first observation of a strange nonchaotic attractor in this model is

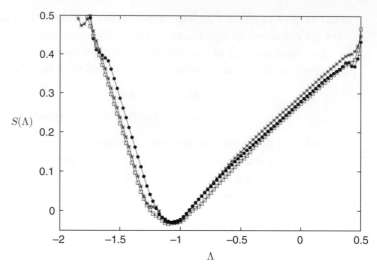

Fig. 5.11 The large deviation function for the system (5.37) with $f(U, \theta) = 3 \tan U \cos(2\pi\theta)$. Squares, stars and filled circles refer to three different values of the observation times τ: 40, 50 and 60, respectively. The oscillations in the tails are due to insufficient statistics.

due to Romeiras et al. (1987)). The minimum of $S(\Lambda)$ is around -1; i.e. the Lyapunov exponent is negative, but positive finite-time Lyapunov exponents can be observed up to $\Lambda \lesssim 0.5$.

Stable chaos

A perhaps more intriguing example of mixed dynamics is the so-called *stable chaos*, an irregular regime characterised by a negative Lyapunov exponent (Politi and Torcini, 2010). This phenomenon, first observed in a chain of coupled maps (Crutchfield and Kaneko, 1988) and later discovered also in neural networks (Zillmer et al., 2006), is, strictly speaking, transient, but it is exponentially long in large systems.

The coupled chain of piecewise linear maps (A.14) is an appropriate testbed for this regime. For a relatively wide range of parameter values, a lattice of such maps converges to more or less structured stable periodic states, but the transient time is exponentially long with the lattice length. Moreover, in spite of the fact the LE is negative, the transient itself is irregular and seemingly stationary (there is no obvious precursor of the final collapse onto some order orbit). In practice, this behaviour is akin to that of (chaotic) cellular automata (Wolfram, 1986), the main difference being that, here, the variables are continuous.

Some light has been shed on the nature of stable chaos by monitoring the evolution of perturbations. By following small but finite perturbations (i.e. studying finite-amplitude Lyapunov exponents; see Chapter 7), it is found that they undergo a sudden amplification, becoming of order 1. The analysis of large deviations, instead, reveals a positive tail.

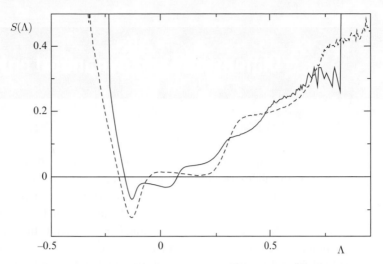

Fig. 5.12 Large deviation function for the model (A.13, A.14) for $\eta = 10^{-4}$ and $\varepsilon = 1/3$. Dashed and solid lines correspond to an observation time $t = 20$ and 40, respectively. The chain length is $L = 200$.

A typical instance is plotted in Fig. 5.12 for two different observation times. Although finite-size corrections are still fairly large, one can notice that the domain of definition of $S(\Lambda)$ extends to positive exponents, suggesting that the rare expansion events are sufficiently important to trigger and sustain the irregular dynamics.

6 Dimensions and dynamical entropies

The Lyapunov exponents of a deterministic system allow quantifying not only its stability and the predictability of the resulting dynamics, but also other important properties such as the fractal dimension of the underlying attractor and its dynamical entropy.

Let us briefly recall the definition of fractal dimension. Given a set $\mathcal{A}$ (which is, for us, a generic attractor), we introduce a covering $\mathcal{C}(\varepsilon)$, as a collection of balls of size ε, such that any point of the attractor belongs to at least one ball. The fractal dimension can be thereby defined from the scaling behaviour of the number $N(\varepsilon)$ of balls of size ε needed to cover the attractor,

$$\mathcal{D} = - \lim_{\varepsilon \to 0} \min \frac{\ln N(\varepsilon)}{\ln \varepsilon}, \tag{6.1}$$

where the minimum over all possible covering procedures has the role of ensuring a negligible overlap among the different balls. This definition is the so-called box dimension. Other definitions are possible which yield, in general, different values. The difference among various definitions follows from the existence of fluctuations in the local density of the invariant measure over arbitrarily small scales, which make the measure "multifractal". As we will see in the final section, there is a strong analogy between these fluctuations and those exhibited by the finite-time LEs that are responsible for the existence of a finite range of generalised LEs. A similar scenario arises also for the dynamical entropy.

In order to make the key elements more understandable, we first neglect the existence of fluctuations, deferring the general treatment to the final section.

6.1 Lyapunov exponents and fractal dimensions

In this section we derive the basic Kaplan-Yorke formula (Kaplan and Yorke, 1979), which expresses the fractal dimension of a given dynamical system in terms of its Lyapunov exponents. We start by illustrating the connection in a simple two-dimensional chaotic system, namely the Hénon map, as it allows an approach that will be straightforwardly extended to open systems in Chapter 11.

Given an attractor $\mathcal{A}$, consider a compact region S_0 of size $\mathcal{O}(1)$ which contains the attractor ($\mathcal{A} \in S_0$) and it is fully contained in its basin of attraction. S_0 can be used to build a proper covering of $\mathcal{A}$, as its forward iterates S_t provide an increasingly refined covering of $\mathcal{A}$ (the almost triangular region ABC in Fig. 6.1a represents one such example for the Hénon map).

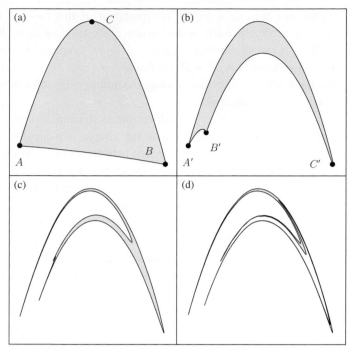

Fig. 6.1 The initial region S_0 covering the Hénon attractor and its first three iterates. The primed letters denote the images of the non-primed ones.

As a result of the expansion along the unstable direction and of the contraction along the stable one, the forward iterates S_t of S_0 resemble increasingly thin and long cylinders (see panels (b)–(d) in Fig. 6.1, where the first three iterates are represented).

Upon cutting S_t transversally into thin slices of size equal to the cylinder thickness, one obtains a partition of the attractor into boxes of size

$$\varepsilon = \exp\{\lambda_2 t\},$$

where λ_2 is the second (negative) Lyapunov exponent (as anticipated, here we neglect the fluctuations of the finite-time Lyapunov exponent which is assumed to be everywhere equal to the asymptotic value – see Section 6.4.1 for a more accurate analysis). The crux of the argument is that such a partition becomes increasingly accurate as the time t evolves (since $\lambda_2 < 0$). By then recalling the definition (6.1) of the fractal dimension, one can conclude that

$$\mathcal{D}_{KY} = -\frac{\ln N}{\ln \varepsilon} = -\frac{\ln(e^{\lambda_1 t}/\varepsilon)}{\ln \varepsilon} = 1 - \frac{\lambda_1}{\lambda_2}, \qquad (6.2)$$

where λ_1 is the positive Lyapunov exponent. Eq. (6.2) is nothing but the well-known Kaplan-Yorke formula (Kaplan and Yorke, 1979) for two-dimensional maps. Although its derivation is rather sketchy, this equation is exact, provided that $\mathcal{D}_{KY}$ is interpreted as the information dimension (Ledrappier, 1981; Young, 1982; Eckmann and Ruelle, 1985) (see also Section 6.4.1).

One can provide a simple interpretation of this formula by introducing the partial dimensions $\mathcal{D}^i$: these are the dimensions of the attractor along the directions oriented along the local invariant directions (i.e. the covariant Lyapunov vectors; see Chapter 4). One can in practice write $\mathcal{D} = \mathcal{D}^1 + \mathcal{D}^2$; in this specific case $\mathcal{D}^1 = 1$, while $\mathcal{D}^2 = |\lambda_1/\lambda_2|$, meaning that the attractor is continuous along the unstable direction, while it is fractal (Cantor-like) along the stable direction.

Let us now consider an N-dimensional dynamical system and let S_0 be a cuboid with edge $\mathcal{O}(1)$. At variance with the simple two-dimensional model, it is no longer obvious how to establish a connection between the elapsed time t and the accuracy of the corresponding partition. The fractal structure of the attractor is progressively uncovered by letting S_t evolve in time; at time t, we do not want to choose a box size ε that is smaller than the scales that spontaneously emerged.

In order to shed light on this problem, it is convenient to adopt a different point of view. Let us consider a small i-dimensional region R_0^i and its forward iterates R_t^i. Under the hypothesis of an ergodic evolution, various initial points will spread over the entire attractor. The volume of the region grows according to the sum of the first i Lyapunov exponents $\mathcal{S}_i$ (cf. Eq. (2.17)). If $\mathcal{S}_i > 0$, the volume of R_t^i grows exponentially while the attractor is being covered. One can thus conclude that the attractor dimension $\mathcal{D}_{KY} > i$; this process is tantamount to determining the "length" of a square: an infinitely long curve is needed to cover the full attractor. Analogously, an exponentially shrinking i-dimensional volume would signal that $\mathcal{D}_{KY} < i$. In practice, the volume neither expands nor contracts only when i corresponds to the true dimension. Since the number of directions is a discrete variable, in general there exists a value $m - 1$ such that $\mathcal{S}_{m-1} > 0$ while $\mathcal{S}_m < 0$. By linearly interpolating between such two limits, one finds

$$\mathcal{D}_{KY} = m - 1 + \frac{\lambda_{m-1}}{\lambda_m}. \tag{6.3}$$

This is the general form of the Kaplan-Yorke formula (see Fig. 6.2 for a geometric construction). The relationship between dimensions and Lyapunov exponents can be further clarified by expressing the volume dynamics in terms of the partial dimensions $\mathcal{D}^k$,

$$V(t, \{\mathcal{D}^k\}) = e^{\sum_k \mathcal{D}^k \lambda_k t}. \tag{6.4}$$

A meaningful value of the dimension $\mathcal{D}_{KY} = \sum_k \mathcal{D}^k$ must be such that the corresponding volume neither expands nor contracts, i.e.

$$\sum_k \mathcal{D}^k \lambda_k = 0. \tag{6.5}$$

It is obvious that there exist different sets $\{\mathcal{D}^k\}$ of partial dimensions, which can satisfy these constraints. The Kaplan-Yorke formula (6.3) is obtained by conjecturing that $\mathcal{D}^k = 1$ for $k < m$, $\mathcal{D} = 0$ for $k > m$, and that the mth partial dimension is the only one possibly having a non-integer value ($0 < \mathcal{D}^m \leq 1$). This formulation makes it transparent the reason why the Kaplan-Yorke formula provides, in general, an upper bound to the (information) dimension. Any other distribution of weights $\mathcal{D}^k$ would indeed give lower values. This is also the reason why the so-defined $\mathcal{D}_{KY}$ is often referred to as the *Kaplan-Yorke* dimension,

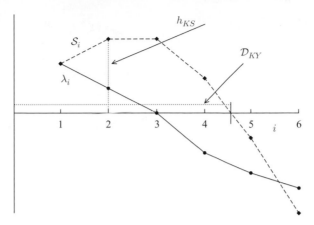

Fig. 6.2 Sketch of the Kaplan-Yorke dimension $\mathcal{D}_{KY}$ and of the Kolmogorov-Sinai entropy h_{KS}. The solid line refers to the Lyapunov exponents λ_i, while the dashed line corresponds to the partial sums $\mathcal{S}_i$.

to underline its possible difference with respect to the effective dimension. One example where there is a finite discrepancy is that of two (or more) uncoupled attractors, since different contracting directions separately contribute to compensate for the expansion along the expanding directions. It has been proved that $\mathcal{D}_{KY}$ is exact for two-dimensional mappings (where there is no ambiguity about the folding directions) and in random attractors, where the nonlinear dynamical rule changes stochastically in time. It is still unclear, however, under which conditions the Kaplan-Yorke formula holds exactly for generic high-dimensional models.

We can now return to the original approach and use the convergence of the region S_t to generate increasingly accurate coverings of an attractor. The identification of the mth direction as the only fractal one is equivalent to defining the proper box size as

$$\varepsilon = \exp(\lambda_m t). \tag{6.6}$$

One can easily see that the number of boxes of size ε, which is needed to cover the corresponding N-dimensional cylinder, scales according to the Kaplan-Yorke dimension $\mathcal{D}_{KY}$.

Finally, notice that the Kaplan-Yorke formula also provides useful information on the number of active degrees of freedom. In fact, in typical dissipative models, the number of variables that are necessary to uniquely identify the different points of an attractor is smaller (in some cases much smaller) than the phase-space dimension (i.e. the number of variables that are used to define the model itself). The Kaplan-Yorke formula can be viewed as a general prescription to estimate the number of "active" variables.

6.2 Lyapunov exponents and escape rate

In the case of chaotic attractors the invariant measure is typically smooth along the unstable directions, while it is assumed to be fractal along one of the stable directions.

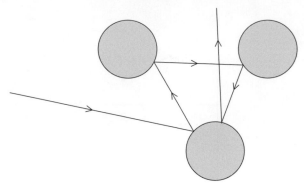

Fig. 6.3   Sketch of a multiple scattering process

There are, however, physical setups where no attractor is present in a given region of phase space, and yet an effective irregular dynamics can be sustained for arbitrarily long times. This is the case of scattering problems, such as the one depicted in Fig. 6.3, where an incoming particle may be trapped because of repeated collisions with the three disks. The invariant set, composed of all non-escaping trajectories, is a chaotic saddle also called repeller or, in mathematical terms, a Smale horseshoe (Katok and Hasselblatt, 1995). This object is fractal along both the stable and the unstable directions.

In practice a chaotic saddle can be generated by "digging holes" (which let trajectories escape) in an otherwise stable attractor. These may be physical holes in real space, such as in the previous example (the open space in between the three disks), or holes in phase space, which arise as a result of a crisis transition (Grebogi et al., 1983). In all such cases the chaotic dynamics is transient (see Lai and Tél (2011) for more details).

In a repeller, the number N of randomly chosen initial trajectories which have not typically escaped after a time t decays exponentially, $N \sim \exp[-\gamma t]$; the rate γ is called the *escape* rate. One can build increasingly accurate approximations of the invariant measure by recursively removing the preimages of the holes (i.e. those initial conditions which enter the holes in a number n of iterates). It is natural to expect that this removal process introduces cuts along the unstable manifold; i.e. it makes the distribution fractal along those directions, too. As the sizes of preimages are related by the growth rates of distances along the unstable direction, one can expect the existence of a relationship between Lyapunov exponents, fractal dimension and escape rate. This is indeed known as the Kantz-Grassberger formula (Kantz and Grassberger, 1985).

We derive such a formula by following the same kind of arguments developed to prove the Kaplan-Yorke formula (6.2) in a two-dimensional setup. We first focus on the unstable direction: the nth preimage of the hole is composed of many intervals of typical size ε_n, which have not yet escaped after n iterates. If the fractal dimension of such a set of points is $\mathcal{D}^1$, then the number of intervals is $\varepsilon_n^{-\mathcal{D}^1}$ whose total length sums up to $\varepsilon_n^{1-\mathcal{D}^1}$. This quantity defines the measure of initial conditions that do not escape before time n. By now taking into account that $\varepsilon_n = e^{\lambda_1} \varepsilon_{n+1}$ and that the ratio between the mass in ε_n and ε_{n+1} is equal to $e^{-\gamma}$, we find that

$$\frac{(e^{-\lambda_1}\varepsilon_n)^{1-\mathcal{D}^1}}{\varepsilon_n^{1-\mathcal{D}^1}} = e^{-\gamma},$$

from which the Kantz-Grassberger formula follows:

$$\mathcal{D}^1 = 1 - \frac{\gamma}{\lambda_1}. \qquad (6.7)$$

By finally using (6.5), we obtain an expression for $\mathcal{D}^2$ and then for the full dimension of the chaotic saddle,

$$\mathcal{D}^2 = \frac{\lambda_1}{|\lambda_2|}\left(1 - \frac{\gamma}{\lambda_1}\right), \qquad \mathcal{D}_{KY} = \mathcal{D}^1 + \mathcal{D}^2 = (\lambda_1 - \gamma)\left(\frac{1}{\lambda_1} + \frac{1}{|\lambda_2|}\right).$$

In Hamiltonian systems, where $\lambda_1 = |\lambda_2|$, the two partial dimensions $\mathcal{D}^1$ and $\mathcal{D}^2$ are equal to one another.

6.3 Dynamical entropies

Another dynamical invariant that is connected to the Lyapunov exponents is the *Kolmogorov-Sinai entropy*, which measures the growth rate of the number of different trajectories that can be generated by a dynamical system and quantifies the overall instability of the underlying dynamics (Sinai, 2009).

For the sake of simplicity, here we refer to discrete-time systems, but the final results apply to continuous-time models as well. Let $\mathbb{B}_\varepsilon$ denote a partition composed of nonoverlapping cells (atoms) $\mathcal{C}_m$ of size ε, which cover the phase space visited by the attractor. One can thus map a generic trajectory $\mathbf{U}(\tilde{t})$ ($1 \le \tilde{t} \le t$) onto a symbol sequence $\{m\}_t = (m(1), m(2), \dots, m(t))$, where $m(\tilde{t})$ identifies the atom containing the configuration at time $\tilde{t}$ ($\mathbf{U}(\tilde{t}) \in \mathcal{C}_{m(\tilde{t})}$). Upon introducing the probability $P(\{m\}_t)$ to observe the symbol sequence $\{m\}_t$, one can define the corresponding entropy

$$H(t, \varepsilon) = -\sum_{\{m(t)\}} P(\{m\}_t) \ln P(\{m\}_t). \qquad (6.8)$$

In chaotic systems the diversity of trajectories grows exponentially with time; therefore the entropy $H(t, \varepsilon)$ can be expected to increase linearly with time. The Kolmogorov-Sinai entropy h_{KS} is defined as the average growth rate,

$$h_{KS} = \lim_{\varepsilon \to 0} \lim_{t \to \infty} \frac{1}{t} H(t, \varepsilon). \qquad (6.9)$$

If the partition is generating, the limit $\varepsilon \to 0$ is not necessary (see Appendix D for a brief discussion and Eckmann and Ruelle (1985) for a more detailed treatment). From now on we assume that this is the case, since this assumption allows for a simplified mathematical analysis.

A relationship between h_{KS} and the Lyapunov exponents can be established by identifying all trajectories characterised by the same given sequence $\{m\}_t$ of cells, as those sitting at time 0 in the set $\mathcal{C} = \mathcal{C}_{m(1)} \cap T^{-1}\mathcal{C}_{m(2)} \cap \ldots \cap T^{t-1}\mathcal{C}_{m(t)}$. As a result, the probabilities in (6.8) can be expressed as

$$P(\{m\}_t) = \mathcal{P}(\mathcal{C}), \tag{6.10}$$

where $\mathcal{P}(\mathcal{C})$ is the measure of $\mathcal{C}$. $\mathcal{C}$ is basically an ellipsoid: let ε_i denote the semi-axis length along the direction of the ith Lyapunov direction (here identified as from the QR decomposition). In the case of a stable direction, we expect ε_i to be approximately equal to the width of $\mathcal{C}_{m(1)}$, i.e. $\varepsilon_i \approx \mathcal{O}(1)$. Since the mutual distance between any two trajectories converges exponentially to zero, this condition is sufficient to ensure that all the forward iterates fall within the same atoms (at the same time). In the case of an unstable direction, the mutual distance increases exponentially in time, and it is therefore necessary to require that $\varepsilon_i \approx e^{-\lambda_i t}$ to ensure that it is always smaller than the typical atom size. As a result, by invoking the smoothness of the invariant measure along the unstable directions, one can write

$$\mathcal{P}(\mathcal{C}) \approx \prod_{i:\lambda_i>0} \varepsilon_i = e^{-\sum_{\lambda_i>0} \lambda_i t},$$

which, with the help of Eqs. (6.8, 6.9, 6.10), implies

$$h_{KS} = \sum_{i:\lambda_i>0} \lambda_i. \tag{6.11}$$

This is the standard expression of the Pesin formula (Pesin, 1977) for generic chaotic attractors. Rigorously speaking, this formula provides an upper bound for the Kolmogorov-Sinai entropy. In the graphical representation of Fig. 6.2 the Kolmogorov-Sinai entropy corresponds to the maximal value of $\mathcal{S}_i$.

Generally, having in mind chaotic saddles as well, it is necessary to take into account that the density along the expanding directions may be singular (i.e. characterised by a partial dimension $\mathcal{D}^i < 1$). In practice, this is tantamount to assuming that

$$\mathcal{P}(\mathcal{C}) \approx \prod_{i:\lambda_i>0} \varepsilon_i^{\mathcal{D}^i} = e^{-\sum_{\lambda_i>0} \lambda_i \mathcal{D}^i t}, \tag{6.12}$$

which implies the general Pesin formula (Pesin, 1977),

$$h_{KS} = \sum_{i:\lambda_i>0} \mathcal{D}^i \lambda_i. \tag{6.13}$$

In chaotic attractors, $\mathcal{D}^i = 1$ and this equation reduces to (6.11). In two-dimensional repellers, the Kantz-Grassberger formula (6.7) implies that

$$h_{KS} = \lambda_1 - \gamma, \tag{6.14}$$

i.e. the dynamical entropy is obtained by subtracting the escape rate from the positive Lyapunov exponent. This is quite intuitive, as the escaped points do not contribute to the complexity of the stationary dynamics.

6.4 Generalised dimensions and entropies

So far, we have treated the dynamical systems as if all trajectories were characterised by the same Lyapunov exponent, so that there is no ambiguity in the identification of the various observables. As we have learned in Chapter 5, however, the presence of fluctuations induces an entire range of (generalised) Lyapunov exponents. In this chapter we extend both the Kaplan-Yorke and Pesin formulas to this more general case.

6.4.1 Generalised Kaplan-Yorke formula

Given a covering of a (fractal) set with boxes of size ε, the generalised dimensions are defined as

$$\mathcal{D}(\beta) = \lim_{\varepsilon \to 0} \frac{1}{\beta - 1} \frac{\ln \sum_i (\mathcal{P}_i)^\beta}{\ln \varepsilon}, \tag{6.15}$$

where $\mathcal{P}_i$ is the probability of the ith box. This formula is similar to the definition of generalised Lyapunov exponents, as it is better seen, by introducing the concept of pointwise dimension

$$\alpha_i = \frac{\ln \mathcal{P}_i}{\ln \varepsilon},$$

which is indeed analogous to the finite-time Lyapunov exponent Λ assigned to a given initial condition (see Eq. (5.1)) with $\ln \varepsilon$ playing the role of the time τ. For $\beta = 0$, this definition reduces to the box dimension defined in (6.1), which is independent of how probable a box is (provided that the probability is strictly larger than zero). For $\beta \to 1$, one obtains the information dimension

$$\mathcal{D}(1) = \lim_{\varepsilon \to 0} \frac{\sum_i \mathcal{P}_i \ln \mathcal{P}_i}{\ln \varepsilon} = \langle \alpha \rangle,$$

which coincides with the average pointwise dimension. Finally, $\mathcal{D}(2)$ is the so-called correlation dimension that can be computed by implementing the Grassberger-Procaccia algorithm (Grassberger and Procaccia, 1983; Beck and Schlögl, 1995).

The definition of dimension can be made more general, allowing for boxes of different sizes (this extension is necessary to carry on the next steps). We proceed by introducing the local observable,

$$V_i(D) = \varepsilon_i^D / \mathcal{P}_i. \tag{6.16}$$

By noticing that $1/\mathcal{P}_i$ can be interpreted as an effective number of boxes, $V_i(D)$ can be seen as an effective D-volume of the set as extrapolated from the parameters of a single box. In particular, if D is larger (smaller) than the local pointwise dimension, then $V_i(D) \ll 1$ ($V_i(D) \gg 1$). D can thereby be estimated by averaging suitable powers of $V_i(D)$. More specifically, let us introduce

$$\langle V \rangle_{\beta,D} = \sum_i \mathcal{P}_i (\varepsilon_i^D / \mathcal{P}_i)^{1-\beta} = \left\langle (V_i(D))^{1-\beta} \right\rangle.$$

In the limit of increasingly fine partitions (i.e. for the average box-size going to zero), $\langle V \rangle_{\beta,D}$ either diverges or vanishes; $D(\beta)$ can be defined as the limit value separating the two regimes. When all boxes have the same size ε, this definition reduces to (6.15), as the common factor $\varepsilon^{(1-\beta)D}$ can be brought out of the average.

If we now go back to Eq. (6.4) and replace the asymptotic value λ_i with the finite-time observable Λ_i, we see that $V(t, \{\mathcal{D}^k\})$ is by all means analogous to $V_i(D)$ in Eq. (6.16). We can thereby consider the power $(1 - \beta)$, averaged over all initial conditions, and take the limit $t \to \infty$ (which is equivalent to refining the partition), thus obtaining the implicit formula

$$\left\langle e^{(1-\beta)\sum_k \mathcal{D}^k(\beta)\Lambda_k t} \right\rangle = \mathcal{O}(1), \tag{6.17}$$

where $\mathcal{D}^k(\beta)$ are the generalised partial dimensions.

The expression in angular brackets in (6.17) is of the type (5.4), the observable Υ being, in this case,

$$\mathcal{V} = \sum \mathcal{D}^i \Lambda_i,$$

so that the condition (6.17) can be equivalently expressed as

$$\mathcal{L}_\mathcal{V}(1 - \beta) = 0, \tag{6.18}$$

where $\mathcal{L}_\mathcal{V}$ is the generalised Lyapunov exponent associated with the observable $\mathcal{V}$. This equation, being a single constraint, does not allow for determining several partial dimensions. However, upon assuming that all but one of the partial dimensions are either equal to 1 or zero (i.e. arguing as in Section 6.1), this equation reduces, for $\beta \to 1$, to the Kaplan-Yorke formula (6.3). It is sufficient to identify $\mathcal{D}_{KY}$ with $\mathcal{D}(1)$ (and interpret λ_m, λ_{m-1} as the standard LEs). In other words, we can conclude that the Kaplan-Yorke formula allows one to express the information dimension via the LEs. As for the other dimensions, including $\mathcal{D}(0)$ and $\mathcal{D}(2)$, the exact relationship is implicit.

In the case of two-dimensional maps, no extra assumption is needed to determine the generalised dimension ($\mathcal{D}^1 = 1$). Whenever, moreover, the volume contraction does not fluctuate, i.e. $\lambda_1 + \lambda_2 = B$, equation (6.17) simplifies to

$$\left\langle e^{(1-\beta)(1-\mathcal{D}^2(\beta))\lambda_1 t} \right\rangle = e^{(1-\beta)\mathcal{D}^2(\beta)Bt},$$

so that

$$\mathcal{L}_1((1 - \beta)(1 - \mathcal{D}^2)) = \mathcal{D}^2(\beta)B,$$

where $\mathcal{L}_1$ is the generalised maximum Lyapunov exponent. Although this equation is simpler than (6.18), it is still an implicit equation that requires some work for it to be solved. Moreover, the reader should be warned that we do not expect this equation to hold in the non-hyperbolic phase (i.e. for sufficiently positive β values), as it is no longer true that $\mathcal{D}^1 = 1$. In full generality we also expect that more than one stable direction may be characterised by a fractal structure (Paoli et al., 1989); this is an additional reason why the formalism needs further refinements.

6.4.2 Generalised Pesin formula

Here, we discuss the case of dynamical entropies. The first step is to generalise the definition of entropy (6.8), by introducing the order-β Renyi entropy

$$H_\beta(t, \varepsilon) = \frac{1}{1 - \beta} \ln \sum_{\{m(t)\}} P(\{m\}_t)^\beta = \frac{1}{1 - \beta} \ln \left\langle P(\{m\}_t)^{\beta - 1} \right\rangle .$$

In the limit case $\beta = 1$, this formula reduces to the standard definition (6.8). As a result, one can define the generalised Kolmogorov-Sinai entropy,

$$h(\beta) = \lim_{t \to \infty} \frac{1}{t} H_\beta(t, \varepsilon),$$

where, for the sake of simplicity, we have dropped the $\varepsilon \to 0$ limit, implicitly assuming that we refer to a generating partition.

From Eqs. (6.10, 6.12), considering that the Lyapunov exponents do fluctuate (i.e. replacing λ_i with Λ_i), it follows that

$$P(\{m\}_t) \approx e^{\mathfrak{S} t}$$

where

$$\mathfrak{S} = \sum_i^{D_u} \mathcal{D}^i \Lambda_i,$$

while D_u denotes the number of unstable directions. Accordingly, from Eq. (5.7),

$$h(\beta) = \mathcal{L}_\mathfrak{S}(1 - \beta).$$

This equation generalises the Pesin formula (6.13) to β-values different from 1. Notice that $\mathcal{L}_\mathfrak{S}(1 - \beta) \neq \sum_{i=1}^{D_u} \mathcal{L}_{\Lambda_i}(1 - \beta)$, for $\beta \neq 1$, as the fluctuations of the single exponents are typically correlated with each other.

7 Finite-amplitude exponents

Standard Lyapunov exponents refer to the evolution of infinitesimal perturbations. A possible strategy for the study of finite perturbations consists in progressively including higher-order corrections to the linear dynamics of the Lyapunov exponent. This route was suggested and explored by Farmer and Sidorowich (1987) (see Pikovsky (1984a) for using corrections to linear dynamics in the context of synchronisation transition), who introduced the concept of *higher order* Lyapunov exponents to characterise the growth rate of higher-order derivatives. As a result, it was found that such exponents can be expressed as linear combinations of the standard Lyapunov exponents (Dressler and Farmer, 1992; Taylor, 1993). Therefore, it was concluded that this strategy does not lead to new tools nor to new indicators.

It proves more fruitful to deal directly with finite perturbations. In the literature, starting from the original papers (Aurell et al., 1996, 1997), the term *finite-size exponent* is typically adopted to refer to the corresponding indicator. Since in this book we deal also with spatially extended systems, where "size" naturally refers to the spatial extension (see the so-called finite-size effects), we prefer to use the name *finite-amplitude* exponent (FAE), which alludes to the "amplitude" of a perturbation, rather than to its "size". A comprehensive review of this concept can be found in Cencini and Vulpiani (2013). Here, we focus on those general aspects that help to clarify the differences and analogies with the standard linear analysis.

7.1 Finite vs. infinitesimal perturbations

Given two neighbouring trajectories $\mathbf{U}_0(t)$ and $\mathbf{U}_1(t)$, let $\Delta(t) = |\mathbf{U}_0(t) - \mathbf{U}_1(t)|$ denote their distance at time t. As long as $\Delta(t)$ is finite, its growth rate varies with the value of Δ itself; this dependence can be captured by a properly defined FAE. Let us fix a series of thresholds θ_n and measure the corresponding (first-passage) crossing times t_n. This way, when different measurements are compared (and possibly averaged), it is clear that all refer to the same instantaneous value of the distance Δ. As for small Δ the perturbation evolution is exponential, and it is natural to space the thresholds exponentially, according to some prescribed multiplicative factor r: $\theta_n = r\theta_{n-1}, r > 1$ (see Fig. 7.1). One can then define the local FAE as

$$\mathbb{L}(\theta_n) = \frac{\ln r}{t_n - t_{n-1}}.$$

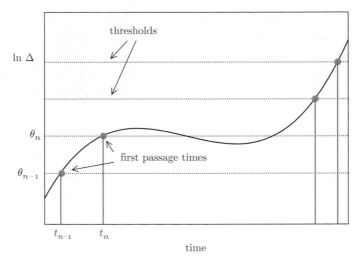

Fig. 7.1 A sketch of the procedure to determine the finite-amplitude Lyapunov exponent.

By then averaging the time interval over an ensemble of M different initial conditions,

$$\langle t_n - t_{n-1} \rangle = \frac{1}{M} \sum_{i=1}^{M} (t_n^{(i)} - t_{n-1}^{(i)}),$$

one obtains the true finite-amplitude exponent (Aurell et al., 1996, 1997)

$$\ell(\theta_n) = \frac{\ln r}{\langle t_n - t_{n-1} \rangle}. \tag{7.1}$$

The reason why the denominator of Eq. (7.1), rather than the local FAE itself, has been averaged in Eq. (7.1) follows from the need of consistency with the definition of the usual Lyapunov exponent. For $\theta_n \to 0$, if a perturbation is amplified by a factor r^N over a time t_N, i.e. if $\Delta(t_N) = r^N \Delta(t_0)$ (with $\Delta(t_N) \ll 1$), one can write

$$\lambda = \lim_{t_N \to \infty} \frac{\ln r^N}{t_N} = \lim_{N \to \infty} \frac{\ln r}{\frac{1}{N} \sum_{1}^{N} (t_i - t_{i-1})} = \frac{\ln r}{\langle t_i - t_{i-1} \rangle}.$$

The last term in the r.h.s. of this equation is equivalent to the r.h.s of Eq. (7.1).

Altogether, finite-amplitude and finite-time exponents differ not only because of the amplitudes of the perturbations that are being considered, but also because of the protocol adopted to determine them. The role of this latter difference has been thoroughly discussed in the analysis of Lagrangian coherent structures by Karrasch and Haller (2013) (see also Section 12.4), but its relevance is more general.

In order to better clarify the point, it is convenient to consider the FAE in the limit of infinitesimal perturbations (Karrasch and Haller, 2013); the resulting indicator will be referred to as the first-passage time Lyapunov exponent. Superficially, the finite-time and the first-passage-time Lyapunov exponent are defined in the same way, as $(\ln r)/\tau$, where τ is the elapsed time and r is the corresponding expansion factor. In the former case, however, τ is an independent variable (fixed a priori), while $r(\tau)$ is being measured, so as to obtain

$\Lambda(t_0, \tau) = (\ln r(\tau))/\tau$.[1] In the latter case, the expansion factor r is an independent variable, while the time-interval $\tau(r)$ is the measured observable; the first-passage-time Lyapunov exponent is thereby defined as $\mathbb{L}(t_0, r) = (\ln r)/\tau(r)$.

As far as averages are being considered, we have already seen in the first part of this section that the two definitions are equivalent. The instantaneous observables, however, may have different properties. Whenever $\Delta(t)$ increases (or decreases) monotonously, no striking differences are expected and one can choose whichever definition. The situation is different in the presence of strong fluctuations, especially if a non-monotonic behaviour is observed (as, e.g., shown in Fig. 7.1). In this case the dependence of Λ on τ is continuous, while that of $\mathbb{L}(t_0, r)$ on r may have jumps due to the local maxima of $\Delta(t)$ as a function of time. This difference makes the finite-time LE a better choice at least in situations where the (local) time stability of particular states is to be characterised, as, e.g., in Lagrangian coherent structures.

7.2 Computational issues

Several problems may arise when the general FAE definition (7.1) is to be implemented. Here, we summarise the most important ones.

First of all, Eq. (7.1) applies only to continuous-time dynamics, where first-passage times can be sharply identified. It can be easily implemented, however, in discrete-time systems, by extending the definition of $\Delta(t)$ to non-integer times,

$$\ln \Delta(t) = \ln \Delta(m(t)) + \ln(\Delta(m(t) + 1)/\Delta(m(t)),$$

where $m(t)$ is the largest integer smaller than t. This formula amounts to assuming a uniform exponential growth in between consecutive time instants. It is easy to check that this definition gives the correct averaging in the linear regime.

A more serious problem arises when averaging has to be implemented to obtain a robust indicator. In the case of infinitesimal perturbations, it is sufficient to select a reference trajectory and thereby average in time (with a sporadic rescaling of the distance to avoid numerical overflows). In fact, the amplitude of the perturbation is immaterial, and contributions obtained at all times can be used to determine the average value. In the FAE case, rescaling of the perturbation is incorrect, as the local orientation of the perturbation is expected to depend on its amplitude. Accordingly, when the maximal distance is reached (which corresponds to the attractor size), it is necessary to start over with a new initial condition. This protocol does not entirely solve the computational problem; a randomly selected initial condition is not typically aligned along the most expanding direction. The only context where this is not a problem is that of one-dimensional maps, where perturbations are scalar quantities. The difficulty can be overcome, nevertheless, by selecting the initial perturbation along the direction of the first covariant Lyapunov vector (as obtained with the help of a standard linear analysis); see Letz and Kantz (2000).

---

[1] The dependence on t_0 is introduced to specify that Λ refers to a trajectory starting at time t_0.

Furthermore, some care should be taken in the presence of a multistable dynamics. Since a finite perturbation may induce a jump towards a different attractor, it is necessary to restrict the average to the perturbed trajectories which converge to the same attractor visited by the unperturbed one.

For small enough amplitude, the FAE ℓ is expected to coincide with the largest Lyapunov exponent, while it saturates to zero for large amplitudes, since perturbations cannot be larger than the size of the accessible phase space. In the intermediate range, ℓ tells us how the growth of a perturbation is affected by nonlinearities.

This is illustrated in Fig. 7.2, with reference to the Lorenz model, where one can notice a long plateau, which corresponds to the maximal Lyapunov exponent, followed by a drop to zero. The same data reveal, however, an additional feature: an initial peak that is obtained when the initial condition is chosen randomly in time (see the circles in Fig. 7.2). This is a spurious effect due to the sampling procedure adopted to select the initial conditions; points characterised by a smaller local Lyapunov exponent are selected with a larger probability, and as a result, the statistics of the average first passage time are biased, at least for the first few crossings. On the other hand, if one starts the calculation of the FAE exponent exactly when the linear perturbation crosses a given threshold, then the statistics are unbiased (see the squares in Fig. 7.2).

An unavoidable problem is that the FAE definition involves neither an infinite-time limit (the trajectories that are being followed are not only finite but even rather short) nor that of infinitesimal perturbations. As a result, the FAE is not mathematically well defined: its value depends on the selection of the variables and on the metric used for the computation of the distance Δ. Nevertheless, it can be profitably used to extract useful information on the stability, whenever it is strongly scale-dependent. We discuss some applications in the next section.

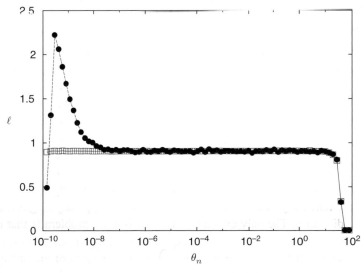

Fig. 7.2 The finite-amplitude exponent for the Lorenz model. The amplitude levels are scaled by a factor $r = \sqrt{2}$. Circles: initial perturbations are chosen randomly in time; squares: initial perturbations are chosen according to the unbiased statistics.

So far, we have assumed the underlying dynamics to be unstable. One can apply the same approach also to stable systems, by starting with large perturbations and following the threshold-crossing downwards towards increasingly fine scales. This has been done, for instance, by Ginelli et al. (2003) to identify the percolation threshold in a chain of stochastic maps. A weakness of this backward approach is the lack of an objective protocol to define the initial direction of the perturbation.

It may seem desirable to extend the definition of the FAE to cover further directions in phase space, obtaining a spectrum as for the usual Lyapunov exponents. Unfortunately, the difficulty of extending the scalar product to a nonlinear environment has prevented any further steps in this direction.

7.2.1 One-dimensional maps

For maps $F(U(t))$ of the unit interval, the direction of the perturbation is not an issue, and one can determine the FAE as a phase-space average of the local expansion rate

$$\ell(\delta) = \langle \ln |\mu(U, \delta)| \rangle, \tag{7.2}$$

where

$$\mu(U, \delta) = \frac{F(U + \delta/2) - F(U - \delta/2)}{\delta}.$$

Here, we discuss an application of this formula in the case of the tent map (A.3) and for the asymmetric Bernoulli map $[F(U) = U/a\,((U - a)/(1 - a))$ if $U < a\ (U > a)]$. Since the invariant measure of both maps is flat, one can define the FAE as

$$\ell(\delta) = \frac{1}{1 - \delta} \int_{\delta/2}^{1 - \delta/2} \ln |\mu(U, \delta)| dU, \tag{7.3}$$

where the integral is restricted to the interval $I_T = [\delta/2, 1 - \delta/2]$, so as to ensure that both $U_1 = U - \delta/2$ and $U_2 = U + \delta/2$ lie within the unit interval, where the map is defined. Moreover, the multiplicative factor has been included to rescale properly the probability to 1 within I_i.

For both maps, the integral can be performed by splitting I_T into the union of $I_1 = [\delta/2, a - \delta/2]$, $I_2 = [a - \delta/2, a + \delta/2]$ and $I_3 = [a + \delta/2, 1 - \delta/2]$. In I_1 (I_3), both U_1 and U_2 are iterated according to the left (right) branch of the map, while in I_2, U_1 falls inside the left branch, while U_2 falls inside the right branch.

The resulting dependence of the FAE on δ is illustrated in Fig. 7.3a, where one can make a comparison with the numerical results obtained using the definition (7.1). In the case of the tent map, the analytic formula (7.3) (dashed line) reproduces almost exactly the numerical data (circles). For the Bernoulli map (see triangles and the solid line), the agreement is less satisfactory in the large amplitude region. The reason is that the distribution of the midpoint $U_1 + U_2$, assumed to be flat in Eq. (7.3), is no longer such for larger amplitudes, when U_1 and U_2 become increasingly decorrelated. The problem would not arise in the case of U being an angle variable (i.e. if the maps were defined on an interval with periodic boundaries).

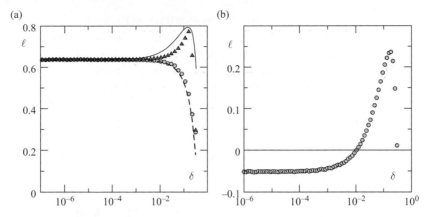

Fig. 7.3 Finite-amplitude exponent for one-dimensional maps. (a) Tent (circles) and Bernoulli (triangles) maps. The absolute value of the slopes of the two map branches is 3 and 3/2 in both cases. The dashed and solid lines correspond to the theoretical predictions. The circles in panel (b) correspond to a numerical simulation of the map (A.14) for $\eta = 0$ (the other parameters are as in (A.14)) with additive noise defined as a sequence of i.i.d. random numbers uniformly distributed in $[-0, 0.1]$.

An interesting phenomenon appears if the one-dimensional map has both a discontinuity and a negative (standard) Lyapunov exponent. In the purely deterministic case, the latter property would imply that the map has a stable periodic orbit (or, more generally, a stable invariant set) and the invariant measure is singular. For a noisy map, however, the invariant density is continuous and the averaging in (7.2) is nontrivial. Of course, in the case of noisy dynamics, the finite-amplitude Lyapunov exponent should be measured from the separation of two trajectories under the action of the same noise (see also Chapter 8). We illustrate a possible nontrivial dependence of ℓ by referring to the map (A.14) for $\eta = 0$ (when it is discontinuous). Simulations are performed by assuming a uniformly distributed additive noise $-d \leq \xi(t) \leq d$ and taking the absolute value of U to enforce a confinement of the dynamics within the unit interval. The numerical results from Eq. (7.2) are reported in Fig. 7.3b, where we see that while $\ell < 0$ for small δ, it becomes positive for moderate amplitudes. This is an example of how a strong nonlinearity (here mimicked by a discontinuity in the map) can generate a large-scale instability. Notice that in this case, the approximate formula Eq. (7.2) is preferable to the general definition; the first-passage-time approach would not be able to capture the mixture of stable/unstable properties of the dynamics.

7.3 Applications

The main applications of FAEs concern the stability of systems with a variety of temporal and amplitude scales, as well as spatially extended systems. Here, we discuss a couple of examples, starting from two coupled Lorenz systems (A.9) with different temporal and amplitude scales,

$$c\dot{x}_1 = \sigma(y_1 - x_1)$$
$$c\dot{y}_1 = rx_1 - y_1 - x_1z_1$$
$$c\dot{z}_1 = -bz_1 + x_1y_1 + \varepsilon z_2$$
$$\dot{x}_2 = \sigma(y_2 - x_2)$$
$$\dot{y}_2 = rx_2 - y_2 - dx_2z_2$$
$$\dot{z}_2 = -bz_2 + dx_2y_2 + d^{-1}\varepsilon z_1.$$

The parameter $c > 1$ is responsible for the time-scale separation: the subsystem 1 is c times slower than subsystem 2. The parameter $d > 1$ controls the separation between the scales of the variables themselves: all the amplitudes in subsystem 2 are d times smaller than in subsystem 1. Correspondingly, perturbations in the subsystem 2 grow faster, but they saturate at amplitudes smaller than the slowly growing perturbations in subsystem 1. This separation of scales is clearly revealed by the evolution of the FAE (see Fig. 7.4). At small amplitudes, the FAE is equal to the Lyapunov exponent of the standard Lorenz system (subsystem 2); after a sharp drop, a second, lower plateau arises at large amplitudes, characterising the perturbation growth within subsystem 1.

The second application is the so-called stable chaos phenomenon (Crutchfield and Kaneko, 1988; Politi et al., 1993), where an irregular dynamics in spatially extended systems is observed despite the negative value of the largest Lyapunov exponent. The simplest model where such a dynamics can occur is a lattice of coupled maps of the type (A13,A14). In the purely deterministic case, each map possesses a stable periodic orbit of period three (for the parameters given in caption of Fig. 7.3). Thus, the dynamics of

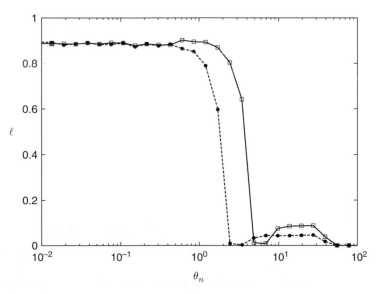

Fig. 7.4 The finite-amplitude exponent for the two-scale Lorenz model, with $c = 10$ and standard parameters ($\sigma = 10, r = 28, b = 8/3$). The threshold levels are scaled by a factor $r = \sqrt{2}; \varepsilon = 0.1$. Squares: ratio of amplitude scales $d = 10$; circles: ratio of scale $d = 20$.

a homogeneous initial state is periodic and linearly stable. However, an inhomogeneous state may remain "turbulent" for extremely long times (exponentially long with the system length). This "turbulence" is sustained by the finite-amplitude instability already seen in Fig. 7.3b (the main difference being that in the coupled-map lattice the effective noise is generated by the spatial coupling). In practice, localised finite perturbations can propagate as "nonlinear" waves (Cencini and Torcini 2001) in spite of the small-amplitude stability (see also Chapter 11).

8 Random systems

The problem of computing Lyapunov exponents arises not only in the study of dynamical systems but also in the context of noisy or disordered linear systems. In the latter case, the fluctuating properties of the Jacobian (or, more generally, of the matrices which determine the linear dynamics) are given beforehand, when the structure of the noise is postulated, rather than emerging self-consistently from the evolution of a deterministic dynamical system. This simplification allows for the development of powerful analytic approaches, especially when the noise is assumed to be δ-correlated.

In the first section we discuss linear discrete-time systems, where the problem of determining the Lyapunov exponents can be formulated in terms of products of random matrices. We present various setups, starting from the case of weak disorder (noise), where the matrices are nearly equal to each other, and show that the variation of the Lyapunov exponents is generically proportional to the square of the disorder amplitude. For an arbitrary amplitude of noise, it is hard to extract analytic information on the entire spectrum of LEs, unless the dynamics is highly symmetric (essentially isotropic), as discussed in Section 8.1.2; otherwise, only the largest LE can be typically determined (in a more or less approximate way). The sparse matrices discussed in Section 8.1.3 provide one such setup, where it is even possible to detect a phase transition upon increasing the amount of disorder. Another setup where powerful semi-analytic techniques have been developed is when the disorder manifests itself as a selection among a few different options (polytomic noise).

The second section of this chapter is devoted to the analysis of continuous-time systems, i.e. of linear stochastic differential equations. One- and two-dimensional linear systems are first discussed in the presence of multiplicative noise. The general Khasminskii theory is then briefly reviewed in Section 8.2.3. Unfortunately, closed expressions of the LEs can be hardly obtained if the noise is not δ-correlated and, even less so, in high-dimensional spaces. A remarkable exception is the fully coupled setup discussed in Section 8.2.4, where the evaluation of the largest LE can be reduced to the computation of an eigenvalue of a Schrödinger operator.

The last section is devoted to a discussion of systems where noise and nonlinearities are simultaneously present. First, we briefly outline a stimulating analogy with supersymmetric quantum mechanics and then discuss the weak noise limit, which, not unexpectedly, reveals analogies with the purely linear case presented in Section 8.1. The role of LEs in noisy nonlinear systems is finally discussed in Section 8.3.3.

8.1 Products of random matrices

The analysis of random matrices deserves special attention. Their study helps to clarify discrete-time dynamical systems and also several physical problems such as the propagation of waves in random linear systems and the statistical properties of disordered systems. An application of the former type (i.e. Anderson localisation) is extensively discussed in Section 12.1, while an instance of the latter type, namely random polymers, is illustrated later in this section. The reader interested in other setups is invited to consult Crisanti et al. (1993).

Here, we discuss various techniques which allow us to derive (approximate and exact) analytical expressions. For later convenience, we first recall that given the linear evolution rule $\mathbf{u}_k(t+1) = \mathsf{A}(t)\mathbf{u}_k(t)$ and the product of matrices $\mathsf{P}(t) = \prod_{j=1}^{t} \mathsf{A}(t)$, the object of study is the sum $\mathcal{S}_k$ of the first k Lyapunov exponents,

$$
\begin{aligned}
\mathcal{S}_k &= \lim_{t\to\infty} \frac{1}{t} \ln \frac{\|\mathsf{P}(j)\mathbf{u}_1(1) \wedge \ldots \wedge \mathsf{P}(j)\mathbf{u}_k(1)\|}{\|\mathbf{u}_1(1) \wedge \ldots \wedge \mathbf{u}_k(1)\|} \\
&= \lim_{t\to\infty} \frac{1}{t} \sum_{j=1}^{t} \ln \frac{\|\mathsf{A}(j)\mathbf{u}_1(j) \wedge \ldots \wedge \mathsf{A}(j)\mathbf{u}_k(j)\|}{\|\mathbf{u}_1(j) \wedge \ldots \wedge \mathbf{u}_k(j)\|},
\end{aligned}
\tag{8.1}
$$

where the vectors $\mathbf{u}_k(1)$ are linearly independent. The time average in Eq. (8.1) can, in principle, be determined as an ensemble average, but this is a formidable task, as the addenda depend on the orientation of the parallelepiped $\mathbf{u}_1(j) \wedge \ldots \wedge \mathbf{u}_k(j)$ and, in the case of time correlated processes, on the past matrices as well.

8.1.1 Weak disorder

We start by analysing matrices A of the type

$$
\mathsf{A}(t) = \mathsf{A}_0 + \varepsilon \mathsf{B}(t),
$$

where A_0 is constant and ε is a small parameter which gauges the amplitude of noise/disorder, while $\mathsf{B}(t)$ is a fluctuating matrix with zero-average entries (if not, the average contributions could be absorbed into A_0). Here, we show that the variation induced by the disorder on the Lyapunov exponents is generally of order $\mathcal{O}(\varepsilon^2)$, with exceptions in some critical cases.

Non-degenerate spectra

We first consider matrices A_0 characterised by distinct eigenvalues (the modules of the eigenvalues are assumed to be all mutually different). For reasons that will soon become clear, it is convenient to work in the basis which diagonalises A_0 and where the eigenvalues are ordered from the most to the least expanding one.

We shall closely follow the approach outlined by Derrida et al. (1987). From Eq. (8.1), it is clear that the Lyapunov exponents can be determined from the matrix $\mathsf{P}(t) = \prod_{j=1}^{t} \mathsf{A}(t)$, which, up to second order in ε, can be written as

$$P(t) \approx \left[1 + \varepsilon C + \varepsilon^2 D \right] A_0^t,$$

where A_0 is diagonal and

$$C = \sum_{i=1}^{t} A_0^{i-1} B(i) A_0^{-i}, \qquad D = \sum_{1 \le i < j \le t} A_0^{i-1} B(i) A_0^{j-i-1} B(j) A_0^{-j}.$$

From the first equality in Eq. (8.1), it follows that

$$S_k = \lim_{t \to \infty} \frac{1}{t} \ln(\det_k P(t)) = \lim_{t \to \infty} \frac{1}{t} \left\{ \ln(\det_k A_0^t) + \ln[\det_k(1 + \varepsilon C + \varepsilon^2 D)] \right\},$$

where $\det_k M$ denotes the determinant of the matrix obtained by considering the first k rows and k columns of M. By then recalling that $\ln(\det_k M) = \mathrm{tr}_k(\ln M)$, one finds

$$S_k = \lim_{t \to \infty} \frac{1}{t} \left\{ \mathrm{tr}_k \ln A_0^t + \mathrm{tr}_k[\ln(1 + \varepsilon C + \varepsilon^2 D)] \right\},$$

and, expanding up to second order,

$$S_k = \lim_{t \to \infty} \frac{1}{t} \left\{ \mathrm{tr}_k \ln A_0^t + \varepsilon \mathrm{tr}_k C + \varepsilon^2 \left[\mathrm{tr}_k D - \frac{1}{2} \mathrm{tr}_k C^2 \right] \right\}.$$

It is easy to see that the first-order contribution in ε vanishes, since the elements of B have zero time-average. By determining all the other elements, one finally obtains

$$S_k = \sum_i^k \ln \mu_i + \varepsilon^2 \sum_{i<j}^k \sum_{1 \le \tau} \frac{\overline{B_{ij}(t) B_{ji}(t+\tau)}}{\mu_i \mu_j} \left(\frac{\mu_j}{\mu_i} \right)^\tau - \frac{\varepsilon^2}{2} \sum_{i,j}^k \frac{\overline{B_{ij} B_{ji}}}{\mu_i \mu_j}. \qquad (8.2)$$

Here μ_i represents the ith eigenvalue of A_0. B_{ij} are the matrix elements expressed in the basis which diagonalises A_0, and the overbar denotes a time average.

The ordering of the variables from the most to the least expanding one is required by the need of self-consistency: S_k is properly defined if and only if it is computed using the most expanding directions. This restriction obliges us to exclude the presence of degeneracies and those nearly critical cases where the coupling itself might change the order of the exponents.

Notice that the convergence of the first correction to the LE is ensured by the proper ordering of the eigenvalues of A_0. The very same correction vanishes for the first Lyapunov exponent and, in general, in the case of δ-correlated disorder.

Complex-conjugate eigenvalues

The key points of the perturbative analysis can already be captured by the study of 2×2 matrices. Moreover, in spite of its simplicity, this setup covers interesting physical problems such as Anderson localisation in the tight-binding approximation (see also Chapter 12) or the emergence of chaos in symplectic maps. It is convenient to formulate the problem in a basis where the action of the unperturbed matrix A_0 corresponds to the multiplication by a factor μ accompanied by a rotation by an angle ϕ (cf. a similar approach in the continuous-time case in Section 8.2.3),

$$\mathsf{A}(t) = \mathsf{A}_0 + \varepsilon \mathsf{B}(t) = \mu \begin{pmatrix} \cos\phi & -\sin\phi \\ \sin\phi & \cos\phi \end{pmatrix} + \varepsilon \begin{pmatrix} B_{11} & B_{12} \\ B_{21} & B_{22} \end{pmatrix}. \tag{8.3}$$

With reference to polar coordinates

$$q(t) = \sqrt{u_1^2(t) + u_2^2(t)}, \qquad \theta(t) = \tan^{-1}(u_2(t)/u_1(t)),$$

the Lyapunov exponent can be expressed as

$$\lambda = \int d\theta \, \overline{\ln \left| \frac{q(t+1)}{q(t)} \right|} \rho(\theta), \tag{8.4}$$

where $\rho(\theta)$ is the (stationary) distribution of the angle θ, while the overline denotes an average over the noise realisations (or, equivalently, a time average).

From (8.3), it is found that $q(t)$ satisfies the recursive relation

$$q^2(t+1) = q^2(t) \left\{ \mu^2 + \varepsilon\mu N_1 + \varepsilon^2 N_2 + \left[\varepsilon\mu C_1 + \varepsilon^2 C_2 \right] \cos 2\theta(t) \right.$$
$$\left. + \left[\varepsilon\mu S_1 + \varepsilon^2 S_2 \right] \sin 2\theta(t) \right\}, \tag{8.5}$$

where

$$\begin{aligned} N_1 &= (B_{11} + B_{22})\cos\phi + (B_{21} - B_{12})\sin\phi \\ C_1 &= (B_{11} - B_{22})\cos\phi + (B_{21} + B_{12})\sin\phi \\ S_1 &= (B_{12} + B_{21})\cos\phi - (B_{11} - B_{22})\sin\phi \\ N_2 &= (B_{11}^2 + B_{12}^2 + B_{21}^2 + B_{22}^2)/2 \\ C_2 &= (B_{11}^2 - B_{12}^2 + B_{21}^2 - B_{22}^2)/2 \\ S_2 &= B_{11}B_{12} + B_{21}B_{22}, \end{aligned} \tag{8.6}$$

By substituting (8.5) in (8.4) and expanding the logarithm, one obtains

$$\lambda = \ln|\mu| + \frac{\varepsilon^2}{4\mu^2} \int d\theta \left\{ 2\overline{N_2} + 2\overline{C_2} \cos 2\theta + 2\overline{S_2} \sin 2\theta \right.$$
$$\left. - \overline{[N_1 + C_1 \cos 2\theta + S_1 \sin 2\theta]^2} \right\} \rho(\theta), \tag{8.7}$$

where we have exploited the zero-average property of the perturbation, so that $\overline{N_1} = \overline{C_1} = \overline{S_1} = 0$. Thus, the deviation of the Lyapunov exponent is quadratic in the noise amplitude also in this case.

Next, it is necessary to determine $\rho(\theta)$. This is the most subtle but eventually easiest step. In the case of an irrational rotation angle ϕ, $\rho(\theta) = 1/(2\pi)$; i.e. all angles are visited with the same probability (before introducing the perturbation). Accordingly, one expects $\rho(\theta)$ to be nearly uniform in the case of small perturbations as well. In fact, analytic calculations reveal that the deviations from a nonuniform behaviour are of order ε^2 (Politi, 2014a). As a result,

$$\lambda = \ln|\mu| + \frac{\varepsilon^2}{8\mu^2}\left[2\overline{(B_{11}+B_{12})(B_{12}-B_{21})}\sin 2\phi \right. \tag{8.8}$$

$$\left. -\left(\overline{B_{11}^2}-\overline{B_{12}^2}-\overline{B_{21}^2}+\overline{B_{22}^2}+2\overline{B_{11}B_{22}}+2\overline{B_{12}B_{21}}\right)\cos 2\phi\right].$$

When ϕ is rationally related to π, the dynamics is not erogidic in the zero-noise limit: there exists a continuum of different periodic orbits. In such cases one expects a nonanalytic behaviour of λ, which nevertheless affects higher-order terms in the expansion, with the remarkable exception of the angles multiple of $\pi/2$ (see Derrida and Gardner (1984) and Campanino and Klein (1990)). At such angles, the leading correction to $\rho(\theta)$ diverges (Politi, 2014a) and the leading term of $\rho(\theta)$ is no longer constant. The reader interested in further details can look at the simple analysis by Izrailev et al. (1998).

Critical case

At last, we discuss the Lyapunov exponent emerging from the product of $\mathbf{A}$ matrices,

$$\mathbf{u}(t+1) = \mathbf{A}(t)\mathbf{u}(t) = [\mathbf{A}_0 + \varepsilon\mathbf{B}(t)]\mathbf{u}(t),$$

when $\mathbf{A}_0$ is at the border between the hyperbolic and the elliptic phase. In the case of 2×2 matrices, it is always possible to select a basis where $\mathbf{A}_0$ has the following representation

$$\mathbf{A}_0 = \begin{pmatrix} 0 & \mu \\ -\mu & 2\mu \end{pmatrix}, \tag{8.9}$$

while we make no assumption on the components of $\mathbf{B}$, apart from being δ-correlated in time. The action of $\mathbf{A}$ can be represented as a one-dimensional map for $R(t) = u_2(t)/u_1(t)$,

$$R(t+1) = \frac{2\mu + \varepsilon B_{22} + (\varepsilon B_{21} - \mu)/R(t)}{\mu + \varepsilon B_{12} + \varepsilon B_{11}/R(t)},$$

which is basically the evolution equation for the phase θ of the vector $\mathbf{u}(t)$. It is easy to see that for $\varepsilon = 0$ the map has a marginally stable fixed point $R = 1$. For small ε, one can expand the recursive equation, obtaining at the first order in ε

$$R(t+1) = 2 - \frac{1}{R} + \varepsilon\frac{B_{22}-2B_{12}}{\mu} + \varepsilon\frac{B_{21}-2B_{11}+B_{12}}{R\mu} + \varepsilon\frac{B_{11}}{R^2\mu}.$$

Since for a small ε, the density of the R values is concentrated around the fixed point $R = 1$, it is convenient to introduce $r = R - 1$. Upon expanding to second order in r, one obtains

$$r(t+1) - r(t) = \varepsilon\frac{-B_{11}-B_{12}+B_{21}+B_{22}}{\mu} + \varepsilon\frac{B_{21}+B_{12}}{\mu}r + r^2,$$

where we have also neglected order-ε corrections to the quadratic term. In view of the smallness of r, one can neglect also the multiplicative noisy term and thereby approximate the evolution equation with the following stochastic equation

$$\dot{r} = r^2 + \varepsilon\xi(t),$$

where $\xi(t)$ is a white noise with intensity

$$\Delta = \overline{(-B_{11} - B_{12} + B_{21} + B_{22})^2}/\mu^2. \tag{8.10}$$

This Langevin equation (which is essentially the same as Eq. (8.27) discussed later) can be analysed by virtue of the Fokker-Planck equation

$$\frac{\partial P}{\partial t} = -\frac{\partial r^2 P}{\partial r} + \frac{\varepsilon^2 \Delta}{2} \frac{\partial^2 P}{\partial r^2},$$

whose stationary solution is

$$\frac{\varepsilon^2 \Delta}{2} \frac{dP}{dr} - r^2 P = j_0,$$

where j_0 represents the current (i.e. the average rotation frequency) to be determined self-consistently, once the probability density has been suitably normalised. If one now introduces the rescaled variables

$$x = \left(\frac{2}{\varepsilon^2 \Delta}\right)^{1/3} r, \qquad Q = \left(\frac{\varepsilon^2 \Delta}{2}\right)^{1/3} P,$$

this equation is rewritten as

$$\frac{dQ}{dx} - x^2 Q = J, \tag{8.11}$$

where $J = j_0 (2/\varepsilon^2 \Delta)^{1/3}$. As a result, we see that the current j_0 is expected to scale as $\varepsilon^{2/3}$. Eq. (8.11) can be solved, yielding

$$Q(x) = J \int_x^\infty e^{(x^3 - y^3)/3} dy,$$

where the current J is determined by imposing the normalisation of the probability density. We can now go back to the definition of the Lyapunov exponent,

$$\lambda = \frac{1}{2} \left\langle \ln \frac{u_1^2(t+1) + u_2^2(t+1)}{u_1^2(t) + u_2^2(t)} \right\rangle.$$

By introducing the variable $R = 1 + r$ and recalling that $r \ll 1$, it is easily found that to the leading order,

$$\lambda = \ln \mu + \langle r \rangle.$$

By then taking into account the variable rescaling, one can write

$$\lambda = \ln \mu + \left(\frac{\varepsilon^2 \Delta}{2}\right)^{1/3} \frac{\int_{-\infty}^{+\infty} x \int_x^\infty e^{(x^3 - y^3)/3} dy}{\int_{-\infty}^{\infty} \int_x^\infty e^{(x^3 - y^3)/3} dy}.$$

Upon computing the integrals (see Derrida and Gardner (1984) for the technical details), one finally obtains

$$\lambda = \ln \mu + \left(\frac{3\varepsilon^2 \Delta}{4}\right)^{1/3} \frac{\sqrt{\pi}}{\Gamma(1/6)} = \ln \mu + 0.28930 \ldots (\varepsilon^2 \Delta)^{1/3}. \tag{8.12}$$

This expression, complemented by Eq. (8.10), is the final result. This approach can be extended to matrices $\mathbf{A}_0$ that are quasi-critical (Derrida and Gardner, 1984; Izrailev et al., 1998).

Perturbative studies can be performed also in the case of degenerate spectra where two directions are accidentally equal (see Zanon and Derrida (1988) and Markoš (1993) for an analysis of two-fold degeneracies). The main message is that the leading correction is typically quadratic in the perturbation amplitude; exceptions are the highly symmetric matrices discussed later in this section, where the correction grows more slowly than any power of ε (Zanon and Derrida, 1988).

Numerical checks

Here, we test the perturbative results, by discussing a setup with 2×2 volume-conserving matrices,

$$\mathbf{A}(t) = \mathbf{A}_0 + \varepsilon \mathbf{B}(t) = \begin{pmatrix} C & -1 \\ 1 & 0 \end{pmatrix} + \varepsilon \begin{pmatrix} b(t) & 0 \\ 0 & 0 \end{pmatrix}. \tag{8.13}$$

This problem emerges within the study of the Schrödinger equation in the tight binding approximation, where C plays the role of the energy, εb corresponds to a random on-site potential and, finally, the time t corresponds to the spatial position along a one-dimensional lattice (see also Chapter 12). For $|C| > 2$ the matrix $\mathbf{A}_0$ is hyperbolic, while for $-2 < C < 2$ it has two complex conjugate eigenvalues. Here, we present the results for three different parameter values, namely $C = 2.5$, $C = 1.8$ and $C = 2$. The outcomes of numerical simulations are reported in Fig. 8.1, where $\Delta\lambda(\varepsilon, C) := \lambda(\varepsilon, C) - \lambda(0, C)$ is plotted vs. ε for a dichotomic noise $\pm \varepsilon$ with variance ε^2.

For $C = 2.5$, $\mathbf{A}_0$ is hyperbolic and the two eigenvalues are distinct (equal to 2 and $1/2$, respectively) so that one can apply Eq. (8.2). In the basis which diagonalises $\mathbf{A}_0$, the

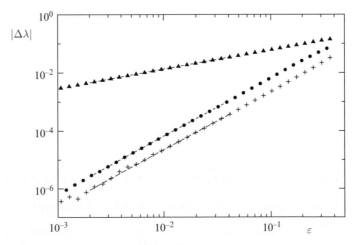

Fig. 8.1 $\Delta\lambda$ versus ε for maps of the type (8.13), with $C = 1.8$ (circles), $C = 2$ (triangles) and $C = 2.5$ (plusses). The three dashed lines result from the perturbative analysis (see the text).

perturbation matrix B is written as

$$\mathsf{B} = \begin{pmatrix} 4/3 & 1/3 \\ -1/3 & -1/3 \end{pmatrix}.$$

As a result Eq. (8.2) implies that $\lambda_1 = \ln 2 - 2\varepsilon^2/9$. The validity of the theoretical prediction is confirmed in Fig. 8.1 (see the dashed line which overlaps with the plusses obtained from numerical simulations). Considering that there are no time correlations, the theoretical formula (8.2) predicts that the sum of the first two exponents remains equal to zero, as it should, given the structure of the matrices.

The circles correspond to $C = 1.8$. For this parameter value, the effect of noise is qualitatively important, as it induces an exponential growth where the unperturbed dynamics would just induce oscillations ($\lambda(0, 1.8) = 0$). By selecting a basis where A_0 has the structure defined in Eq. (8.3), the matrix B takes the form

$$\mathsf{B} = \begin{pmatrix} b(t) & b(t)/\tan \phi \\ 0 & 0 \end{pmatrix},$$

where $\tan \phi = \sqrt{4 - C^2}/C$. From Eq. (8.8) one finds that

$$\lambda = \frac{\varepsilon^2}{8 \sin^2 \phi},$$

so that the numerical coefficient in front of ε^2 is equal to $0.657\ldots$. The corresponding curve reproduces perfectly the results of the numerical simulations (see the dashed line which overlaps the circles).

Finally, we have considered the critical case $C = 2$ (see the triangles), where the theory predicts a universal coefficient (see Eq. (8.12)). Once again the numerical results are in agreement with the theoretical predictions.

8.1.2 Highly symmetric matrices

In the presence of strong disorder, the problem of determining the Lyapunov spectrum is very difficult. A relevant simplification arises when the $N \times N$ matrices $\mathsf{A}(t)$ are nonsingular, statistically independent and their entries identically distributed. When this is the case, it is "sufficient" to average the local expansion rate $R(\mathsf{A}, \{\mathbf{u}\}_k)$ implicitly defined in (8.1) (see the argument of the logarithm in the last term) over all possible orientations of the vectors and realisations of the matrix A. Moreover, since $R(\mathsf{A}, \{\mathbf{u}\}_k)$ does not vary when the vectors $\mathbf{u}_1, \ldots, \mathbf{u}_k$ are replaced by any orthonormal basis which spans the same subset, one can set the denominator in Eq. (8.1) equal to 1.

The computation of the average expansion rate still requires determining the probability distribution of the angles. We have already encountered an example of this difficulty in the previous section, while describing the perturbative approach in a simple two-dimensional setup.

A second simplifying hypothesis may help to obtain analytical expressions:

Property I: *given any orthogonal matrix* Q, *the matrix* $\mathsf{Q}^\mathsf{T} \mathsf{A}^\mathsf{T} \mathsf{A} \mathsf{Q}$ *has the same distribution of angles as* $\mathsf{A}^\mathsf{T} \mathsf{A}$.

This property ensures invariance under rotations: if both the volume $\|\mathbf{u}_1(j-1) \wedge \ldots \wedge \mathbf{u}_k(j-1)\|$ and (accordingly) the matrix $\mathsf{A}(j)$ are simultaneously rotated, the new matrix has the same distribution as the original one. As a result, it turns out that

$$\mathcal{S}_k = \langle \ln \|\mathsf{A}\mathbf{e}_1 \wedge \ldots \wedge \mathsf{A}\mathbf{e}_k\| \rangle, \tag{8.14}$$

where the angular brackets denote an average over all possible realisations of the matrix A, while the $\mathbf{e}_j$s are the vectors of the standard Euclidean basis.

Gaussian ensemble

A simple setup where this property is satisfied is that of matrices A whose entries are i.i.d. Gaussian variables with zero mean and variance σ (besides being uncorrelated in time). Such full matrices are instances of the Jacobians that may arise in mean field models, where all the variables are mutually coupled (see Section 10.4.3 for a discussion of some properties thereof and, more specifically, the neural network models studied by Cessac et al. (1994) and Curato and Politi (2013)).

As shown in Appendix C (see also Newman (1986a)),

$$\lambda_k = \ln \sigma + \frac{1}{2}\left[\ln 2 + \psi\left(\frac{N-k+1}{2}\right)\right],$$

where $\psi(y) = \Gamma'(y)/\Gamma(y)$ is the digamma function, while $\Gamma(y)$ is the standard gamma function.

In the thermodynamic limit ($N \to \infty$), one can approximate the digamma function as $\psi(u) = \ln u$. Upon further introducing the integrated density $\rho = k/N$ (see Section 10.1 for the definition of ρ in the general context of spatially extended systems), one obtains

$$\lambda(\rho) = \ln(\sigma\sqrt{N}) + (1/2)\ln(1-\rho). \tag{8.15}$$

Therefore, the Lyapunov exponents remain finite only if the variance σ^2 scales as $1/N$. This is a trivial consequence of the central limit theorem. The amplitude of the generic component u_i' of $\mathbf{u}' = \mathsf{A}\mathbf{u}$ is of the same order as that of u_i only if the (zero-average) matrix entries are of order $1/\sqrt{N}$. In fact, a well-known theorem states that the eigenvalues of A are uniformly distributed on a circle of radius σ (Girko, 1984).

It is instructive to compare Eq. (8.15) with the distribution of "Lyapunov" exponents resulting from the repeated application of the same matrix. For the sake of simplicity, we assume $\sigma = s/\sqrt{N}$. The circular law for the eigenvalues of a single matrix implies that the probability distribution $P(\mu)$ of the moduli μ is linear $P(\mu) = 2\mu/s^2$ for $\mu \le s$, while it is zero above s. The Lyapunov spectrum can be determined by expressing the value λ of the Lyapunov exponent vs. the integrated density ρ (defined in such a way that $\rho = 0$ corresponds to the maximum exponent). It is therefore convenient to pass from $P(\mu)$ to its integral $\rho(\mu) = \int_m^s d\mu P(\mu) = 1 - (\mu/s)^2$. By recalling that $\lambda = \ln \mu$, it is found that

$$\rho(\lambda) = 1 - e^{2\lambda}/s^2.$$

Upon inverting this equation, one obtains the expression

$$\lambda = \ln s + \frac{1}{2}\ln(1-\rho),$$

which is identical to Eq. (8.15). This means that the instability of a "fixed point", resulting from the repeated application of the same random matrix, is the same as in the "dynamical" case, when the matrices are mutually independent. The result is not too surprising since the high symmetry of the matrices prevents the orientations of vectors to play any role.

This result can be extended to non-Gaussian distributions, under the condition that the fourth moment is finite and the density itself everywhere bounded (Isopi and Newman, 1992). This is in a sense a consequence of the combined action of the thermodynamic limit and the central limit theorem.

If one restricts one's interest to the largest Lyapunov exponent, the determination of the probability density of the angles is less of a difficulty, and analytical results can be obtained for Gaussian, mutually correlated entries A_{ij}. In fact, given the covariance matrix Σ of A, Kargin (2014) has proved that in the thermodynamic limit, the largest Lyapunov exponent is

$$\lambda = \frac{1}{2} \ln \lim_{N \to \infty} \text{tr}(\Sigma).$$

In the case of i.i.d. and δ-correlated entries, this formula reduces to Eq. (8.15) for $\rho = 0$, as all diagonal entries are equal.

Finally, under the general condition that the distribution of matrices $A^T A$ is rotationally invariant plus some technical requisites, Newman (1986b) proved that the Lyapunov density spectrum $\rho(\lambda)$ (arising from the product of a random sequence of A-matrices) is determined implicitly by the formula

$$\int d\mu \frac{S(\mu)\mu^2}{(1-\rho)e^{2\lambda} + \rho\mu^2} = 1, \qquad (8.16)$$

where the integrated density ρ is defined in a way that $\rho(\lambda_{max}) = 0$, while $\rho(\lambda_{min}) = 1$. The proof of this formula requires a series of technical conditions which prevent it from being applied to, e.g., the previously discussed Gaussian random matrices. One can, nevertheless, verify a posteriori that Eq. (8.16) holds also in such a case. In fact, the eigenvalues μ of a matrix of the type $\sqrt{A^T A}$ are distributed according to the so-called quarter-circle law (Marčenko and Pastur, 1967a, b; Wigner, 1967)

$$S(\mu) = \begin{cases} \frac{2}{\pi s}\sqrt{1 - \frac{\mu^2}{4s^2}} & 0 \le \mu \le 2s, \\ 0 & \text{otherwise.} \end{cases}$$

Upon replacing this equation into Eq. (8.16) and computing the integral, one obtains

$$\frac{1}{2\rho^2}\left[(1-\rho)\frac{e^{2\lambda}}{s^2} + 2\rho - \sqrt{(1-\rho)^2\frac{e^{4\lambda}}{s^4} + 4\rho(1-\rho)\frac{e^{2\lambda}}{s^2}}\right] = 1.$$

Besides $\rho = 1$, this equation admits the physically relevant solution $\rho = 1 - e^{2\lambda}/s^2$, which coincides with Eq. (8.15) (remember that $s = \sqrt{N}\sigma$). Accordingly, one expects Eq. (8.16) to cover a range of models wider than that for which it has been proven.

8.1.3 Sparse matrices

Full matrices arise in contexts where all variables are mutually coupled. In the physical world it is more likely to encounter setups where the coupling is restricted to small subsets of variables. In such a case one has to deal with *sparse* matrices, where most of the entries are equal to zero, as, for instance, in a chain of oscillators. Here, we consider $N \times N$ matrices A where each row contains K randomly chosen non-zero elements. This matrix structure naturally emerges in networks characterised by a connectivity K. The number K is, in principle, an integer, but one could consider setups where the connectivity fluctuates (either in space or in time), in which case K should be interpreted as an average and could thereby take non-integer values as well. The entries of A are assumed to be positive, i.i.d. variables, with a distribution $g(a)$ (for a discussion of the more general case with random signs, see Cook and Derrida (1990)). Assessing the stability properties of a product of such sparse matrices is a formidable task. So far, it has been possible to determine only the largest Lyapunov exponent in the thermodynamic limit.

Free energy in a random environment

Before determining the expression of the Lyapunov exponent, here we illustrate the equivalence with a statistical-mechanics problem: the determination of the free energy for a directed polymer in a random environment. Consider a sequence of one-dimensional layers, each composed of N points ($1 \le i \le N$), and suppose that each point in layer t is connected with K randomly chosen points in layer $t+1$. Furthermore, a random energy $\varepsilon_{ij}(t)$ is attached to the bond connecting the site i (in layer t) with site j (in layer $t+1$) according to some distribution. One then wants to consider all paths (polymers) ω of length t, which start from a specific point in the first layer. By denoting with E_ω the energy of a generic path ω (it is the sum of the ε_{kj} energies of the single bonds), one can define the partition function

$$Z_t(i) = \sum_\omega e^{-E_\omega \beta}, \tag{8.17}$$

where the sum runs over all walks ω of length t, which start from the point i, while β is the inverse temperature (with a unit Boltzmann constant). It is easy to show that Z_t satisfies the recursive relation

$$Z_{t+1}(i) = \sum_{m=1}^{N} e^{-\varepsilon_{mi}(t)\beta} Z_t(m).$$

The iteration of this equation consists in applying a sparse matrix $A(t)$ ($A_{mi} = e^{-\varepsilon_{mi}\beta}$) to the vector $\mathbf{Z}_t$. By recalling that the free energy F (per bond) is determined by the asymptotic growth rate of the partition function [$Z_t = \exp(-t\beta F)$] one can conclude that $\lambda = -\beta F$, so that the problem of determining the Lyapunov exponent is equivalent to that of determining the free energy. The equivalence goes beyond the specific case herein discussed, but it requires a specific mathematical formalisation that depends on the case of interest.

Here, we determine the Lyapunov exponent by following an intuitive microcanonical approach. For the sake of simplicity we keep referring to $\mathbf{Z}_t$ as the vector to be iterated for the computation of λ. The ith component of the vector $\mathbf{Z}_t$ is the sum of K^t terms, each being the product of t matrix elements (see (8.17) for the polymer interpretation). It is convenient to express the amplitude of each addendum (walk) by referring to its average growth rate, $\prod_{n=1}^{t} \ln a_{ij}(n) = \exp(\tilde{\lambda}t)$. As a result, one can express the Lyapunov exponent as an integral,

$$\lambda = \lim_{t\to\infty} \frac{1}{t} \ln \int \mathcal{N}(\tilde{\lambda}, t) \exp(\tilde{\lambda}t) d\tilde{\lambda}, \qquad (8.18)$$

where $\mathcal{N}(\tilde{\lambda}, t)$ denotes the typical number of addenda characterised by the growth rate $\tilde{\lambda}$. In order to determine $\mathcal{N}(\tilde{\lambda}, t)$, it is necessary to estimate the probability $P(\tilde{\lambda}, t)$ that a single walk is characterised by the growth $\tilde{\lambda}$. This can be done by introducing the generating function for the distribution of the entries of the matrix,

$$G(q) = \ln \left[\int_0^{+\infty} g(a)a^q da \right]. \qquad (8.19)$$

In fact, it is easily seen that $P(\tilde{\lambda}, t) \approx \exp(tf(\tilde{\lambda}))$, where $f(\tilde{\lambda})$ is the Legendre transform of $G(q)$:

$$f(\tilde{\lambda}) = \min_q (G(q) - q\tilde{\lambda}).$$

Notice that $f(\tilde{\lambda})$ has typically a parabolic shape with a maximum equal to zero for some value of $\tilde{\lambda}$.

We can now determine the typical number of $\tilde{\lambda}$ values by multiplying the probability $P(\tilde{\lambda}, t)$ by the number K^t of walks,

$$\mathcal{N}(\tilde{\lambda}, t) = \exp(H(\tilde{\lambda})t) := \exp[(f(\tilde{\lambda}) + \ln K)t].$$

In general, there exists a finite interval $\Delta\tilde{\lambda} = [\tilde{\lambda}_m, \tilde{\lambda}_M]$ where $H(\tilde{\lambda})$ is larger than zero (i.e. $H(\tilde{\lambda}_m) = H(\tilde{\lambda}_M) = 0$). This implies that only the $\tilde{\lambda}$-values belonging to $\Delta\tilde{\lambda}$ are exponentially abundant, while the other ones are practically absent.

From Eq. (8.18), it follows that the Lyapunov exponent is given by

$$\lambda = f(\tilde{\lambda}^*) + \tilde{\lambda}^* + \ln K,$$

where $\tilde{\lambda}^*$ is the value which maximises the r.h.s., i.e. where $f'(\tilde{\lambda}) = -1$. This is true, however, only if $\lambda^* \in \Delta\tilde{\lambda}$. In such a regime (hereafter called phase I), by recalling that the Legendre transform implies $f' = -q$, one obtains the expression

$$\lambda = G(1) + \ln K.$$

If $\tilde{\lambda}^*$ falls outside $\Delta\tilde{\lambda}$ (phase II), the leading contribution to the Lyapunov exponent comes from the left border of $\Delta\tilde{\lambda}$, where $\mathcal{N} \approx 1$,

$$\lambda = \tilde{\lambda}_m.$$

The proper phase is selected depending on the sign of

$$H(\tilde{\lambda}^*) = G(1) + \ln K - G'(1).$$

A positive (negative) $H(\tilde{\lambda}^*)$ implies that we are in phase I (II).

We now discuss a simple model which nicely shows that the relevant control parameter is the amount of disorder. First of all, let us notice that since a uniform rescaling of the matrix entries induces a shift of the Lyapunov exponents, one can, without loss of generality, fix the average amplitude by imposing $\langle \ln a_{ij} \rangle = 0$. We then consider a Gaussian distribution of $v_{ij} = \ln a_{ij}$, with zero mean and variance β^2

$$P(v) = \frac{1}{\sqrt{2\pi}\beta}e^{-v^2/(2\beta^2)}.$$

With this assumption, the integrals in Eq. (8.19) are standard Gaussian integrals that can be easily computed. Upon increasing β, the system passes from phase I to phase II,

$$\lambda = \begin{cases} \ln K + \beta^2/2 & \text{for } \beta < \beta_c, \\ \sqrt{2\ln K}\beta & \text{for } \beta > \beta_c, \end{cases}$$

where $\beta_c = \sqrt{2\ln K}$. The comparison with the polymer setup shows that β is equivalent to the inverse temperature, so that the high-temperature phase corresponds to the weak disorder regime, while the low-temperature phase corresponds to the strong disorder regime.

In the weak disorder limit, λ is the sum of two contributions: the first addendum $\ln K$ takes into account the "amplification" arising from the sum of K contributions; the second addendum coincides with the generalised Lyapunov exponent $\mathcal{L}(1)$ ($\mathcal{L}(q) = G(q)/q$) associated with the product of the scalars a_{ij} (cf. Chapter 5). The reason why the q-value 1 appears in the expression of the Lyapunov exponent instead of the standard one $q = 0$ is that each component of the vector $\mathbf{Z}_t$ is the *sum* of an increasing number of multipliers (see also the discussion on weakly coupled dynamical systems in the next chapter). In fact, the case $K = 1$ is singular: this formula does not apply, as there is not an exponential number of walks to be summed over. In the language of statistical mechanics, the low-temperature phase arising above β_c is equivalent to a glassy phase characterised by a replica-symmetry breaking (Derrida and Spohn, 1988).

The results of direct simulations performed for $K = 4$ and different system sizes are plotted in Fig. 8.2 and compared with the theoretical predictions (see the solid and dashed lines in panel (a)). There, one can see an overall agreement, but also that the Lyapunov exponent converges quite rapidly to the theoretical curve below $\beta_c = \sqrt{2\ln 4}$, while the convergence is much slower in phase II (crosses correspond to $N = 10^4$). Additional information can be extracted by looking at the average structure of the Lyapunov vectors. Given the normalised vector $\mathbf{Z}_t$ (the Euclidean norm set equal to 1), its components are ordered from the largest to the smallest one and then averaged over time to obtain $\bar{Z}$. In panels (b) and (c) of Fig. 8.2 two distributions below and above the transition are reported. In both cases one can notice a power-law decay of the amplitudes. The crucial difference is that in the weak disorder phase, the decay is slow, and this implies an extended nature of

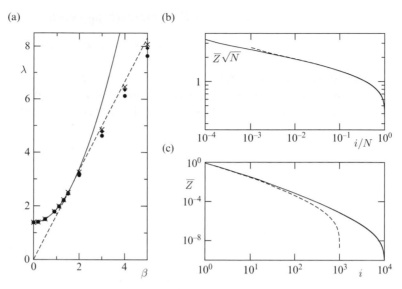

Fig. 8.2 (a) The maximum Lyapunov exponent for a sparse network with connectivity $K = 4$ and a Gaussian distribution of the logarithms of the multipliers, with a standard deviation β. Solid and dashed lines correspond to phases I and II, respectively. Circles, diamonds and crosses correspond to $N = 100, 1000$ and 10000, respectively. The average vector structure is reported for $\beta = 0.5$ (panel (b)) and $\beta = 3$, (panel (c)).

the vectors (the curves obtained for different system sizes indeed overlap after multiplying the amplitudes by $\sqrt{N}$). On the other hand, in the strong disorder phase, the vector is localised, and the presence of new variables leads to the addition of smaller and smaller components.

Finally, it is important to notice that nowhere in the theoretical analysis have we made use of a specific spatial arrangements of the bonds; so long as the amplitudes $a_{ij}(t)$ are independent of each other, the spatial structure may remain the same at all times, and yet all of the previous results still apply. This strengthens the connection with physical setups such as (quenched) sparse networks.

8.1.4 Polytomic noise

In some cases, the disorder manifests itself as a random selection of matrices out of a finite set of options; this is what may be referred to as *polytomic* noise. It is a relatively simple setup where one has the chance of deriving analytical results and may serve as an approximation for continuous distributions.

Here, we first discuss the microcanonical appproach (Deutsch and Paladin, 1989), which allows the derivation of simple approximate expressions of the Lyapunov exponent. Then, we illustrate an application of zeta-function formalism which leads to a series of exponentially converging estimates (Mainieri, 1992). Finally a faster-than-exponentially converging approach by Pollicott (2010) is presented, which applies to strictly positive matrices.

Microcanonical approach

The microcanonical approach can, in principle, be applied to a generic polytomic noise; for the sake of simplicity we restrict our analysis to the dichotomic case and denote with A and B the two possible matrices. We illustrate the method by following a different approach from that of the original paper (Deutsch and Paladin, 1989), as it is amenable to a more transparent interpretation. The starting point is the exact equality

$$\left\| \left(A + xB \right)^N \right\| = e^{\nu(x)N}, \tag{8.20}$$

where x is an unknown parameter, while $\nu(x)$ is the logarithm of the largest (in modulus) eigenvalue of $(A + xB)$. The l.h.s. of this equation can be rewritten, by using the binomial expansion, as

$$\left(A + xB \right)^N = \sum_m \binom{N}{m} x^{N-m} e^{\lambda(p)N}, \tag{8.21}$$

where we have introduced the major assumption that for large N, the contributions of a generic term, composed of m matrices A and $N - m$ matrices B, is independent of the matrix arrangement and can be thereby interpreted as the Lyapunov exponent for a random product with a fraction $p = m/N$ of matrices A. The very fact we restrict the calculations to products of matrices where the fraction of elements of one species is fixed is the motivation for the name, *microcanonical*, attributed to the procedure. By now equating Eqs. (8.20, 8.21) and transforming the sum into an integral, one can write

$$e^{\nu(x)N} = \int \exp\left([-p \ln p - (1-p) \ln(1-p) + (1-p) \ln x + \lambda(p)] N \right) dp,$$

where we have also made use of the Stirling formula to approximate the binomial coefficient. The integral can be performed by invoking the steepest descent method to obtain

$$\nu(x) = \lambda(p^*) - p^* \ln p^* - (1 - p^*) \ln(1 - p^*) + (1 - p^*) \ln x$$
$$0 = \lambda'(p^*) - \ln p^* + \ln(1 - p^*) - \ln x.$$

Given x, the second equation allows determining the correct value of the fraction p^*, which contributes to $\nu(x)$. In practice, this is the inverse formula of the one we need, since it is easy to determine $\nu(x)$ by diagonalising $xA + (1-x)B$, while the difficult task is to determine $\lambda(p)$. As for standard Legendre transforms, however, one can invert this formula, obtaining

$$\lambda(p) = \nu(x^*) + p \ln p + (1-p) \ln(1-p) - (1-p) \ln x^*,$$
$$0 = \nu'(x^*) - \frac{1-p}{x^*}.$$

Once the probability p has been assigned, this formula yields the value of x to be inserted in the expression of $\lambda(p)$.

If the matrices reduce to scalar quantites ($|A|$ and $|B|$), this approach is exact: $\nu(x) = \ln(|A| + x|B|)$, which then implies $x^* = A(1-p)/(Bp)$, so that the Lyapunov exponent is $\lambda = p \ln A + (1-p) \ln B$.

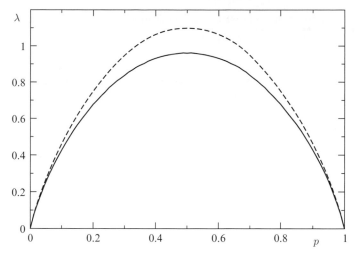

Fig. 8.3 The maximum Lyapunov exponent for a product of the 2×2 matrices (8.22) with $b = 4$ and varying probability p of the matrix $\mathbf{A}$. The solid line refers to the numerical simulations, while the dashed curve corresponds to the microcanonical approach.

An instructive test of the microcanonical approach can be made by referring to the following pair of matrices

$$\mathbf{A} = \begin{pmatrix} 1 & 0 \\ b & 1 \end{pmatrix}; \qquad \mathbf{B} = \begin{pmatrix} 1 & b \\ 0 & 1 \end{pmatrix}. \tag{8.22}$$

Both are characterised by two unit eigenvectors, i.e. by a zero Lyapunov exponent, so that any deviation from 0, for $0 < p < 1$, is entirely due to the non-commutativity of the two matrices. In Fig. 8.3 we compare the outcome of numerical simulations (solid line) with the microcanonical approach (dashed line) for $b = 4$.

We see that the microcanonical approach is able to capture the increase of the exponent due to the non-commutativity of the matrices. Notice that the matrix non-commutativity may induce counter-intuitive phenomena such as the onset of instabilities from the combination of otherwise stable dynamics. For instance, upon scaling down the amplitude of all matrix elements by a factor 2, the two matrices $\mathbf{A}$ and $\mathbf{B}$ would correspond to a stable dynamics (with an exponent $-\ln 2$), while the Lyapunov exponent of a random product with equal probabilities would be positive (≈ 0.27). In practice, we see that noise may turn a stable into an unstable dynamics, not because it allows the overcoming of some energy barrier, but because it qualitatively affects the evolution of arbitrarily small perturbations. On the one hand, this phenomenon can be seen as the counterpart of noise stabilisation discussed in Section 8.2.3. On the other hand, it is the dynamical system multiplicative version of the Parrondo's paradox (Parrondo and Dins, 2004; Harmer and Abbott, 1999), i.e. a possibility to generate a winning game out of a suitable combination of losing games.

In general, the accuracy of the method depends on the fluctuations of the eigendirections between the two matrices and worsens in the presence of negative entries, as they induce partial cancellations which affect the quality of the theoretical estimates. In some instances arising in the context of statistical mechanics, the microcanonical approach proves to be rather accurate (see, e.g., Crisanti et al. (1993)).

Zeta-function formalism

Given a polytomic noise, where the $N \times N$ matrix A_j can assume k different values, a generic sequence of t such matrices is uniquely identified by $\{i\} = \{i_1, i_2, \ldots, i_n\}$, where $i_j \in \{1, 2, \ldots, k\}$, while $|\{i\}| = n$ denotes its length and $p_{\{i\}} = p_{i_1} p_{i_2} \ldots p_{i_n}$ the probability of the given sequence. One can therefore define the Ruelle zeta-function (see Eq. (5.25)) as

$$\zeta(z, q) = \exp\left(\sum_{t=1}^{\infty} \frac{z^t}{t} \sum_{|\{i\}|=t} p_{\{i\}} \mu_{\{i\}}^q \right),$$

where $\mu_{\{i\}}$ denotes the leading eigenvalue of the product of matrices identified by the $\{i\}$ sequence, i.e. a specific periodic realisation of the noise. Notice also that, at variance with dynamical systems, where the probability of an orbit is determined by its multipliers, here it is just a given property of the underlying stochastic process.

Let us then recall that the generalised Lyapunov exponent $\mathcal{L}(q)$ is related to the radius of convergence $\hat{z}$, by the equality $\hat{z} = \mathrm{e}^{-\mathcal{L}(q)q}$. Moreover, it is easily seen that $\zeta^{-1}(z, 0) = 1 - z$, so that $\hat{z}(0) = 1$. As a result, $\partial_z \zeta^{-1}(1, 0) = -1$ and Eq. (5.28) thereby simplifies to

$$\lambda = -\partial_q \zeta^{-1}(1, 0). \tag{8.23}$$

It is now convenient to introduce the notion of a prime cycle as that of a sequence that cannot be expressed as a repetition of shorter blocks (in the case of a dichotomic A/B process, BAA is primary, while AAA is not). Therefore, one can group together into a single term the contribution of each prime cycle and that of its "multiples" (e.g. BAA, $BAABAA$, $BAABAABAA$, etc.). The contribution to $\zeta(z, q)$ of such a term is, for the prime sequence $\{i\}$,

$$\zeta_{\{i\}}(z, q) = \frac{1}{1 - p_{\{i\}} z^{|\{i\}|} \mu_{\{i\}}^q},$$

where we have also taken into account the contribution of the cyclic permutations of $\{i\}$. As a result, the inverse ζ-function can be written as

$$\zeta^{-1}(z, q) = \prod_{\{i\} \in \mathcal{C}} \left(1 - p_{\{i\}} z^n \mu_{\{i\}}^q \right),$$

where the product goes over all prime cycles of any length (cf. (5.26, 5.27)). The next step consists in expressing the infinite product as a power series in z,

$$\zeta^{-1}(z, q) = \sum_{\{i\} \in \mathcal{Z}} (-1)^l p_{\{i\}} z^n \mu_{\{i\}}^q,$$

where the sum now runs over the ensemble $\mathcal{Z}$ of sequences $\{i\}$ that can be expressed as the juxtaposition of distinct prime sequences, and l is their number. Notice that the multiplicity of each sequence depends on the different ways it can be decomposed. The sequence AB will appear twice: as a single block of length 2 or the juxtaposition of A followed by B;

the sequence AA does not appear at all, as the two blocks A are not distinct. Now, from Eq. (8.23),

$$\lambda = -\sum_{\{i\}} (-1)^l p_{\{i\}} \ln \mu_{\{i\}}.$$

By truncating this formula to sequences of length n, one can obtain increasingly accurate estimates of the Lyapunov exponent that converge exponentially to the asymptotic value (Mainieri, 1992).

Super-exponential convergence

The zeta-function formalism can be further improved to accelerate the convergence at least for positive matrices (Pollicott, 2010). Due to the complicated technicalities, here we limit ourselves to describing the final formula (the reader interested in the proof is invited to consult Pollicott (2010)). We anticipate that the faster convergence is ensured by the use of information contained in the eigenvectors of finite products of matrices (in the standard application of the zeta-function formalism, only the eigenvalues are considered). So, let $\mu_{\{i\}}$ and $\mathbf{u}_{\{i\}}$ denote the maximal (non-degenerate) eigenvalue and the corresponding eigenvector for the sequence $\{i\}$ of matrices.

Moreover, $\mathbf{u}_{\{i\}}(l)$ is the lth iterate of $\mathbf{u}_{\{i\}}$, i.e. the eigenvector obtained by cyclically shifting $\{i\}$ by l sites (a sequence that is obviously characterised by the same probability). Finally, without loss of generality, it is also assumed that each matrix has a unit determinant (if not, one can rescale all the entries and eventually sum the contribution arising from the matrix rescaling).

We can now proceed to the introduction of the auxiliary function

$$T_{j,t}^{(n)} = \sum_{|\{i\}|=n} p_{\{i\}} \frac{\exp\left(t \sum_{l=0}^{n-1} \tilde{\lambda}_j \left(\mathbf{u}_{\{i\}}(l)\right)\right)}{1 - \mu_{\{i\}}^{-2}},$$

where $\tilde{\lambda}_j(\mathbf{u})$ is the Lyapunov exponent measured over a single time step, by applying the matrix A_j to the vector $\mathbf{u}$:

$$\tilde{\lambda}_j(\mathbf{u}) \equiv \ln \frac{|\mathsf{A}_j \mathbf{u}|}{|\mathbf{u}|}.$$

One can now define the following functional coefficients

$$a_n^{(j)}(t) = \sum_{n_1 k_1 + \ldots n_l k_l = n} \frac{(-1)^l}{k_1 \ldots k_l!} \frac{(T_{j,t}^{(n_1)})^{k_1} \ldots (T_{j,t}^{(n_l)})^{k_l}}{n_1^{k_1} \ldots n_2^{k_2}},$$

where the sum is over all linear combinations of $n_i \times k_i$ products that sum to n (with $n_i \neq n_j$ when $i \neq j$). As an example, the possible decompositions of $n = 6$ are: (6×1), (3×2), (2×3), (1×6) for $l = 1$; $(5 \times 1 + 1 \times 1)$, $(4 \times 2 + 2 \times 1)$, $(4 \times 1 + 1 \times 2)$, $(3 \times 1 + 1 \times 3)$, $(2 \times 2 + 1 \times 2)$, $(2 \times 1 + 1 \times 4)$ for $l = 2$ and finally $(3 \times 1 + 2 \times 1 + 1 \times 1)$ for $l = 3$. The Lyapunov exponent can be finally expressed as

$$\lambda = \sum_{j=1}^{k} \frac{\sum_{i=1}^{\infty} c_i^{(j)}}{\sum_{i=1}^{\infty} b_i^{(j)}},$$

where $b_i^{(j)} = n a_i^{(j)}(0)$ and $c_i^{(j)} = (\partial a_i^{(j)}/\partial t)(0)$. The external sum over k components expresses a well-known fact: the Lyapunov exponent is the average over k possible options (given the k-polytomic noise). The inner sums express the averaging over the distribution of the possible angles. One can obtain a series of approximations of the Lyapunov exponent by truncating the two inner sums:

$$\lambda_n = \sum_{j=1}^{k} \frac{\sum_{i=1}^{n} c_i^{(j)}}{\sum_{i=1}^{n} b_i^{(j)}}.$$

Since the computation of $b_n^{(j)}$ and $c_n^{(j)}$ involves only the knowledge of eigenvalues and eigenvectors of products of at most n matrices, this equation can be used as a way to approximate the true Lyapunov exponent. Pollicott (2010) has proved that the convergence is faster than exponential. More precisely, he found that the truncation error decreases at least as $\exp(-Cn^{N/(N-1)})$, where N is the dimension of the matrix.

8.2 Linear stochastic systems and stochastic stability

In this section we turn our attention to continuous-time stochastic processes.

8.2.1 First-order stochastic model

We start with the relatively simple, but rather instructive, example of a one-dimensional stochastic equation,

$$\frac{du}{dt} = (a + \xi(t))u, \tag{8.24}$$

where a stationary, multiplicative noise $\xi(t)$ drives the variable u. At first, we make no assumptions about the noise except that its average vanishes $\langle \xi \rangle = 0$ (in fact, the average value can be always absorbed in the parameter a). From the solution

$$u(t) = u(0) \exp \left(at + \int_0^t \xi(t')dt' \right),$$

one easily obtains the Lyapunov exponent

$$\lambda = \left\langle \frac{d}{dt} \ln \frac{u(t)}{u(0)} \right\rangle = a.$$

The computation of the generalised Lyapunov exponents (cf. Chapter 5; in the context of stochastic systems they are often called moment Lyapunov exponents) is slightly more involved:

$$\mathcal{L}(q) = \frac{1}{q} \lim_{t \to \infty} \frac{1}{t} \ln \left\langle u^q(t) \right\rangle = a + \frac{1}{q} \lim_{t \to \infty} \frac{1}{t} \ln \left\langle \exp(q \int_0^t \xi(t')dt') \right\rangle.$$

The problem reduces to the determination of the characteristic function of the random variable $\int_0^t \xi(t')dt'$. If $\xi(t)$ is Gaussian, then the integral is also Gaussian, and for large t its variance is Dt, where $D = \int_{-\infty}^{\infty} \langle \xi(t)\xi(t+\tau) \rangle \, d\tau$ is the diffusion constant of a random walk for which ξ is the velocity. Thus, for Gaussian noise,

$$\mathcal{L}(q) = a + \frac{qD}{2}. \tag{8.25}$$

Such a linear dependence of the generalised Lyapunov exponents on the index q is often called Gaussian approximation (although it may hold also for non-Gaussian noise, and as we will see in the next section, may not hold exactly for Gaussian noise in setups more complex than (8.24)).

8.2.2 Noise-driven oscillator

A simple but important two-dimensional model is a linear oscillator with parametric noise

$$\frac{d^2 u}{dt^2} + (E + \xi(t))u = 0, \tag{8.26}$$

where we assume again $\langle \xi \rangle = 0$. A closed description can be obtained if one further assumes that the noise is Gaussian and δ-correlated, $\langle \xi(t)\xi(t') \rangle = 2\sigma^2 \delta(t - t')$. One can apply the powerful Fokker-Planck formalism and also perform an effective averaging (here, we follow the approach by Zillmer and Pikovsky (2003)).

This simple, but not exactly solvable – in the statistical sense – model finds applications in various fields of physics:

1. In the theory of one-dimensional Anderson localisation, the spatial coordinate can be interpreted as a time variable. Accordingly, Eq. (8.26) can be read as a one-dimensional stationary Schrödinger equation for a single particle in a δ-correlated potential ξ. The eigenenergy E can be either positive or negative – negative values correspond to a band gap. Although, strictly speaking, one should consider (8.26) with two boundary conditions $|u| \to 0$ at $t \to \pm\infty$, it is customary to treat (8.26) as an initial-value problem and to look at the growth rate of the variable u as $t \to \infty$ (Lifshits et al., 1988; Crisanti et al., 1993). The growth rate gives the localisation length, while the fluctuations of $u(t)$ are important for the description of conductance fluctuations in finite samples (see Section 12.1 for more details on the Anderson localisation problem).
2. In the theory of parametric resonance of a linear oscillator, the parameter E is assumed to be positive and is interpreted as the square of the oscillator's frequency, while ξ accounts for the frequency fluctuations. The oscillations $u(t)$ grow as a result of the noisy pump; it can be monitored from the growth rate of the different moments. Usually the oscillator is also characterised by the presence of a linear damping of the type $-2\gamma \dot{u}$. The equation with this damping term can be nevertheless given the form (8.26) by virtue of the transformation $u \to e^{-\gamma t}\tilde{u}$, $E \to E - \gamma^2$.

3. The computation of the largest LE in some high-dimensional Hamiltonian systems can be reduced to an equation such as (8.26), where $u(t)$ is the local perturbation amplitude and E is a suitable curvature (Pettini, 2007). This correspondence is further discussed in Chapter 10.

Remarkably, the generalised Lyapunov exponents for $q = 0, 2, 4, 6, \ldots$ can be determined analytically. The case $q = 0$ corresponds to the usual Lyapunov exponent, which can be computed as follows. With the ansatz $v = \dot{u}/u = d\ln u/dt$, Eq. (8.26) can be reduced to the first-order nonlinear Langevin-type equation

$$\dot{v} = -v^2 - \xi(t) - E. \tag{8.27}$$

Here v, after reaching $-\infty$, is reinjected at $+\infty$, which corresponds to a zero-crossing of $u(t)$. One can then write the Fokker-Planck equation for the probability distribution $P(t, v)$ of v and determine the stationary solution (which, of course, is a solution with a constant probability flow),

$$P(v) = \frac{1}{\sqrt{\pi} \, \sigma^{4/3}} \frac{e^{-\frac{v^3}{3\sigma^2} - \frac{Ev}{\sigma^2}} \int_{-\infty}^{y} e^{\frac{z^3}{3\sigma^2} + \frac{Ez}{\sigma^2}} dz}{\int_0^\infty z^{-1/2} \, e^{-\frac{z^3}{12} - \frac{Ez}{\sigma^{4/3}}} dz}.$$

The Lyapunov exponent is finally given by the average of v (see, e.g., Lifshits et al. (1988)),

$$\lambda = \langle v \rangle = \left\langle \frac{d\ln u}{dt} \right\rangle = \frac{\sigma^{2/3}}{2} \frac{\int_0^\infty z^{1/2} \, e^{-\frac{z^3}{12} - \frac{Ez}{\sigma^{4/3}}} dz}{\int_0^\infty z^{-1/2} \, e^{-\frac{z^3}{12} - \frac{Ez}{\sigma^{4/3}}} dz}. \tag{8.28}$$

This expression is essentially the same as (8.12), and this is not a pure coincidence. Indeed, the oscillator is a two-dimensional system, and its evolution over finite time is described by a 2×2 matrix. In the case $E = \sigma = 0$, the evolution of the oscillator (8.26) over time T can be expressed as

$$u(T) = u(0) + Tu'(0), \qquad u'(T) = u'(0).$$

In the variables $x = u$, $y = u + Tu'$, this evolution is described by the matrix (8.9) with $\mu = 1$. Thus, the continuous-time approximation described in Section 8.1.1 leads to a description that is equivalent to the one developed here for the noisy continuous-time linear oscillator.

The exact expression (8.28) reveals that the Lyapunov exponent is a scaling function of the parameters E and σ: $\lambda(E, \sigma) = \sigma^{2/3}\bar{\lambda}(E\sigma^{-4/3})$. Of course, this follows directly also from the renormalisation of time in (8.26): by rescaling $t = |E|\sigma^{-2}t'$ one obtains an equation which contains only the parameter $E\sigma^{-4/3}$. The scaling is then obtained by expressing the Lyapunov exponent in terms of the original time. The same scaling extends to the generalised Lyapunov exponents.

For $q = 2, 4, 6, \ldots$ a different analytical approach can be used. As Eq. (8.26) is a linear stochastic equation, the evolution of the moments of order q of the type $\langle u^{q-k}\dot{u}^k \rangle$ leads to a closed system of linear equations. By using Eq. (8.26), one can explicitly write

$$\frac{d}{dt}\langle u^{q-k}\dot{u}^k \rangle = (q - k)\langle u^{q-k-1}\dot{u}^{k+1} \rangle - kE\langle u^{q-k+1}\dot{u}^{k-1} \rangle - k\langle \xi(t)u^{q-k+1}\dot{u}^{k-1} \rangle.$$

The average of the last term can be performed by using the Furutsu-Novikov-formula,

$$k\langle \xi(t)u^{q-k+1}\dot{u}^{k-1}\rangle = -k(k-1)\sigma^2\langle u^{q-k+2}\dot{u}^{k-2}\rangle, \tag{8.29}$$

which thus allows deriving a closed set of equations that can be written in a matrix form,

$$\dot{\mathbf{w}} = \mathbf{C}\mathbf{w},$$

where $\mathbf{w} = \{\langle u^q\rangle, \langle u^{q-1}\dot{u}^1\rangle, \ldots, \langle \dot{u}^q\rangle\}$, while $\mathbf{C}$ is the band diagonal matrix

$$\mathbf{C} = \begin{pmatrix} 0 & q & 0 & 0 & & & \\ -E & 0 & q-1 & 0 & \ddots & & \\ 2\sigma^2 & -2E & 0 & q-2 & \ddots & & \\ 0 & 6\sigma^2 & -3E & 0 & \ddots & & \\ & \ddots & \ddots & \ddots & \ddots & & \\ & & & & 0 & 2 & 0 \\ & & & & (1-q)E & 0 & 1 \\ & & & & (q-1)q\sigma^2 & qE & 0 \end{pmatrix}.$$

The exponential growth of the moments of order q is determined by the eigenvalue of $\mathbf{C}$ with the largest real part. By definition (see Chapter 5), the generalised LE $\mathcal{L}(q)$ (for even q) is thus equal to this eigenvalue divided by q (for odd q, $\langle u^q\rangle \neq \langle |u^q|\rangle$, and this approach does not provide the correct exponent).

In the simplest case, $q = 2$, $\mathcal{L}(2)$ is a solution of the cubic equation $\mathcal{L}(2)^3 + E\mathcal{L}(2) - \sigma^2/2 = 0$, i.e.

$$\mathcal{L}(2) = \begin{cases} \left(\frac{\sigma^2}{4} + \sqrt{\frac{\sigma^4}{16} + \frac{E^3}{27}}\right)^{1/3} - \frac{E}{3}\left(\frac{\sigma^2}{4} + \sqrt{\frac{\sigma^4}{16} + \frac{E^3}{27}}\right)^{-1/3}, & \text{if } \frac{E}{3} \geq -\left(\frac{\sigma}{2}\right)^{4/3}, \\ 2\sqrt{\frac{|E|}{3}}\cos\left(\frac{1}{3}\arctan\sqrt{\frac{|E|^3}{27} - \frac{\sigma^4}{16}}\right), & \text{if } \frac{E}{3} < -\left(\frac{\sigma}{2}\right)^{4/3}. \end{cases} \tag{8.30}$$

For larger q-values one has to find the roots of the corresponding polynomial of order $q+1$ numerically; this is technically straightforward.

Generally, the Lyapunov exponents obtained for different even q-values differ from each other, indicating that the finite-time growth rate fluctuates, as discussed in Chapter 5. Within a Gaussian approximation, the generalised Lyapunov exponents grow linearly with q (see Eq. (8.25)) and can be thereby expressed as a function of two of them ($q = 0, 2$ in this case). The detailed analysis performed by Zillmer and Pikovsky (2003) reveals that the Gaussian approximation is valid for large positive and negative values of the relevant parameter $E\sigma^{-4/3}$, while strong deviations are found when $E\sigma^{-4/3}$ is close to zero (in fact, for a given index q the deviations are significant for $|E|^{3/2}\sigma^{-2} \lesssim 2q$).

We now discuss the dynamics of Eq. (8.26) in a more general setup, where the noise is neither Gaussian nor white. Here, we briefly describe the results derived in Pinsky (1986) and Arnold et al. (1986) for a noise generated by a Markov process, in the limit of small noise intensity. More precisely, the evolution equation is

$$\frac{d^2u}{dt^2} + (E + \sigma F(\xi(t)))u = 0,$$

where $\xi(t)$ is a Markov process and F is a function with zero mean. In Arnold et al. (1986) a continuous-time Markov process was considered, which can be represented as a white-noise-driven stochastic process – the Ornstein-Uhlenbeck process belongs to this class. In Pinsky (1986) a Markov process with a finite number of states (described by a suitable master equation) was considered – a telegraph process with two states is of this type. In such contexts, an essential ingredient is represented by the correlation properties of the noise term $F(\xi(t))$ that are quantified by the autocorrelation function $C(t) = \langle F(t)F(0) \rangle$, while the corresponding power spectrum is $f(\omega) = (2\pi)^{-1} \int_{-\infty}^{\infty} \cos \omega t C(t) dt$. In the weak noise limit, $\sigma \to 0$, the expression for the leading term of the LE is, for positive E (Arnold et al., 1986),

$$\lambda = \sigma^2 \frac{\pi f(2E^{1/2})}{4E}, \tag{8.31}$$

while for negative E,

$$\lambda = \sqrt{-E} + \sigma^2 \frac{1}{4E} \int_0^{\infty} e^{-2\sqrt{-E}t} C(t) dt.$$

The relation (8.31) has an insightful physical interpretation. The amplitude of the Lyapunov exponent depends solely on the Fourier component of the noise at twice the frequency of free oscillations. This can be interpreted as a parametric resonance by noise. Indeed, in the case of a periodic $F(t)$, one has a Mathieu equation with a set of instability regions, where the oscillations grow exponentially. The largest unstable region is around twice the bare frequency and has a width (i.e. a frequency mismatch) $\sim \sigma$. As for Eq. (8.31), this region only contributes to the overall instability, when all frequencies are simultaneously present.

In the case F is a telegraph process with two states ± 1 and a transition rate a, the autocorrelation function is $C(t) = e^{-2a|t|}$ and formula (8.31) yields

$$\lambda = \frac{a\sigma^2}{8E(a^2 + E)},$$

in accordance with Pinsky (1986). Arnold et al. (1986) also explored the very slow noise limit (the characteristic correlation time τ tends to infinity) and fixed noise intensity σ (see theorem 6.2 in Arnold et al. (1986)). For large and positive values of E, the Lyapunov exponent scales as $\lambda \sim \tau^{-1}$.

Finally, we briefly mention some further results related to noise-driven oscillators. Mallick and Peyneau (2006) presented some analytic considerations of the model (8.26) with a Gaussian Ornstein-Uhlenbeck noise $\xi(t)$ that is characterised by a purely exponential autocorrelation function. The authors are able to derive approximate expressions for the Lyapunov exponents both in the limit of short correlation times and as expansions for small noise amplitudes. As for the generalised Lyapunov exponents in the presence of a correlated noise, several approximation schemes are discussed by Vallejos and Anteneodo (2012), together with a comparison with numerics.

8.2.3 Khasminskii theory

In this section we consider generic multi-dimensional linear stochastic systems. A mathematically rigorous presentation can be found in Khasminskii (2012) and Arnold (1998). Here we illustrate the theory on a physical level (following chapter 6 of Khasminskii (2012)), shifting from the usual Ito interpretation of the stochastic differential equations, typically adopted in the mathematical literature, to the Stratonovich one.

Let us introduce a general system of N linear stochastic differential equations for the variables $u_1, \ldots, u_N$ with L noise terms $\xi_1(t), \ldots, \xi_L(t)$,

$$\dot{u}_k = a_{kl}u_l + b^n_{kl}\xi_n(t)u_l, \qquad k = 1, \ldots, N, \quad n = 1, \cdots, L, \tag{8.32}$$

with the convention of an implicit sum over repeated indices. The Lyapunov exponent is the average growth rate of the norm $r = (u_1^2 + \cdots + u_N^2)^{1/2}$,

$$\lambda = \left\langle \frac{d}{dt} \ln r \right\rangle = \left\langle \frac{\dot{r}}{r} \right\rangle. \tag{8.33}$$

It is convenient to introduce the normalised variables $y_k = u_k/r$, for which one obtains from (8.32) a closed system of equations

$$\dot{y}_k = a_{kl}y_l + b^n_{kl}\xi_n(t)y_l - y_k y_l a_{lm}y_m - b^n_{lm}\xi_n(t)y_m y_k y_l.$$

This system describes the stochastic evolution of the variables y_k lying on a sphere of unit radius. Under quite general assumptions, one can assume that, at large times, these variables are stationary functionals of the noises $y_k = F_k[\xi_n]$ (there can be situations where these functionals are not unique, meaning that ergodicity is broken). On the other hand, the evolution of r can be expressed using (8.32) as

$$\dot{r} = \frac{u_l \dot{u}_l}{r} = r[y_l a_{lm}y_m + b^n_{lm}\xi_n(t)y_m y_l].$$

Thus, the Lyapunov exponent (8.33) can be expressed as the average over the statistics of noise terms

$$\lambda = \langle y_l a_{lm}y_m + b^n_{lm}\xi_n(t)y_m y_l \rangle. \tag{8.34}$$

The sign of the Lyapunov exponent (8.34) determines the stochastic stability of the linear system (8.32): for $\lambda < 0$ any initial state eventually vanishes with probability one; for $\lambda > 0$ any non-zero initial state diverges to infinity with probability one. Unfortunately, explicit analytic formulas for the LE are known in very simplified setups only. For instance, in the presence of a Gaussian white noise $\langle \xi_n(t)\xi_s(t') \rangle = \delta_{ns}\delta(t - t')$, a powerful method based on the Fokker-Planck equation (Risken, 1989; Gardiner, 2009) can be used to determine the statistical properties of y_k. Even in this special case, however, general closed expressions can be obtained only for $N = 2$ (the noise-driven oscillator of Section 8.2.2 is one such example). As the variables y_1, y_2 lie on the unit circle, one can introduce the angle $\phi = \arctan(y_2/y_1)$, thus reducing the problem to that of finding the stationary distribution of one-dimensional Fokker-Planck equation; general solutions on the circle are well known (Stratonovich, 1967; Risken, 1989). They allow one to calculate the averages required in (8.34), including $\langle \xi_n(t)y_m y_l \rangle$, which can be determined with the help of the Furutsu-Novikov formula as in (8.29). We will not present these general lengthy expressions but rather consider a particular illustrative example adapted from Khasminskii (2012),

$$\dot{u}_1 = b_1 u_1 - \sqrt{2}\sigma u_2 \xi(t),$$
$$\dot{u}_2 = b_2 u_2 + \sqrt{2}\sigma u_1 \xi(t).$$

By introducing the variables $\phi = \arctan(u_2/u_1)$ and $r = \sqrt{u_1^2 + u_2^2}$ one can rewrite the evolution equations as

$$\dot{\phi} = (b_2 - b_1)\sin\phi\cos\phi + \xi(t) \tag{8.35}$$

$$\dot{r} = r(b_1\cos^2\phi + b_2\sin^2\phi). \tag{8.36}$$

The stationary probability density for the Langevin equation (8.35) is

$$W(\phi) = \frac{\exp[2\kappa\cos^2\phi]}{2\pi e^\kappa I_0(\kappa)}, \qquad \kappa = \frac{b_1 - b_2}{4\sigma^2},$$

where I_0 is the modified Bessel function. By averaging the growth rate of r in Eq. (8.36) over this distribution, one obtains the Lyapunov exponent

$$\lambda = \left\langle b_1\cos^2\phi + b_2\sin^2\phi \right\rangle = \frac{1}{2}\left[b_1 + b_2 + (b_1 - b_2)\frac{I_1(\kappa)}{I_0(\kappa)}\right], \tag{8.37}$$

where I_1 is the Bessel function of order 1. Remarkably, expression (8.37) reveals an interesting example of noise stabilisation. Suppose that $b_1 > 0 > b_2$ and $b_1 + b_2 < 0$. Then, in the limit of small noise, where $\kappa \to \infty$, we have $I_1(\kappa)/I_0(\kappa) \to 1$ and $\lambda = b_1 > 0$, i.e. as expected the LE coincides with the largest eigenvalue of the linear deterministic system. For large noise $\kappa \to 0$, $I_1(\kappa)/I_0(\kappa) \to 0$ and $\lambda = (b_1 + b_2)/2 < 0$, i.e. the system is stable. The physical reason for this stabilisation is obvious: noise "rotates" the angle ϕ, mixing unstable (u_1) and stable (u_2) directions, so that the resulting exponent becomes the average of the eigenvalues of the deterministic problem.

As shown by Arnold and Imkeller (1995), the Khasminskii approach can be extended to the computation of all generalised Lyapunov exponents.

8.2.4 High-dimensional systems

Some analytical results can be obtained in globally coupled, high-dimensional systems, when the single matrix elements are statistically identical. In the case of δ-correlated noise, it is possible to determine the entire Lyapunov spectrum. In more general conditions, only the largest LE is known.

δ-correlated noise

Let the evolution be determined by the stochastic differential equation

$$\dot{u}_i = \sum_{j=1}^{N} K_{ij}(t)u_j, \tag{8.38}$$

where $\langle K_{ij}(t)K_{hl}(t')\rangle = \sigma^2\delta_{ih}\delta_{jl}\delta(t-t')$. Upon integrating the equations over a small time interval $1/m$, this equation can be formally rewritten as

$$\mathbf{u}(t+\frac{1}{m}) = \mathbf{A}\mathbf{u}(t),$$

where $\mathbf{A} = \exp(\mathbf{K}/\sqrt{m})$. Since $\mathbf{A}$ satisfies the property I listed in Section 8.1.2, Eq. (8.14) can be rewritten as (Newman, 1986a)

$$\mathcal{S}_k = \lim_{m\to\infty}\frac{m}{2}\left\langle\ln\|e^{\mathbf{K}/\sqrt{m}}\mathbf{e}_1\wedge\ldots\wedge e^{\mathbf{K}/\sqrt{m}}\mathbf{e}_k\|^2\right\rangle. \tag{8.39}$$

As shown in Appendix C, the Lyapunov exponents are

$$\lambda_k = \frac{\sigma^2}{2}(N-2k+1). \tag{8.40}$$

This is a perfectly symmetric spectrum ($\lambda_k = -\lambda_{N-k+1}$), as expected for a time-reversible system such as (8.38). The result is, however, less obvious than it might appear. It has been obtained by following Newman's approach, which implicitly adopts the Stratonovich interpretation, which preserves the time reversibility of the stochastic equation. Had one followed Ito's approach, all Lyapunov exponents would have had to be shifted down by an amount $\sigma^2/2$, thus losing the symmetry of the Lyapunov spectrum.

In the thermodynamic limit, assuming $\sigma^2 = s/N$, the spectral density is given by the linear expression

$$\lambda = \frac{s}{2}(1-2\rho). \tag{8.41}$$

As for the discrete-time case discussed in Section 8.1.2, it is instructive to compare the Lyapunov spectrum obtained in the dynamical regime with that resulting from the repeated application of a same random matrix $\mathbf{K}$. On the one hand, the eigenvalue spectrum of $\mathbf{K}$ obviously follows the circular law, and the stability is determined by the real part of the eigenvalues (rather than by their moduli, as in the discrete time case). On the other hand, by expanding the density in the vicinity of the largest real part one finds

$$\lambda = s[1-(3\rho\pi)^{2/3}].$$

This spectrum is substantially different from (8.41), starting from the maximum exponent which, in the dynamical regime, is half of that for the fixed point. Moreover, the latter spectrum exhibits a singular behaviour instead of the linear shape of the former one.

Correlated noise

In the presence of a correlated noise, the determination of the entire Lyapunov spectrum is problematic. One can, nevertheless, determine the largest LE. We assume the same model structure as in Eq. (8.38), except for the normalisation of the matrix elements and their correlations, namely

$$\dot{u}_i = \frac{1}{\sqrt{N}}\sum_j K_{ij}u_j \tag{8.42}$$

and

$$\langle K_{ij}(t)K_{km}(t+\tau)\rangle = \delta_{ik}\delta_{jm}C(\tau).$$

Given the tangent vector $\mathbf{u}$, it is convenient to introduce its two-time autocorrelation

$$\chi^2(t,\tau) = \xi^2(t_1,t_2) = \left\langle \sum_i u_i(t_1)u_i(t_2) \right\rangle,$$

where $t = (t_1+t_2)/2$ and $\tau = t_2-t_1$, while the average is performed over all realisations of the noise. Since $\chi^2(t,0)$ is the square Euclidean norm of the vector, it is readily understood that

$$\chi^2(t,\tau) = \hat{\chi}^2(\tau)e^{2\mathcal{L}(2)t}, \tag{8.43}$$

where $\mathcal{L}(2)$ is the generalised LE for $q = 2$, since the divergence rate is obtained by averaging the second moment.

From the equations of motion (8.42) it follows that

$$\frac{\partial^2 \xi^2}{\partial t_1 \partial t_2} = \left\langle \frac{1}{N} \sum_{i,j,k} K_{ij}(t_1)K_{ik}(t_2)u_j(t_1)u_k(t_2) \right\rangle.$$

Because of the statistical independence of the matrix elements, and assuming also that the vector components are independent of the K_{ij} entries, we can rewrite this equation as

$$\frac{\partial^2 \xi^2}{\partial t_1 \partial t_2} = C(t_2 - t_1)\xi^2(t_1,t_2). \tag{8.44}$$

Upon then introducing the variables t and τ, Eq. (8.44) can be rewritten as

$$\frac{1}{4}\frac{\partial^2 \chi^2}{\partial t^2} - \frac{\partial^2 \chi^2}{\partial \tau^2} = C(\tau)\chi^2(t,\tau),$$

which, after making use of Eq. (8.43), reduces to the functional eigenvalue equation,

$$\frac{\partial^2 \hat{\chi}^2}{\partial \tau^2} + C(\tau)\hat{\chi}^2(\tau) = \mathcal{L}^2(2)\hat{\chi}^2. \tag{8.45}$$

This is a Schrödinger-like equation, where τ plays the role of a spatial variable, while the correlation function $C(\tau)$ corresponds to a pseudo-potential and $\mathcal{L}^2(2)$ is a pseudo-energy. The generalised LE corresponds to the square root of the largest eigenvalue of the Schrödinger equation (the sign of the potential is inverted with respect to the usual case encountered in quantum mechanics). If the elements K_{ij} are time-independent random numbers (i.e. the disorder is quenched), then $C(\tau) = C(0)$ is a time-independent constant. In this case $\mathcal{L}^2(2)$ is equal to the variance $C(0)$ of the Jacobian elements. By recalling that the theory of Gaussian random matrices predicts that $\lambda^2 = C(0)$ (Girko, 1984) (see also Section 8.1.2), this result reveals that fluctuations are negligible in this context and the dependence of $\mathcal{L}(q)$ on q is at least partially absent. In general physical contexts, where correlations vary with τ, $C(0)$ can be still used as a proxy of $\mathcal{L}^2(2)$; in such cases, it gives an upper bound to the true generalised exponent.

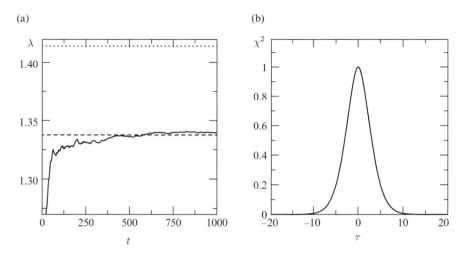

Fig. 8.4 (a) Time convergence of the LE of model (8.42) for $\gamma = 0.05$ and $\sigma^2 = 0.2$, as obtained from a simulation with $N = 800$ variables. The dotted line corresponds to the naive estimate of $\mathcal{L}(2)$, obtained as the standard deviation of the Jacobian entries; the dashed line corresponds to the exact value of $\mathcal{L}(2)$, derived from the largest eigenvalue of Eq. (8.45); (b) the corresponding eigenfunction.

We now briefly illustrate an application of this formalism in the case of a purely exponential decay of the correlations, i.e.

$$\dot{K}_{ij} = -\gamma K_{ij} + \eta_{ij}(t),$$

where η_{ij}s are independent white-noise signals,

$$\langle \eta_{ij}(t)\eta_{km}(t \mid \tau)\rangle = \sigma^2 \delta_{ik}\delta_{jm}\delta(\tau).$$

In this case, $C(\tau) = (\sigma^2/2\gamma)\exp(-\gamma|\tau|)$. The "heaven" state of the Schrödinger equation for $\gamma = 0.05$ and $\sigma^2 = 0.2$ is plotted in Fig. 8.4b. The corresponding energy yields $\mathcal{L}(2)$ (see the dashed line in Fig. 8.4a). The comparison with a direct computation of λ (see the solid curve) confirms the correctness of the theoretical estimate and suggests that fluctuations are negligible, as $\mathcal{L}(2)$ is numerically indistinguishable from λ.

This approach is quite suited for the study of neural networks such as

$$\dot{U}_i = -U_i + \frac{g}{\sqrt{N}}\sum_j G_{ij}\tanh U_j, \tag{8.46}$$

where G_{ij} are i.i.d. Gaussian variables with zero average and unit variance. In fact, this model is the setup where a formula of the type (8.45) had been first derived by Sompolinsky et al. (1988).

The tangent space dynamics of model (8.46) is described by the equation

$$\dot{v}_i = \frac{g}{\sqrt{N}}\sum_j \frac{G_{ij}}{\cosh^2 U_j}v_j, \tag{8.47}$$

where $v_i = u_i \exp(t)$ and $\mathbf{u}$ is the usual tangent vector. Eq. (8.47) is of the type (8.42), with $K_{ij} = gG_{ij}/\cosh^2(U_j)$. In this context, the fluctuations of the Jacobian elements are the result of the underlying chaotic motion in real space. With the help of a sophisticated dynamical mean field approach, Sompolinsky et al. (1988) were able to derive a self-consistent equation for the correlation function of U_j and thereby determine the Lyapunov exponent.

8.3 Noisy nonlinear systems

For noise-driven dynamical systems the definition of the Lyapunov exponents is essentially the same as for purely deterministic ones. Let $\mathbf{U}(t)$ denote a generic trajectory of the stochastic system, while $\mathbf{u}(t)$ denotes the evolution of a perturbation of $\mathbf{U}(t)$ under the action of the same noise forces $\xi(t)$. In the case of continuous time, they satisfy the equations

$$\dot{\mathbf{U}} = \mathbf{F}(\mathbf{U}, \xi, t), \qquad \dot{\mathbf{u}} = \mathsf{J}(t)\mathbf{u}, \qquad \mathsf{J} = \frac{\partial \mathbf{F}}{\partial \mathbf{U}}.$$

The Lyapunov exponents are then defined from the asymptotic growth rate of $\mathbf{u}(t)$ in the usual way, as described in Chapter 2. A completely analogous definition applies to the discrete time case (noise-driven or random mappings), too.

At variance with the purely deterministic case, the computation of the Lyapunov exponents is generally nontrivial already in simple setups, since constant or strictly periodic solutions do not exist. One can, nevertheless, apply the same numerical methods defined in Chapter 3, complemented by the implementation of suitable numerical algorithms for the integration of stochastic differential equations.

8.3.1 LEs as eigenvalues and supersymmetry

In Sections 5.5 and 8.2.4 we have already seen that the problem of computing the LEs can be "reduced" to that of determining the eigenvalues of a suitable evolution operator. This is quite a general strategy that is further strengthened by the existence of a remarkable analogy with supersymmetric quantum mechanics.

Here, we illustrate the relationship by discussing a simple one-dimensional dynamical system with additive noise (Benzi et al., 1985; Graham, 1988),

$$\dot{x} = f(x) + \xi(t), \qquad f(x) = -\frac{dV(x)}{dx}, \tag{8.48}$$

where $\xi(t)$ is a Gaussian δ-correlated process, and, in order to make the link more transparent, we have changed notation, denoting the noisy nonlinear process with $x(t)$.

Since the system is one-dimensional, its LE can be determined by averaging the derivative of the force,

$$\lambda = \langle f'(x) \rangle = -\left\langle \frac{d^2 V(x)}{dx^2} \right\rangle, \tag{8.49}$$

while the evaluation of the generalised exponents $\mathcal{L}(q)$ involves the following limit (see Chapter 5),

$$q\mathcal{L}(q) = \lim_{\tau \to \infty} \frac{1}{\tau} \ln \left\langle \left(\frac{\delta x(\tau)}{\delta x(0)} \right)^q \right\rangle = \lim_{\tau \to \infty} \frac{1}{\tau} \ln \left\langle e^{q \int_0^\tau f'(x(t))dt} \right\rangle, \tag{8.50}$$

where the computation of the average requires knowledge of the probability density $Q[x(t)]$ of a generic path. $Q[x(t)]$ can be determined by setting up a suitable path-integral formalism. Let us start by rewriting (8.48) as $\dot{w} = (\dot{x} - f(x))$, so that $\xi(t) = \dot{w}(t)$, where w is a Wiener process (random walk). The probability P to observe a path $w(t)$ for a time τ can be then expressed as

$$P[w(t)][dw(t)] = \frac{1}{N} \exp \left\{ -\frac{1}{2} \int_0^\tau (\dot{w})^2 dt \right\} [dw(t)], \tag{8.51}$$

where N is a suitable normalisation constant. As we are eventually interested in the distribution $Q[x(t)][dx(t)]$ of the variable x, it is necessary to perform a change of variable from w to x. The corresponding Jacobian is (Horsthemke and Bach, 1975)

$$[dw(t)] = \exp \left\{ -\frac{1}{2} \int_0^\tau \left(\frac{df}{dx} \right)^2 dt \right\} [dx(t)],$$

which, upon insertion into Eq. (8.51), leads to

$$Q[x(t)] = \frac{1}{N} \exp \left\{ -\frac{1}{2} \int_0^\tau \left[(\dot{x} - f(x))^2 + \frac{df}{dx} \right] dt \right\}.$$

By then neglecting boundary effects (at times 0 and τ), $Q[x(T)]$ can be finally expressed in terms of the potential V introduced in (8.48),

$$Q[x(t)] = \frac{1}{N} \exp \left\{ -\int_0^\tau \left[(\dot{x})^2 + \left(\frac{dV}{dx} \right)^2 - \frac{d^2V}{dx^2} \right] dt \right\}.$$

As a result,

$$\left\langle e^{q \int_0^\tau f'(x(t))dt} \right\rangle = \int [dx(t)] Q[x(t)] e^{q \int_0^\tau f'(x(t))dt}$$

$$= \frac{1}{N} \int [dx(t)] \exp \left\{ -\int_0^\tau \left[\frac{1}{2} (\dot{x})^2 + U(x) + q \frac{d^2V}{dx^2} \right] dt \right\}, \tag{8.52}$$

where

$$U(x) = \frac{1}{2} \left[\left(\frac{dV}{dx} \right)^2 - \frac{d^2V}{dx^2} \right].$$

On the other hand, the standard path-integral formulation of quantum mechanics (Feynman and Hibbs, 2010) implies that for large enough τ,

$$\frac{1}{N} \int [dx(t)] \exp \left\{ -\int_0^\tau \left[\frac{1}{2} (\dot{x})^2 + W(x) \right] dt \right\} \sim e^{-\tau E_0}, \tag{8.53}$$

where E_0 is the ground state of the Hamiltonian

$$\hat{H} = -\frac{1}{2}\frac{\partial^2}{\partial x^2} + W(x), \tag{8.54}$$

where $W(x)$ is some generic potential. By now combining Eqs. (8.50, 8.52 and 8.53), it is found that

$$\mathcal{L}(q) = -E_0(q)/q, \tag{8.55}$$

where $E_0(q)$ is the ground state of the Hamiltonian (8.54), with

$$W(x) = U(x) + q\frac{d^2 V}{dx^2}. \tag{8.56}$$

Accordingly, the computation of the LEs is traced back to the determination of the (lowest) eigenvalue of a suitable operator (the Hamiltonian (8.54)). The problem is, however, a bit more tricky. For $q = 0$, the ground state has zero energy (the corresponding eigenstate is nothing but the stationary distribution of the process $x(t)$). In the small-q limit, a standard perturbation theory, nevertheless, implies that

$$E_0(q) \approx q\left\langle\frac{d^2 V}{dx^2}\right\rangle,$$

which, if combined with Eq. (8.55), yields the expected expression for the usual Lyapunov exponent from $\lambda = (d\mathcal{L}/dq)|_{q=0}$ (cf. Eq. (8.49)).

Now the supersymmetry comes into play. By setting $q = 0$ in Eq. (8.56), one obtains the "bosonic" Hamiltonian

$$\hat{H}_b = -\frac{1}{2}\frac{\partial^2}{\partial x^2} + \frac{1}{2}\left[\left(\frac{dV}{dx}\right)^2 - \frac{d^2 V}{dx^2}\right],$$

which can be written as as $\hat{H}_b = A^\dagger A$, where

$$A^\dagger = \frac{1}{\sqrt{2}}\left(-\frac{d}{dx} + \frac{dV}{dx}\right), \qquad A = \frac{1}{\sqrt{2}}\left(\frac{d}{dx} + \frac{dV}{dx}\right).$$

On the other hand, by setting $q = 1$ in Eq. (8.56), one obtains the "fermionic" Hamiltonian

$$\hat{H}_f = AA^\dagger = -\frac{1}{2}\frac{\partial^2}{\partial x^2} + \frac{1}{2}\left[\left(\frac{dV}{dx}\right)^2 + \frac{d^2 V}{dx^2}\right].$$

$\hat{H}_b$ and $\hat{H}_f$ are an instance of a pair of partner Hamiltonians that may be encountered in supersymmetric quantum mechanics (see Cooper et al. (1995) and Feigel'man and Tsvelik (1982) for a more general treatment). The two operators share the same set of eigenvalues E^n, with the exception of the ground state $E^0 = 0$ that is present only in the bosonic Hamiltonian. The eigenfunctions are related as follows: given the nth bosonic eigenfunction ψ_b^n, $A\psi_b^n = \psi_f^{n-1}$ identifies the $(n-1)$th fermionic partner; similarly, $A^\dagger\psi_f^n = \psi_b^{n+1}$.

We can thus conclude that the first excited state of the bosonic Hamiltonian (which coincides with the ground state of $\hat{H}_f$) corresponds to the generalised Lyapunov exponent

$-\mathcal{L}(1)$, thus suggesting a strategy to determine, at least, $\mathcal{L}(1)$. An extension of this formalism to higher-dimensional problems and futher references can be found in Gozzi and Reuter (1994) and Tanase-Nicola and Kurchan (2003). Although this formalism is rather elegant, it has not yet led to truly new results.

8.3.2 Weak-noise limit

In the weak-noise limit, one naively expects the LEs of a noisy system to converge to those of the deterministic one (at least when the corresponding attractor is structurally stable). Accordingly, the Lyapunov exponents of a stable steady state perturbed by a small noise remain all negative, while the largest Lyapunov exponent of a noisy chaotic regime is still positive. Of course, larger noise intensities may change the sign of the maximum Lyapunov exponent. In the next section we present some such transitions. Here, we discuss the nontrivial situation where the deterministic dynamics are either periodic or quasiperiodic, i.e. when the maximum Lyapunov exponent in the noise-free case is zero.

If a deterministic autonomous system possesses a stable limit cycle, the largest Lyapunov exponent is zero (it corresponds to the indifferent equilibrium of the phase of the oscillations), while all other exponents are negative. In the presence of a small external forcing, the oscillator dynamics can be reduced to that of the phase (see Kuramoto (1984) for a general reduction procedure and Goldobin et al. (2010) for a general noisy case). Thus, one obtains a one-dimensional equation for the noise-driven phase of the oscillations. Since in one-dimensional driven systems chaos is impossible independent of the nature of driving, the Lyapunov exponent can be only non-positive. Typically, if there is a multiplicative component in the driving term, the Lyapunov exponent is negative. For a white Gaussian noise this can be shown analytically (Goldobin and Pikovsky, 2004; Teramae and Tanaka, 2004). In the simplest case of one component in the force, the equation for the phase can be written as a Langevin equation

$$\dot{\Phi} = \omega + \varepsilon f(\Phi)\xi(t), \qquad (8.57)$$

where $f(\Phi)$ is a 2π-periodic phase sensitivity function and $\xi(t)$ is a δ-correlated, white Gaussian noise. A linear perturbation φ obeys

$$\dot{\varphi} = \varepsilon\varphi f'(\Phi)\xi(t),$$

so that the Lyapunov exponent is obtained by averaging the growth rate

$$\lambda = \varepsilon \left\langle f'(\Phi)\xi(t) \right\rangle = \varepsilon^2 \left\langle f''(\Phi)f(\Phi) \right\rangle,$$

where the last expression follows from the application of the Furutsu-Novikov formula. One can now determine λ, by averaging over the distribution density $\rho(\Phi)$ obtained by solving the Fokker-Planck equation, which corresponds to the Langevin equation (8.57) (Goldobin and Pikovsky, 2004). In the small noise limit, the phase density is uniform (to leading order in ε), and, after a partial integration, it is found that

$$\lambda = \varepsilon^2 \left\langle f''(\Phi)f(\Phi) \right\rangle \approx -\varepsilon^2 \left\langle [f'(\Phi)]^2 \right\rangle.$$

Thus, the Lyapunov exponent is negative and scales as $\sim \varepsilon^2$, analogously to the weak noise limit for products of random matrices (see Section 8.1). If the forcing of the phase is not purely additive (i.e. if $f(\Phi) \neq const$), λ is strictly negative.

When the unperturbed system has two zero Lyapunov exponents, the situation is different. The dynamics reduces to that of a two-dimensional driven system, the two degrees of freedom corresponding to the modes with a zero Lyapunov exponent. As a result, a small noise can induce a positive Lyapunov exponent, i.e. a chaotic dynamics. In fact, we have already considered such a case in Section 8.2.2 when studying a noise-driven oscillator (see Eq. (8.26)). Remarkably, the small-noise asymptotics for the largerst Lyapunov exponent in a stochastically perturbed Hamiltonian dynamics is also the same as in (8.28), namely $\lambda \sim \sigma^{2/3}$ (see, e.g., Pinsky and Wihstutz (1988, 1992), Baxendale and Goukasian (2002) and Lam and Kurchan (2014)).

8.3.3   Synchronisation by common noise and random attractors

We now discuss the physical interpretation of Lyapunov exponents in a noise-driven system. In deterministic dynamical systems, the largest Lyapunov exponent characterises the stability of a typical trajectory. It serves as a criterion to identify the type of the dynamics: a positive Lyapunov exponent implies chaos, while a zero or negative largest LE implies a regular dynamics. In noisy systems, individual trajectories are unavoidably irregular. A proper physical interpretation of the sign of the largest LE follows from its very definition: the Lyapunov exponents describe the stability of a trajectory under the assumption that the noise term is the same for the reference and the perturbed trajectory. Thus, the LEs should be interpreted having in mind identical systems driven by the same *common noise*, but starting from different initial configurations. In the case of a negative largest Lyapunov exponent, slightly different initial states converge to each other. If this property is also valid globally, for typical initial states, one speaks of synchronisation by common noise (Pikovsky, 1984b, c). If the largest Lyapunov exponent is positive, then even close initial states diverge and one speaks of desynchronisation.

In the mathematical literature, the ensemble of phase-points obtained at a given time by evolving identical dissipative systems driven by the same noise, but with different initial conditions, is often called a random attractor (Crauel et al., 1997; Schmalfuß, 1997; Arnold, 1998). In the case of synchronisation, the random attractor is a randomly moving point, while in the case of desynchronisation it is a fluctuating fractal set (Yu et al., 1990). At each time instant this set looks like a strange attractor in a deterministic system, but its form varies in time, according to the noise.

There are two spectacular experiments where (de)synchronisation by common noise can be observed. One involves an ensemble of particles floating on the surface of a fluid. Turbulent fluid motion, restricted to the two-dimensional surface, can be considered as a noisy two-dimensional velocity field. Floating particles all move in the same field but have different initial conditions. If the largest Lyapunov exponent is negative, the particles concentrate in a point, while if the Lyapunov exponent is positive, a fractal set is observed (Sommerer, 1994). In neurosciences, the equivalent concept of reliability has been introduced by Mainen and Sejnowski (1995). Here, one does not consider an ensemble of identical systems, but applies the same noisy force (which is pre-recorded so that the experiment can be repeated many times) to the very same system (neuron) many times. As the initial conditions of the neuron are not controlled, effectively one has an ensemble of identical neurons driven by a common noise. A negative largest Lyapunov exponent means that in all repetitions of the experiment, the neuron fires exactly in the same manner, i.e. it responds reliably to a noisy external force. If the Lyapunov exponent becomes positive, the reliability is lost (Goldobin and Pikovsky, 2006). In the mathematical literature transitions from one to the other type of random attractor are called stochastic D-bifurcations (Arnold, 1998).

Coupled systems

Increasingly complex regimes can be progressively understood by studying systems composed of coupled elementary units. This makes it possible to start from a known "internal dynamics" and thereby pinpoint the effect of the mutual interactions. Such an approach proves to be very useful in several contexts, the most popular one being synchronisation, i.e. the adjustment of the rhythm of interacting oscillatory systems (see Pikovsky et al. (2001) for details).

In this chapter we study the effect of coupling on LEs, starting from the simple Lozi map (A.5), which can be viewed as the two-dimensional transformation

$$\mathbf{U}(t+1) = \mathbf{F}(\mathbf{U}(t)) = \begin{pmatrix} 1 - a|U_1(t)| + bU_2(t) \\ U_1(t) \end{pmatrix}. \tag{9.1}$$

We consider two linearly coupled maps described by the variables $\mathbf{U}$ and $\mathbf{V}$, respectively,

$$\begin{pmatrix} \mathbf{U}(t+1) \\ \mathbf{V}(t+1) \end{pmatrix} = \begin{pmatrix} (1-\varepsilon)I_2 & \varepsilon I_2 \\ \varepsilon I_2 & (1-\varepsilon)I_2 \end{pmatrix} \begin{pmatrix} \mathbf{F}(\mathbf{U}(t)) \\ \mathbf{F}(\mathbf{V}(t)) \end{pmatrix}, \tag{9.2}$$

where ε represents the coupling strength and I_2 is a 2×2 identity matrix. The whole system is four-dimensional; accordingly, it has four Lyapunov exponents $\lambda_m(\varepsilon)$ that all depend on the coupling parameter ε. For vanishing coupling $\varepsilon = 0$, the two maps – each characterised by two LEs – are independent, and, because the mappings are identical, there are two pairs of exponents: $\lambda_1(0) = \lambda_2(0)$, $\lambda_3(0) = \lambda_4(0)$. The dependence of the four LEs on the coupling parameter ε is reported in Fig. 9.1. The LEs seem to vary smoothly with ε, but a closer look reveals a highly singular dependence that goes under the name of coupling sensitivity (see Section 9.1).

Moreover, from Fig. 9.1 one can infer the presence of a qualitative change beyond $\varepsilon = \varepsilon_c \approx 0.085$, when the second exponent λ_2 changes sign, becoming negative (notice also that λ_1 and λ_3 become strictly constant). This is because for $\varepsilon > \varepsilon_c$, the states of the two systems are identical $\mathbf{U}(t) = \mathbf{V}(t)$, although chaotic – i.e. we are in the presence of synchronisation. In Section 9.2 we discuss how the LE analysis can help to characterise the transition to synchronisation. In particular, we will see that this can be conveniently done by introducing special, *transverse* LEs.

9.1 Coupling sensitivity

Coupling sensitivity is the strong repulsion between LEs that is observed when two or more identical chaotic systems are weakly coupled. This effect, first discovered and described

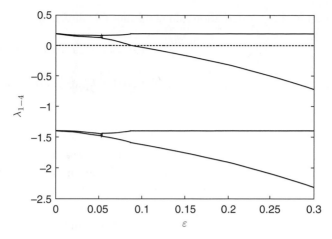

Fig. 9.1 Lyapunov exponents of two coupled identical Lozi maps, for $a = 1.4$ and $b = 0.3$, as functions of the coupling constant ε.

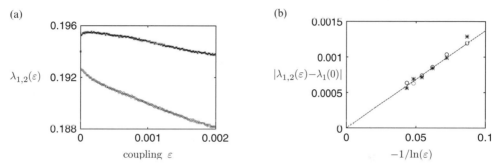

Fig. 9.2 Coupling sensitivity of chaos in the Lozi maps of Fig. 9.1. The actual values of the two largest LEs are plotted in panel (a), while the variations induced by the coupling are plotted in panel (b), revealing scaling $\sim -1/\ln(\varepsilon)$.

by Daido (1984, 1985, 1987), is illustrated in Fig. 9.2. Panel (a) contains an enlarged view of the behaviour of the two largest LEs of Fig. 9.1. There, one can appreciate the highly singular dependence on the coupling strength ε: the separation between the exponents $\lambda_{1,2}$ becomes sizable already for extremely small coupling. The scaling behaviour is revealed in panel (b): the variation of the Lyapunov exponents induced by the coupling is on the order of $\sim 1/|\log(\varepsilon)|$.

This effect is rather universal and can be observed in many setups, including dissipative and Hamiltonian differential equations, symplectic maps, etc.

9.1.1 Statistical theory and qualitative arguments

Here, we present a statistical theory of coupling sensitivity, following Cecconi and Politi (1999) and Zillmer et al. (2000). It is sufficient to consider two coupled linear continuous-time stochastic processes (cf. (8.24) for the uncoupled case) that mimic the dynamics of infinitesimal perturbations along the unstable direction in weakly coupled chaotic systems,

$$\frac{du_1}{dt} = (\chi_1(t) + \lambda_1^0)u_1 + \varepsilon(u_2 - u_1),$$

$$\frac{du_2}{dt} = (\chi_2(t) + \lambda_2^0)u_2 + \varepsilon(u_1 - u_2).$$

(9.3)

The relevant parameters can be grouped as follows:

(1) Lyapunov exponents $\lambda_{1,2}^0$ of the uncoupled systems.
(2) Fluctuations of the local growth rates. They are modeled by the terms $\chi_{1,2}(t)$, which are assumed to be random processes with zero mean values. In order to apply the powerful theory of the one-dimensional Fokker-Planck equation, we furthermore assume these noise terms to be independent, Gaussian and δ-correlated,

$$\langle \chi_i \rangle = 0 \quad \text{and} \quad \langle \chi_i(t)\chi_j(t') \rangle = 2\sigma_i^2 \delta_{ij}\delta(t - t'),$$

where $\sigma_{1,2}^2$ describe the fluctuations of the local expansion rates in the chaotic systems, as discussed in Chapter 5 and Section 8.2.
(3) The coupling is described by the last terms on the r.h.s. Its strength is gauged by the constant ε. A symmetric coupling is initially assumed; asymmetric coupling is considered later.

By assuming that the two variables u_1, u_2 have the same sign (such a condition is established at long but finite times even when the signs are initially different), one can introduce the new variables

$$v_1 = \ln(u_1/u_2) \quad \text{and} \quad v_2 = \ln(u_1 u_2)$$

and rewrite the stochastic model as

$$\frac{dv_1}{dt} = \chi_1 - \chi_2 - 2\varepsilon \sinh(v_1) + \lambda_1^0 - \lambda_2^0,$$

$$\frac{dv_2}{dt} = \chi_1 + \chi_2 + 2\varepsilon \cosh(v_1) + \lambda_1^0 + \lambda_2^0 - 2\varepsilon.$$

(9.4)

Remarkably, the dynamics of the variable v_1 does not depend on that of v_2; thus a standard statistical approach based on the one-dimensional Fokker-Planck equation – see, e.g., Risken (1989) and Gardiner (2009) – can be applied to find the stationary distribution of v_1,

$$\rho(v_1) = C \exp\left(\frac{\lambda_1^0 - \lambda_2^0}{\sigma_1^2 + \sigma_2^2} v_1 - \frac{2\varepsilon}{\sigma_{-1}^2 + \sigma_2^2} \cosh v_1 \right).$$

(9.5)

With the help of the solution (9.5), the largest LE λ_{max} can be obtained (for the sake of simplicity, the subscript max will be omitted),

$$\lambda = \lim_{t\to\infty} \frac{1}{2t} \langle \ln(u_1^2 + u_2^2) \rangle.$$

The norm $u_1^2 + u_2^2$ can be expressed in terms of v_1 and v_2 as

$$\ln(u_1^2 + u_2^2) = v_2 + \ln(2\cosh v_1).$$

Since one is interested in the long-time limit, the stationary distribution of v_1 may be used. Because $\langle \ln(2\cosh v_1) \rangle_{\rho(v_1)}$ is finite and time independent, the only contribution to the largest Lyapunov exponent comes from the growth of v_2. Thus Eq. (9.4) gives the expression for λ,

$$\lambda = \frac{1}{2}\langle \dot{v}_2 \rangle = \varepsilon \langle \cosh v_1 \rangle + \frac{1}{2}(\lambda_1^0 + \lambda_2^0 - 2\varepsilon). \tag{9.6}$$

By averaging according to the stationary distribution $\rho(v_1)$, one obtains

$$\langle \cosh v_1 \rangle = \frac{K_{1-l}(\varepsilon/\sigma^2) + K_{1+l}(\varepsilon/\sigma^2)}{2K_{|l|}(\varepsilon/\sigma^2)},$$

where K_l are the modified Bessel (Macdonald) functions (Abramowitz and Stegun, 1964) $\sigma^2 = (\sigma_1^2 + \sigma_2^2)/2$ and $l = (\lambda_1^0 - \lambda_2^0)/\sigma^2$.

By substituting this expression into Eq. (9.6), an analytical formula for the largest Lyapunov exponent is obtained. We write it as a scaled relationship,

$$\frac{\lambda - \frac{1}{2}(\lambda_1^0 + \lambda_2^0 - 2\varepsilon)}{\sigma^2} = \frac{\varepsilon}{\sigma^2}\frac{K_{1-l}(\varepsilon/\sigma^2) + K_{1+l}(\varepsilon/\sigma^2)}{2K_{|l|}(\varepsilon/\sigma^2)}. \tag{9.7}$$

This expression reveals that the relevant parameters of the problem are the coupling strength and the Lyapunov-exponent mismatch (both normalised to the effective fluctuation σ^2 of the bare exponents).

Simplified expressions can be obtained in the small coupling limit.

Small coupling, equal Lyapunov exponents

From (9.5), if the Lyapunov exponents of two interacting systems are equal, then $\lambda_1^0 = \lambda_2^0 = \lambda^0$ and the parameter l vanishes, leading to

$$\lambda = \varepsilon \frac{K_1(\varepsilon/\sigma^2)}{K_0(\varepsilon/\sigma^2)} + \lambda^0 - \varepsilon.$$

For small coupling ε/σ^2 the leading term in ε is singular, as it follows from the expansions of K_1 and K_0 (Abramowitz and Stegun, 1964),

$$\lambda - \lambda^0 \approx \frac{\sigma^2}{|\ln(\varepsilon/\sigma^2)|}. \tag{9.8}$$

This formula corresponds to Daido's singular dependence of the Lyapunov exponent on the coupling parameter (Daido, 1984, 1985).

Small coupling, different Lyapunov exponents

The expansion (9.8) remains valid for small values of the mismatch $|l|$, if $(\varepsilon/\sigma^2)^{|l|}$ is close to 1. For larger mismatches, when

$$\left(\frac{\varepsilon}{\sigma^2}\right)^{|l|} \ll 1,$$

the largest Lyapunov exponent is

$$\lambda \approx 2\sigma^2 |l| \frac{\Gamma(1-|l|)}{\Gamma(1+|l|)} \left(\frac{\varepsilon}{2\sigma^2}\right)^{2|l|} + \frac{1}{2}(|\lambda_1^0 - \lambda_2^0| + \lambda_1^0 + \lambda_2^0), \tag{9.9}$$

i.e. the singularity becomes of power-law type with the power depending on the system's mismatch. Note also that this is the correction to the largest LE of the two uncoupled systems $\lambda_{1,2}^0$.

Asymmetric coupling

The case of asymmetric coupling, i.e. that of different constants $\varepsilon_1, \varepsilon_2$ in two coupling terms in Eq. (9.3), can be reduced to the symmetric coupling by rescaling $u_1 \to \sqrt{\varepsilon_2} u_1$, $u_2 \to \sqrt{\varepsilon_1} u_2$, so that the effective coupling constant is given by the geometric mean $\sqrt{\varepsilon_1 \varepsilon_2}$ (see Zillmer et al. (2000) for details). Of course, this transformation fails if the coupling is purely unidirectional; i.e. if, say, $\varepsilon_2 = 0$; then the Lyapunov exponents can be easily determined as $\lambda^0, \lambda^0 - \varepsilon_1$ and no singularity is observed.

The main coupling-sensitivity expression (9.8) can be qualitatively understood from the following arguments. Let us, for simplicity, assume that $\lambda_1^0 = \lambda_2^0 = 0$ and rewrite the stochastic model (9.3) in terms of the logarithms $z_{1,2} = \log u_{1,2}$:

$$\begin{aligned}
\frac{dz_1}{dt} &= \chi_1(t) + \varepsilon(\exp(z_2 - z_1) - 1), \\
\frac{dz_2}{dt} &= \chi_2(t) + \varepsilon(\exp(z_1 - z_2) - 1).
\end{aligned} \tag{9.10}$$

The average drift velocity of either z_1 or z_2 can be interpreted as the maximal LE of the coupled system. If $z_1 \approx z_2$, the coupling terms can be neglected and the variables $z_{1,2}$ will perform independent random walks driven by the noise terms $\chi_{1,2}$. When the separation becomes large enough, however, the coupling starts influencing the dynamics. For example, if $z_2 - z_1 \sim \log \varepsilon$, then the variable z_1 is essentially advanced. One can view the dynamics as a random walk in the stripe $|z_1 - z_2| < \log \varepsilon$ with "reflections" from the boundaries, where the "lagged" variable bounces towards the "leading" one (see Fig. 9.3). In practice, whenever the fluctuations drive one of the two perturbations too far ahead, this *acceleration* is transmitted to the other component, thus resulting in an increase of the Lyapunov exponent. The velocity gain can be estimated as follows. Since the random walker diffuses with a constant σ^2, the time to reach the boundary while starting from the diagonal is $(\log \varepsilon)^2 / \sigma^2$ (this follows from the diffusion law $(\Delta z)^2 \sim \sigma^2 t$, considering that $\Delta z \approx \log \varepsilon$). When the boundary is reached, the smallest between z_1 and z_2 is pushed by $\approx |\log \varepsilon|$. Thus, the average velocity increases by $\sim \sigma^2 / |\log \varepsilon|$, which is exactly the LE increase predicted by Eq. (9.8).

Although this theory has been developed for two coupled one-dimensional systems, the effect is very general and arises in generic identical systems. Daido (1985) showed that both Lozi and Hénon maps exhibit coupling sensitivity; Livi et al. (1992) found that it can be observed also in a chain (coupled map lattice) of arbitrary length of one-dimensional maps. In Chapter 11, we show that coupling sensitivity is so strong in globally coupled identical systems as to yield a correction that remains finite even in

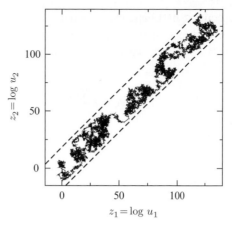

Fig. 9.3 Illustration of the typical dynamics of a perturbation in the stochastic model (9.10). The dashed lines correspond to the "reflectiing barriers" $|z_2 - z_1| = |\log \varepsilon|$ (here for $\sigma = 0.1$ and $\varepsilon = 10^{-8}$).

the thermodynamic limit, when the coupling vanishes. This singular behaviour arises also when different systems are coupled, provided that they are characterised by equal bare Lyapunov exponents (but not necessarily equal fluctuations).

Besides the (near) equality of the bare LEs, a second crucial prerequisite for the onset of coupling sensitivity is the existence of fluctuations in the multiplier evolution of the single uncoupled systems. Accordingly, this phenomenon does not arise in coupled symmetric tent maps (the local multiplier is everywhere the same, except for an irrelevant sign), as well as in logistic maps at the Ulam point, as they are conjugate to the tent maps and no multiplier fluctuations exist in the long time limit.

9.1.2 Avoided crossing of LEs and spacing statistics

The strong LE separation occurring in coupled systems resembles level repulsion in quantum mechanics and other linear systems (see, e.g., Landau and Lifshitz (1958)). Here, we explore the analogy, first referring to a simple case, and then studying (mostly in a numerical way) a more general setup.

Let us consider the 2×2 matrix

$$A = \begin{pmatrix} e^{\lambda_1^0} & \varepsilon \\ \varepsilon & e^{\lambda_2^0} \end{pmatrix}.$$

For $\varepsilon = 0$, A is diagonal, and its eigenvalues, expressed in the form of "Lyapunov" exponents, are λ_1^0 and λ_2^0 (we assume them to be real). As soon as the "coupling" ε is switched on, the eigenvalues become

$$e^{\lambda_{1,2}} = \frac{e^{\lambda_1^0} + e^{\lambda_2^0}}{2} \pm \frac{\Delta}{2}, \qquad \Delta^2 = (e^{\lambda_1^0} - e^{\lambda_2^0})^2 + 4\varepsilon^2. \tag{9.11}$$

This formula tells us that when $\lambda_1^0 = \lambda_2^0$, the coupling induces a splitting of the two "levels" of order $\sim \varepsilon$.

In the context of Lyapunov exponents, one is interested in the singular values of a *product* of matrices of type **A**. If there are no fluctuations, then the LEs coincide with these λ_1 and λ_2 values; otherwise we have to refer to Eq. (9.7), which, assuming the case of a symmetric splitting, yields

$$\Delta = \varepsilon \frac{K_{1-l}(\varepsilon/\sigma^2) + K_{1+l}(\varepsilon/\sigma^2)}{2K_{|l|}(\varepsilon/\sigma^2)}.$$

As shown by Ahlers et al. (2001), this expression is well approximated by the simple formula

$$\Delta^2 = (\lambda_1^0 - \lambda_2^0)^2 - \left(\frac{2\sigma^2}{\ln(\varepsilon/\sigma^2)}\right), \tag{9.12}$$

which is similar to the non-fluctuating case (9.11), but it contains a much stronger dependence on the coupling strength. In practice, very small ε-values (even those emerging as round-off errors in numerical modeling) induce a non-negligible difference between the two LEs.

When the properties of a given quantum system are investigated, the energy-level repulsion manifests itself as avoided crossing. In fact, the presence of a small coupling between "neighbouring" eigenstates implies that the curves obtained by plotting the corresponding eigenvalues as a function of the control parameter cannot cross. The minimal distance between two such curves is, according to (9.11), of order ε. The same effect occurs for Lyapunov exponents but in a much stronger way. This phenomenon is illustrated in Fig. 9.4 with reference to a system composed of $N = 6$ globally coupled Chirikov-Taylor standard maps (cf. (A.6))

$$P_i(t+1) = P_i(t) + K_i \sin Q_i(t) + \frac{\varepsilon}{n}\sum_{j=1}^{n} \sin(Q_j(t) - Q_i(t)) \tag{9.13}$$

$$Q_i(t+1) = Q_i(t) + P_i(t+1).$$

In Fig. 9.4, the positive Lyapunov exponents are reported in the absence (panel (a)) and presence (panel (b)) of a small coupling, as a function of a suitable control parameter τ (see the figure caption for its definition). The comparison between the two panels reveals the existence of a strong repulsion that manifests itself in spite of the smallness of ε ($= 10^{-12}$).

Finally, we address the statistical implications of coupling sensitivity, invoking an analogy with the theory of random matrices. The repulsion among the eigenvalues, expressed by formula (9.11), implies that the probability density $P(\Delta)$ of a spacing $\Delta \ll 1$ between nearby levels is smaller than predicted by the Poissonian distribution, i.e. the distribution expected for independent levels. In fact, it turns out that $P(\Delta) \sim \Delta^\beta$, where the exponent β depends on the type of matrices (e.g. orthogonal, symplectic) (Mehta, 2004).

This argument can be extended to the Lyapunov exponents in random systems. Here we deal with the statistics of singular values of products of random matrices, where there are in fact two sources of randomness: (i) temporal fluctuations of the matrix entries expressed by the parameter σ^2 (they are a manifestation of the randomness along a given trajectory) and (ii) "quenched" randomness of the entries associated with the selection of the system parameters (in the example of globally coupled maps (9.13), the parameters K_i are to

(a)
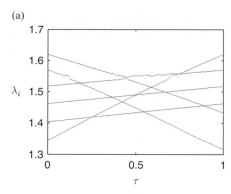

(b)
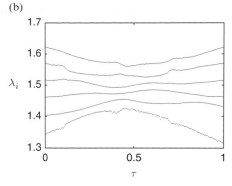

Fig. 9.4 The positive Lyapunov exponents of a set of $N = 6$ globally coupled Chirikov-Taylor maps (cf. Eq. 9.13) as a function of a control parameter τ. $K_i(\tau) = K_i^0 + \tau(K_i^1 - K_i^0)$, with $(K_i^0, K_i^1) = (7.5, 10), (8, 8.5),$ $(8.5, 9), (9, 9.5), (9.5, 7.25)$ and $(10, 8.25)$, respectively, for $i = 1, 6$. Panel (a) refers to the zero-coupling case, while $\varepsilon = 10^{-12}$ in panel (b).

be considered as random quenched ones). If, as in the theory of random matrices, we assume that the effective coupling strengths are uniformly distributed close to zero, then the spacing density of the Lyapunov exponents follows from relation (9.12) for $\lambda_1^0 = \lambda_2^0$,

$$P(\Delta) = \frac{4\sigma^2}{\Delta^3} e^{-2\sigma^2/\Delta^2},$$

so that the probability decrease for $\Delta \to 0$ is much stronger than any power law. This consideration shows that in large chaotic systems, the LEs tend to be widely separated from one another. If one observes equal or close Lyapunov exponents, there must be a special reason for this. For example, systems with continuous symmetries can have several zero Lyapunov exponents, while discrete symmetries may induce degeneracies, as discussed in Section 2.5.6. Another possibility is that the effective fluctuations of different modes are highly or completely correlated, so that these considerations, based on the assumption of independent noises in Eq. (9.3), do not apply and the relation (9.3) is not valid. We discuss such a situation for *transverse* Lyapunov exponents in Section 9.2.

9.1.3 A statistical-mechanics example

The effect of coupling sensitivity can be observed also in statistical contexts such as an Ising chain in a random external field (Derrida and Hilhorst, 1983) or Anderson localisation in weakly coupled, one-dimensional chains (Zillmer et al., 2002).

We illustrate this phenomenon for an Ising chain in a random magnetic field that is characterised by the Hamiltonian

$$H = \sum_m \left[J\sigma_m\sigma_{m+1} + (h + \eta_m)\sigma_m \right],$$

where $\sigma = \pm 1$ describes the spin state, J is the exchange energy and $h + \eta_m$ is the magnetic field, split into a constant and a randomly fluctuating component.

The problem of determining the free energy has been tackled with different approaches (Derrida and Hilhorst, 1983; Mello and Robledo, 1993; Zillmer and Pikovsky, 2005). Here we rely heavily on the relationship between the free energy per spin and the maximal Lyapunov exponent of a product of suitable random (transfer) matrices. We have already seen such a correspondence at work in Section 8.1 while discussing a problem of directed polymers. We start by revisiting the relationship in this context. Let $Z_+(n)$ $(Z_-(n))$ denote the partition function of a chain of n spins, whose nth spin is in the up (down) state. On the one hand, the total partition function is obviously $Z(n) = Z_+(n) + Z_-(n)$. On the other hand, it is easily seen that the spatial evolution of the two Z components satisfies the recursive relation

$$\begin{pmatrix} Z_+(n+1) \\ Z_-(n+1) \end{pmatrix} = \frac{1}{\sqrt{\varepsilon}} \begin{pmatrix} e^{\beta(h+\eta_m)} & \varepsilon e^{\beta(h+\eta_m)} \\ \varepsilon e^{-\beta(h+\eta_m)} & e^{-\beta(h+\eta_m)} \end{pmatrix} \begin{pmatrix} Z_+(n) \\ Z_-(n) \end{pmatrix}, \qquad (9.14)$$

where $\beta = (k_B T)^{-1}$, and we have introduced $\varepsilon = e^{-2\beta J}$ to emphasise that ε plays the role of a "coupling parameter" between Z_+ and Z_-. Paradoxically, such two Z components turn out to be weakly coupled ($\varepsilon \ll 1$) in the so-called strong-exchange coupling limit (i.e. when $J \gg 1$). Due to the standard relation between partition function and free energy $Z(n) = e^{-\beta F n}$, it is transparent that $\lambda = -\beta F$.

With the exception of the multiplicative factor $1/\sqrt{\varepsilon}$, the matrix structure is analogous to that introduced to describe weakly coupled one-dimensional chaotic systems. In fact, Zillmer and Pikovsky (2005) argued that the discrete mapping (9.14) can be considered as a discretisation of the continous-time system (9.3), with the following correspondence of parameters $\lambda_{1,2}^0 = \pm\beta h$, $2\sigma^2 = \beta^2 \langle \eta^2 \rangle$ (the parameter ε is the same in both contexts). Accordingly, by combining the equality $F = -\lambda/\beta$ with the expression (9.7) (and recalling the existence of the multiplicative factor $\sim \varepsilon^{-1/2}$ in (9.14)), one finally obtains

$$F(\varepsilon) = J - \frac{\varepsilon}{2\beta} \frac{K_{1-\kappa}(\varepsilon/2\sigma^2) + K_{1+\kappa}(\varepsilon/2\sigma^2)}{K_1(\varepsilon/2\sigma^2)}, \qquad \kappa = h/2\sigma^2.$$

Expansions (for small κ and ε) similar to Eqs. (9.8, 9.9) yield inverse-logarithmic and power-law dependencies of the free energy on ε, in accordance with the results obtained by Derrida and Hilhorst (1983) and Mello and Robledo (1993). With reference to the original spin-coupling constant J, the following dependencies are obtained in the two cases: (i) $F \approx J - C_1/J$ and (ii) $F \approx J - h - C_2 \exp[-4\kappa J]$, respectively (with suitable constants $C_{1,2}$).

9.1.4 The zero exponent

Degeneracies involving the zero Lyapunov exponent are special, as they are typically associated with the presence of some symmetry. So long as the symmetry is preserved, there is no reason to expect any dependence on the coupling. What happens, however, when the symmetry is broken? In this section we focus on the zero exponent that is associated with the invariance under time translation of autonomous continuous-time dynamical systems. The presence of a zero LE can be interpreted as the marginal stability of a phase-like variable.

Given any two, uncoupled autonomous systems, two zero LEs are expected to occur. In the presence of an arbitrarily weak coupling, however, only one zero exponent is preserved. It is therefore natural to ask what happens to the other exponent: is it affected by coupling sensitivity?

We address this question for a weakly and periodically forced chaotic oscillator, i.e. in the case of a unidirectional coupling from a periodic oscillator to a chaotic one. It is convenient to build a stroboscopic map by monitoring the internal state of the oscillator each time the phase of the forcing term reaches a given value. The simplest meaningful such map is the generalised special flow (Pikovsky et al., 1997a, b; Politi et al., 2006)

$$r(t+1) = f(r(t)) + 2\varepsilon c g(r(t)) \cos(\phi + \alpha),$$
$$\phi(t+1) = \phi(t) + \Delta + ar(t) + \varepsilon \cos(\phi(t)), \tag{9.15}$$

where Δ is the detuning between the average frequency of the chaotic oscillator and the forcing frequency, ε denotes the amplitude of the forcing term, and we have assumed that the forcing period is equal to 1. In the absence of coupling, the map takes a skew product structure: the dynamics of the amplitude r is just described by the function $f(r)$. As a result, there are two exponents: λ_1, associated with the amplitude dynamics, which we assume to be positive; and $\lambda_2 = 0$, associated with the phase evolution.

When the forcing is switched on, both exponents vary. Here we are mostly interested in the variation of λ_2. Upon introducing the tangent vector $\mathbf{v}$ and the Jacobian $\mathbf{J}$ of the transformation, one can formally write

$$\lambda_2 = \int_{-1}^{+1} dr \int_0^{2\pi} d\phi P(r, \phi) \ln |\mathbf{J}(r, \phi)\mathbf{v}_\perp|,$$

where $P(r, \phi)$ is the invariant measure in the presence of the perturbation, and $\mathbf{v}_2$ is a unit vector aligned along the direction of the second covariant Lyapunov vector. Such an equation is not generally solvable. An analytic perturbative procedure can be nevertheless implemented in some special cases. One such example corresponds to the following choice of functions in the map (9.15): (i) $f(r) = 1 - 2|r|$ (with $r \in [-1.1]$), a map which admits a Markov partition (the atoms being the intervals $[-1, 0)$, and $[0, 1]$), and (ii) $g(r) = r^2 - |r|$, which does not destroy the Markov partition, when the coupling is switched on.

Under such assumptions, it has been found by Politi et al. (2006) that

$$P(r, \phi) = \frac{1}{4\pi} + \varepsilon p_1(r, \phi) + \mathcal{O}(\varepsilon^2)$$

and

$$|\mathbf{J}(r, \phi)\mathbf{v}_2| = \varepsilon F_1(r, \phi) + \varepsilon^2 F_2(r, \phi) + \mathcal{O}(\varepsilon^3),$$

with suitable expressions for p_1, F_1 and F_2. (The reader interested in the details is invited to consult Politi et al. (2006).) As $p_1(r, \phi)$, $F_1(r, \phi)$ and $F_2(r, \phi)$ have zero average (do not contain the zero Fourier mode), it is easily concluded that $\lambda_2 = \mathcal{O}(\varepsilon^2)$. This is analogous to the case of weakly disordered matrices and oscillators driven by a weak stochastic force, analysed in Chapter 8. Depending on the choice of the various parameters, λ_2 can be either positive or negative.

This setup typically appears in the context of phase synchronisation studies (Pikovsky et al., 2001). Accordingly, the perturbative calculations show that no two zero LEs exist, even below the onset of phase synchronisation.

9.2 Synchronisation

In this section we consider the case of relatively strong coupling, where Lyapunov exponents help to explain synchronisation properties of chaotic systems.

9.2.1 Complete synchronisation and transverse Lyapunov exponents

Now we turn our attention to the properties of the LEs for strong coupling, when a synchronous dynamics is induced. We start by referring to the simple setup (9.2), but the formalism is presented in such a way that an extension to any two identical maps is straightforward.

The system of two coupled maps possesses a completely synchronous state $\mathbf{U}(t) = \mathbf{V}(t)$, for any coupling strength ε. A standard analysis allows the assessment of its stability. An infinitesimal perturbation $\mathbf{u}, \mathbf{v}$ evolves according to the linear transformation

$$\begin{pmatrix} \mathbf{u}(t+1) \\ \mathbf{v}(t+1) \end{pmatrix} = \begin{pmatrix} (1-\varepsilon)\mathsf{l}_2 & \varepsilon\mathsf{l}_2 \\ \varepsilon\mathsf{l}_2 & (1-\varepsilon)\mathsf{l}_2 \end{pmatrix} \begin{pmatrix} \mathsf{J}(t) & 0 \\ 0 & \mathsf{J}(t) \end{pmatrix} \begin{pmatrix} \mathbf{u}(t) \\ \mathbf{v}(t) \end{pmatrix},$$

where $\mathsf{J}(t) = \frac{\partial \mathbf{F}}{\partial \mathbf{U}}$ is a 2×2 matrix which, for the synchronous state, is the same for both variables $\mathbf{u}, \mathbf{v}$.

It is now convenient to introduce new variables, according to the eigenvectors of the coupling matrix: $\mathbf{w} = \mathbf{u} + \mathbf{v}$ and $\mathbf{r} = \mathbf{u} - \mathbf{v}$. They allow decoupling the equations of motion,

$$\mathbf{w}(t+1) = \mathsf{J}(t)\mathbf{w}(t),$$

$$\mathbf{r}(t+1) = (1 - 2\varepsilon)\mathsf{J}(t)\mathbf{r}(t).$$

The equation for $\mathbf{w}$ is the same as for an infinitesimal perturbation of a single map; thus it yields the Lyapunov exponents of the uncoupled system λ_i^0 $(i = 1, 2)$. The equation for $\mathbf{r}$ differs only by the constant factor $(1 - 2\varepsilon)$; thus the resulting LEs are equal to $\lambda_i^0 + \log(1 - 2\varepsilon)$ $(i = 1, 2)$.

The interpretation of the two sets of LEs is straightforward. As the perturbation $\mathbf{w}$ does not break the symmetry of the synchronous state $\mathbf{U} = \mathbf{V}$, the corresponding Lyapunov spectrum characterises the *internal* stability of the synchronous dynamics (which, in this case, is composed of a positive and a negative LE). On the other hand, the perturbation $\mathbf{r}$ does break the symmetry: the corresponding LEs, called transverse LEs (often denoted as $\lambda_\perp$), characterise the *external* stability of the synchronous trajectory with respect to symmetry-breaking perturbations. The synchronous state is stable if both $\lambda_{1,2}^0 + \log(1 - 2\varepsilon) < 0$. This allows the identification of the critical coupling in the model (9.1, 9.2) $\varepsilon_c = 0.5(1 - \exp(-\lambda_{max})) \approx 0.088$. It is noteworthy that the expression for the largest transverse exponent can be interpreted as a sum of two contributions: (i) the largest LE of uncoupled chaos, λ^0, and (ii) the transverse Lyapunov exponent of the coupling matrix

$$\begin{pmatrix} 1 - \varepsilon & \varepsilon \\ \varepsilon & 1 - \varepsilon \end{pmatrix},$$

which is just the logarithm of the eigenvalue $1 - 2\varepsilon$.

For $\varepsilon > \varepsilon_c$ the synchronous state is stable: a generic trajectory of the full system converges to the manifold $\mathbf{u} = \mathbf{v}$, and the LEs plotted in Fig. 9.1 correspond to these internal and transverse exponents. Below the critical coupling, the largest transverse LE becomes positive, the synchronous state loses its stability and an asymmetric regime establishes itself; its LEs no longer coincide with those of the synchronous regime.

In full generality, the synchronisation transition can be summarised in the following way: let us assume that an N-dimensional phase space contains an invariant M-dimensional manifold $\mathcal{M}$ (in the example (9.2), $\mathcal{M}$ is the two-dimensional diagonal $\mathbf{U} = \mathbf{V}$, embedded in a four-dimensional phase space). The linear perturbations to a generic trajectory in $\mathcal{M}$ can be divided into those which do not leave the manifold itself (there exist M linearly independent components) and those that are transverse ($N - M$ independent components). Accordingly, the LEs can be divided into M internal ones $\lambda_\|$, which characterise the evolution within $\mathcal{M}$ and $N - M$ transverse ones $\lambda_\perp$. The latter ones control the stability of the "synchronous" state: in order for it to be stable, all transverse LEs need to be negative.

The transverse Lyapunov exponent determines the stability of the synchronisation manifold on average but not a uniform stability. A more detailed description of the stability is provided by finite-time transverse exponents (Pikovsky and Grassberger, 1991). Indeed, as discussed in Chapter 5, the distribution of the finite-time exponents is determined by the large deviation function $S(\Lambda_\perp)$ (cf. Eq. (5.8)), which has a minimum at the true exponent but extends over a finite domain. This domain is typically determined by the values taken by the Lyapunov exponent on non-generic invariant sets (fixed points, periodic orbits, etc.). From the point of view of stability of the synchronisation manifold, the domain of the large deviation function of the transverse exponent $[\Lambda_\perp^{min}, \Lambda_\perp^{max}]$ is of major importance. The values of $\Lambda_\perp$ depend on the coupling parameter ε. For large couplings $\Lambda_\perp^{max} < 0$, which means that *all* small perturbations of the synchronous state decrease. In some range of couplings it may happen that $\Lambda_\perp^{max} > 0$ while $\lambda_\perp < 0$. This means that while small perturbations on average decrease, some invariant synchronous sets have unstable transverse directions and the corresponding (atypical) perturbations grow. This makes the synchronous set a Milnor attractor: a set that attracts the typical neighbouring points but not all of them (irrespective of the neighbourhood size). In this regime the synchronous state is very sensitive to perturbations and noise; see Pikovsky and Grassberger (1991) and Pikovsky et al. (2001) for more details.

9.2.2 Clusters, the evaporation and the conditional Lyapunov exponent

In systems of globally coupled, identical chaotic units it is possible to observe highly symmetric synchronisation patterns. The model (A.26) with $\mathbf{F}_k = \mathbf{F}$ and $G_{kj} = g$ identifies a rather general setup for the characterisation of such regimes. The system admits a large variety of synchronous states, because any subset of the local variables can form a completely synchronised cluster: any partition into clusters is invariant. Given a cluster of m elements $\mathbf{U}_1 = \mathbf{U}_2 = \ldots = \mathbf{U}_m = \mathbf{U}_{cl}(t)$,[1] its transverse stability can be assessed by checking whether its elements tend to "evaporate". This can be done by considering perturbations of the type $\mathbf{w} = \mathbf{u}_i - \mathbf{u}_j$, where $i, j \leq m$. It is clear that any such perturbation $\mathbf{w}$ does not affect $\overline{\mathbf{V}}$ and thus $\overline{\mathbf{Z}}$. The evolution of $\mathbf{w}$ is governed by an equation where only a linearisation with respect to $\mathbf{U}_{cl}$ is performed,

$$\dot{\mathbf{w}} = \frac{\partial \mathbf{F}(\mathbf{U}_{cl}, \overline{\mathbf{V}}, \overline{\mathbf{Z}})}{\partial \mathbf{U}_{cl}} \mathbf{w}.$$

This equation yields the transverse LEs of the cluster stability. The largest one is often called the evaporation or the split exponent: if it is negative, the cluster is stable; if it is positive, the cluster dissolves.

Cluster states generically emerge in sufficiently symmetric setups. Unidirectional coupling (i.e. when one or more units are driven by the same master system) offers one such example,

$$\dot{\mathbf{U}} = \mathbf{F}_1(\mathbf{U}),$$
$$\dot{\mathbf{V}}_1 = \mathbf{F}_2(\mathbf{V}_1) + \varepsilon(\mathbf{U} - \mathbf{V}_1),$$
$$\dot{\mathbf{V}}_2 = \mathbf{F}_2(\mathbf{V}_2) + \varepsilon(\mathbf{U} - \mathbf{V}_2),$$

where the three vectors $\mathbf{U}$, $\mathbf{V}_1$ and $\mathbf{V}_2$ all are assumed to have the same dimension. In this setup, complete synchronisation of the two driven systems ($\mathbf{V}_1 = \mathbf{V}_2$) is a natural phenomenon. Whenever this occurs, a more or less complex functional relationship exists between the U- and the V-oscillators. This relationship is referred to as generalised synchronisation (Rulkov et al., 1995). The stability of generalized synchrony is determined by the LEs transverse to the corresponding generalised-synchrony manifold. They are called "conditional" LEs (Carroll and Pecora, 1991; Pecora and Carroll, 1991; Pyragas, 1996, 1997). Remarkably, generalised synchronisation ($\mathbf{U}(t) \neq \mathbf{V}_1(t) = \mathbf{V}_2(t)$) can be observed also when $\mathbf{F}_1 = \mathbf{F}_2$.

9.2.3 Synchronisation on networks and master stability function

In Section 9.2.1 we saw that eigenfunctions and eigenvalues of the coupling matrix play an important role in the determination of the transverse Lyapunov exponents; in the example presented therein, the transverse LE was obtained by just adding the logarithm of the eigenvalue of the coupling matrix to the original (longitudinal) Lyapunov exponent. This is an instance of a rather general property of networks of oscillators with a certain type of coupling, as discussed next. In such cases, the *master stability function* (Pecora and Carroll, 1998, 1999) allows the decomposing of the transverse LE into the contribution arising from the dynamics of the individual units and the contribution of the coupling matrix.

---

[1] The indices of the cluster can be assumed to range from 1 to m, without loss of generality.

We illustrate the method for a network of N identical, m-dimensional dynamical systems of the type (A.26) where

$$\mathbf{F}_k = \mathbf{f}(\mathbf{U}_k) + \overline{\mathbf{V}} = \mathbf{f}(\mathbf{U}_k) + \sum_{j=1}^{N} G_{kj}\mathbf{c}(\mathbf{U_j}), \tag{9.16}$$

where the velocity field $\mathbf{f}$ and the coupling function $\mathbf{c}$ are the same for all units (as the factor $1/N$ is irrelevant for this analysis, it is implicitly included in the definition of the matrix $\mathbf{G}$). The structure of the network is described by the $N \times N$ coupling matrix G_{kj}, which is assumed to satisfy the constraint $\sum_j G_{kj} = 0$ for all k, so that $\mathbf{G}$ mimics a sort of generalised diffusion among the various nodes of the network. Due to this condition, the synchronous state $\mathbf{U}_1 = \ldots = \mathbf{U}_N$ is invariant, no matter which network topology is chosen.

The linearised equation for the perturbations is

$$\dot{\mathbf{u}}_k = \frac{\partial \mathbf{f}}{\partial \mathbf{U}}\mathbf{u}_k + \frac{\partial \mathbf{c}}{\partial \mathbf{U}} \sum_{j=1}^{N} G_{kj}\mathbf{u_j}, \tag{9.17}$$

where $\frac{\partial \mathbf{f}}{\partial \mathbf{U}}$ and $\frac{\partial \mathbf{c}}{\partial \mathbf{U}}$ are $m \times m$ Jacobians of $\mathbf{f}$ and $\mathbf{c}$, determined in the synchronous state. The main idea behind the master stability function approach is to diagonalise the coupling matrix $\mathbf{G}$. Let us assume that $\mathbf{M}^{-1}\mathbf{G}\mathbf{M} = \Gamma$, where Γ is a diagonal matrix. By multiplying (9.17) by $\mathbf{M}^{-1}$ on the left and introducing $\mathbf{v} = \mathbf{M}^{-1}\mathbf{u}$, one obtains a set of mutually independent equations

$$\dot{\mathbf{v}}_k = \left(\frac{\partial \mathbf{f}}{\partial \mathbf{U}} + \Gamma_k \frac{\partial \mathbf{c}}{\partial \mathbf{U}} \right) \mathbf{v}_k, \tag{9.18}$$

where Γ_k denote the eigenvalues of $\mathbf{G}$. The only coupling terms that are left are those linking the m variables which characterise each single unit. By solving these equations for each Γ value, a set of m exponents is obtained. Those corresponding to $\Gamma_1 = 0$ (always present because of the conditions imposed on the matrix $\mathbf{G}$) characterise the internal stability of the synchronous dynamics, while the other LEs are the transverse ones.

From Eq. (9.18), one can see that different coupling configurations lead to the same linearised equations, the only possible difference being the (generally complex) eigenvalue Γ_k. Therefore, one can consider Γ_k as an arbitrary complex parameter $\Gamma_k = \alpha + i\beta$ and study the Lyapunov exponents of the linear complex equation

$$\dot{\mathbf{z}} = \left(\frac{\partial \mathbf{f}}{\partial \mathbf{U}} + (\alpha + i\beta)\frac{\partial \mathbf{c}}{\partial \mathbf{U}} \right) \mathbf{z}.$$

Let now $\lambda_{max}(\alpha, \beta)$ denote the largest transverse Lyapunov exponent for a given pair of α, β values. This function is called *master stability function*. It depends only on the velocity field $\mathbf{f}(\mathbf{U})$ and the coupling function $\mathbf{c}(\mathbf{U})$.

The domain of the (α, β) plane where $\lambda_{max} < 0$, is called the stability region (which may be empty). Given a particular coupling matrix $\mathbf{G}$ (i.e. network topology), the synchronous state is stable if and only if all the eigenvalues Γ_k lie inside the stability region.

As an example, the master stability function for the Lorenz model (A.9) is presented in Fig. 9.5, for $\mathbf{c} = (0, 0, z)$ (i.e. only the third variables are coupled – among themselves).

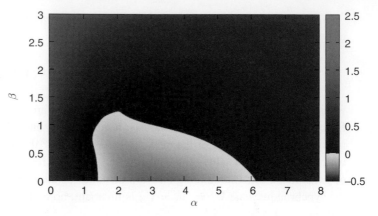

Master stability function for Lorenz systems coupled through the third variable. The grey levels are chosen so as to highlight the borders of the stability domain.

There exists a closed domain in plane (α, β) where the largest Lyapunov exponent is negative. This implies that it is in principle possible to design a network structure that is able to synchronise such chaotic oscillators.

Bloch-Lyapunov exponents

One-dimensional diffusively coupled chains of oscillators represent an interesting and quite general setup where this formalism can be profitably applied. In the case of nearest-neighbour coupling, their evolution is described by an equation of the type

$$\dot{\mathbf{U}}_x = \mathbf{f}(\mathbf{U}_x) + \varepsilon[\mathbf{c}(\mathbf{U}_{x-1}) - 2\mathbf{c}(\mathbf{U}_x) + \mathbf{c}(\mathbf{U}_{x+1})], \tag{9.19}$$

where $\mathbf{c}$ is typically a linear transformation, specifying how the local variables are coupled. For instance, for $m = 2$, the two options

$$\mathbf{c} = \begin{pmatrix} 1 & 0 \\ 0 & 1 \end{pmatrix}, \qquad \mathbf{c} = \begin{pmatrix} 0 & 1 \\ 1 & 0 \end{pmatrix}$$

mean either that each of the two variables is affected by its own diffusion or that they cross interact.

It is easily seen that Eq. (9.19) is a variant of Eq. (9.16), where $\mathbf{G}$ is a tridiagonal matrix with $G_{x,x} = -2$ and $G_{x,x\pm1} = 1$ and all of the other terms are equal to zero. As the sum of all elements in each row is trivially equal to zero, the state $\mathbf{U} = const$ is a possible synchronous solution; its stability can be investigated by diagonalising $\mathbf{G}$, a procedure which, in this case, corresponds to Fourier transforming the equations of motion in tangent space. One has to be careful about the boundary conditions: if periodic conditions are assumed, the usual Fourier transform is to be implemented; otherwise either sine or cosine transforms may be required. In practice, in a finite system of length L, the modes $\exp(2\pi ikx/L)$ for $1 < k < L$ have to be considered to study the stability of the synchronous solution. Their stability provides the transverse Lyapunov exponents, while the stability of the zero mode yields the internal stability of the homogeneous solution.

In spatially uniform systems (either unbounded or in the presence of periodic boundary conditions), the spatially homogeneous state is invariant as well as all the states with arbitrary periodicity in space. Indeed, if the initial condition for model (9.19) satisfies $\mathbf{U}_x = \mathbf{U}_{x+L}$ in an infinite lattice, then this property holds at all times. On the other hand, a linear perturbation $\mathbf{u}_x$ must not have the same periodicity. As in the standard theory of linear waves in a periodic potential, here the Bloch ansatz $\mathbf{u}_x(t) = \tilde{\mathbf{u}}_x(t)e^{i\kappa x}$ allows one to reduce the dynamics of the perturbation to the original domain, because now $\tilde{\mathbf{u}}_x = \tilde{\mathbf{u}}_{x+L}$. The corresponding Lyapunov exponent depends on the quasi-momentum κ, which attains values in the interval $[-\pi/L, \pi/L]$. The value $\kappa = 0$ corresponds to the internal exponent (i.e. to perturbations that do not violate the original periodicity), while the other κ-values correspond to the transverse exponents. For applications of these Bloch-Lyapunov exponents, see Pikovsky (1989) and Straube and Pikovsky (2011).

10 High-dimensional systems: general

In physical systems characterised by a large number of degrees of freedom, it is instructive to investigate the scaling behaviour of the Lyapunov spectrum with the "system size". In this chapter we show that in many setups one can define a Lyapunov density spectrum when the number of degrees of freedom is large enough (i.e. in the so-called thermodynamic limit). Altogether, this is a manifestation of the *extensive* character of the underlying dynamics.

Dealing with infinite-dimensional systems poses a series of questions that have to be carefully investigated. First, it is no longer true that all norms are mutually equivalent; this, in fact, leads to ambiguities in the classification of a given dynamical behaviour as chaotic. Second, a problem of commutativity between the infinite-time limit and the thermodynamic (infinite-size) limit may arise.

In the specific case of spatially extended systems, it is natural to extend the analysis of infinitesimal perturbations to account for their spatial structure and possible propagation phenomena. The chronotopic approach provides a unified framework for a comprehensive characterisation of the space-time properties of tangent-space dynamics. This includes the definition of spatial Lyapunov exponents, convective exponents and a super-invariant entropy density. Occasionally, it is necessary to go beyond the linear stability analysis and to introduce the propagation velocity of finite perturbations, i.e. to study damage-spreading phenomena.

Selected model classes are specifically analyzed. Semi-analytical formulas of the entire Lyapunov spectrum are derived for a chain of Hamiltonian oscillators. Differential-delay equations that show analogies and differences with spatially extended systems are also discussed. Finally, some network structures are discussed to illustrate the shape of the Lyapunov spectrum in the presence of massive as well as sparse coupling.

10.1 Lyapunov density spectrum

In spatially extended systems, the number N of degrees of freedom is proportional to the system size. The reader can, for simplicity, just think of a finite discrete lattice where each site is characterised by d variables. In a one-dimensional chain composed of L elements, one has $N = dL$ (in the case of coupled logistic maps (A.10), $d = 1$ since the local variable is a scalar; in a chain of Rössler oscillators (A.17), $d = 3$).

A first general question concerns the scaling behaviour of the Lyapunov spectrum (i.e. the set of all Lyapunov exponents) in the thermodynamic limit, when $L \rightarrow \infty$.

Heuristic considerations suggest that the evolution of a large system can be thought of as a juxtaposition of many, almost independent, domains. As a result, it is reasonable to expect that the number of unstable directions is proportional to L, or in other words, space-time chaos is extensive. This is indeed the core of a conjecture formulated by Ruelle (1982), who suggested that, in the large-L limit, the dependence of the LEs on their index i and the system size L can be combined into a dependence on the integrated density $\rho = i/L$,

$$\lambda(\rho = i/L) = \lim_{L \to \infty, \rho = \text{const}} \lambda_i(L). \tag{10.1}$$

Asymptotically, the variable ρ ranges from 0 (which corresponds to the maximum LE) to the number d of variables per site. The conjectured existence of a *Lyapunov density spectrum* $\lambda(\rho)$ has been confirmed in several contexts, which include the Kuramoto-Sivashinsky equation (Pomeau et al., 1984; Manneville, 1985), the symplectic Fermi-Pasta-Ulam chain (Livi et al., 1986) and coupled maps (Kaneko, 1985).

In Fig. 10.1a we plot the spectra obtained for a chain of coupled logistic maps (A.10) with periodic boundary conditions (p.b.c.). Notice that in a finite system, the "optimal" definition of the integrated density is $\tilde{\rho} = (i - 1/2)/L$. The $1/(2L)$ shift indeed accounts for first-order finite-size corrections (as also confirmed by the nice overlap in the figure). The advantage of this definition is particularly transparent in symplectic systems of length L, where $\lambda_i = -\lambda_{2L-i+1}$; in fact, the equality $\lambda(\tilde{\rho}_i) = -\lambda(2 - \tilde{\rho}_i)$ holds exactly for any L. Anyway, asymptotically, there is no difference between $\tilde{\rho}$ and ρ.

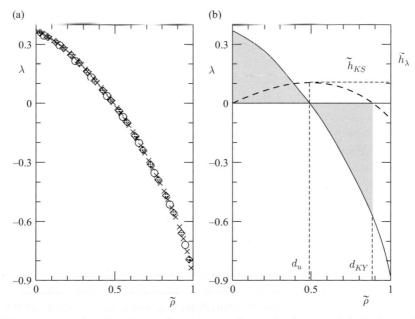

Fig. 10.1 Rescaled Lyapunov spectra for a chain of coupled logistic maps with $a = 2$ and $\varepsilon = 1/6$. Circles, diamonds and crosses in panel (a) refer to $L = 10, 20$ and 40. In panel (b), the geometric interpretation of the density of Kaplan-Yorke dimension and of the Kolmogorov-Sinai entropy is given. The dashed curved line corresponds to the integral $\tilde{H}_\lambda$ of the Lyapunov density spectrum.

Once the very existence of a Lyapunov density spectrum (in the thermodynamic limit, see Eq. (10.1)) has been established, it is natural to analyse the implications of this result for the scaling behaviour of dimension and dynamical entropies in spatially extended systems. We start from the Pesin formula for the Kolmogorov-Sinai entropy (see Section 6.3), which can be rewritten as

$$h_{KS} = \sum_{\lambda_i > 0} \lambda_i = L \int_0^{d_u} d\rho \lambda(\rho),$$

where d_u is implicitly defined as $\lambda(d_u) = 0$ (Nd_u is the number of unstable directions, i.e. it is the dimension of the unstable manifold). Notice that the second equality is obtained by imagining the ρ axis as split into cells of size $1/L$, i.e. by interpreting the sum as the discretisation of an integral.

Analogously, one can express the Kaplan-Yorke formula (Section 6.1) as

$$0 = \int_0^{D_{KY}/L} d\rho \lambda(\rho).$$

Both h_{KS} and D_{KY} are extensive variables; i.e. these observables are proportional to the system size L and one can indeed introduce $\tilde{h}_{KS} = h_{KS}/L$ and $d_{KY} = D_{KY}/L$ as the density of the dynamical entropy and the density of fractal dimension. The meaning of both concepts is illustrated in Fig. 10.1b, where one can see that $\tilde{h}_{KS}$ can be interpreted as the area under the positive part of the density spectrum (and it also coincides with the maximum of $\tilde{h}_\lambda(\rho) = \int_0^\rho \lambda(\rho')d\rho'$), while the density of fractal dimension corresponds to the abscissa where $\tilde{h}_\lambda(\rho) = 0$.

The concept of the Lyapunov density spectrum is rather general and applies to other contexts, some of which are summarised later in this chapter (see, for instance, systems with delay and network structures). One important example that is worth discussing is that of spatially continuous systems. In fact, there are at least two reasons why an extended system may be high-(possibly infinite-)dimensional: (i) a large spatial size and (ii) the extension to extremely tiny scales. This previous analysis refers to the former case, which is also referred to as the thermodynamic limit. The characterisation of spatially continuous systems such as those described by partial differential equations (in a bounded region) requires different considerations. It is convenient to approach the continuum limit starting from a lattice structure and increasing the density of points. In particular, it is convenient to assign a length a to the link between two neighbouring sites and to then define the total length $L = Ma$, where M is the number of points. The continuous limit corresponds to increasing M, while keeping L constant. In this process, the argument $\rho = i/L$ in $\lambda(\rho)$ extends to arbitrarily large values, spaced by a finite amount $1/L$. This relationship is very similar to the correspondence between a time series and its Fourier transform. On the one hand, the frequency resolution increases when longer and longer signals are considered. In contrast, increasingly high frequencies become accessible upon reducing the sampling time; the frequency plays the same role as ρ in the Lyapunov analysis.

The correctness of this way of arguing is illustrated in Fig. 10.2, where the Lyapunov spectra of the Kuramoto-Sivashinsky partial differential equation (A.19) are reported for two different spatial resolutions, without rescaling the index i. Upon looking at the LEs obtained with the higher resolution, one can see that while the first 42 LEs nicely overlap

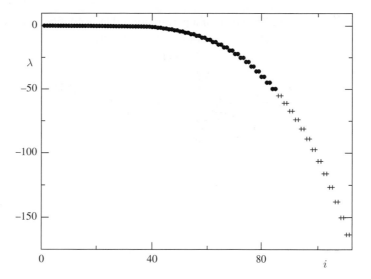

Fig. 10.2 Two Lyapunov spectra for the Kuramoto-Sivashinsky equation in a system of length $L = 96$ and different spatial resolutions: solid circles refer to 42 Fourier modes, plusses to 85. (Data: courtesy of K. A. Takeuchi.)

with the exponents obtained with the lower resolution, the last ones are characterised by very negative values. In practice, the increased spatial resolution unveils further rather stable degrees of freedom. The good overlap on the left of the first part of the spectrum suggests that the spatial resolution of the simulations is sufficiently fine to ensure an accurate reproduction of the dynamical properties.

10.1.1 Infinite systems

When the thermodynamic limit is invoked in the computation of Lyapunov spectra, one should realise that a delicate point is implicitly touched: the commutativity of the limits $t \to \infty$ and $L \to \infty$, which, in turn, calls into play the non-equivalence of norms in infinite-dimensional spaces.

Sinai (1996) showed that the order of the two limits does not matter in some systems of N particles with short-range interactions. This is not, however, always the case. One example is the Hamiltonian mean-field model (A.28). There, in the magnetised phase, if the limit $N \to \infty$ is taken first, all LEs are found to be equal to zero, since the microscopic dynamics is strictly periodic (the single "pendula" either oscillate or rotate). If, however, the order of the two limits is exchanged, a number of finite positive (and negative) LEs emerge; this is an extreme case of the coupling sensitivity discussed in Section 11.2 (see Ginelli et al. (2011) for a more exhaustive analysis).

In spatially extended systems, the problem can be elucidated by introducing a suitable class of norms. Let ΔU_x denote the difference between any two configurations of a chain of dynamical systems defined on a discrete lattice ($x \in \mathbb{I}$). The following s-norm can be introduced,

$$\|\Delta U_x\|_s = \sum_x |\Delta U_x| e^{-s|x|},$$

where $s > 0$. The norm of a perturbation that is confined to sites further than x_0 from the origin, e.g. $\Delta U_x = \theta(x - x_0)$ (where $\theta(x)$ is the Heaviside function), is

$$\|\Delta U_x\|_s = \frac{e^{-x_0 s}}{e^s - 1}.$$

Accordingly, it can be made arbitrarily small, by selecting x_0 sufficiently large, without requiring ΔU_x to be locally small, as explicitly done when the evolution equations are linearised in the standard approach. Let us further assume that the evolution of the semi-infinitely extended perturbation ΔU_x basically consists of a propagation towards smaller x values with a velocity v_f. As a result, the perturbation amplitude increases exponentially in time with a rate

$$\lambda_f = v_f s. \tag{10.2}$$

This formula tells us that besides the amplification of uncertainty from the least to the most relevant digits, another mechanism may be responsible for the generation of a deterministic chaotic dynamics: the spatial propagation of uncertainty from the boundaries towards the region of interest.

This is the mechanism that generates the complex dynamics observed in discrete-state systems, where the local variable can take a finite number of distinct values only. Deterministic and random cellular automata (Ilachinski, 2001) belong to such a class, which includes the Ising model of coupled spins as a prominent physical example. Given the finiteness of the state space, only periodic regimes can be generated in finite discrete-state systems. The spatial extension does, nevertheless, bring enough complexity for non-trivial regimes to be sustained in the thermodynamic limit as, e.g., revealed by the chaotic cellular automata studied by Wolfram (1986) and the rich variety of regimes exhibited by the so-called *game of life* (Berlekamp et al., 1982). The need to consider strictly infinite systems is analogous to the need to use infinite-precision arithmetic in the simulation of a nonlinear system, where a finite precision would necessarily kill any aperiodic dynamics.

Stable chaos (Politi and Torcini, 2010) is a more general instance of spatially induced chaos that emerges in systems characterised by continuous variables, where a standard Lyapunov exponent can be defined and even be negative. A chain of harmonic oscillators such as

$$\ddot{U}_x = U_{x+1} - 2U_x + U_{x-1}$$

may be included in this category. A chain of any finite length is trivially integrable and thereby characterised by zero Lyapunov exponents. However, in an infinite chain, (finite) perturbations travelling from either the left or the right propagate according to the sound velocity v_s, so that $\lambda_f = s v_s > 0$. This is the reason why infinite integrable chains may be considered as ergodic systems (the two infinitely long leads playing the role of suitable heat baths).

As a further remark about Eq. (10.2), one should note that λ_f depends on s, i.e. it depends on the norm. It is fair to recognise that the propagation velocity v_f is a more objective indicator to characterise this regime. A detailed discussion of the different velocities that can be defined in spatially extended systems is presented in Section 10.3, where the convective Lyapunov exponents and damage-spreading phenomena are introduced and analysed.

Unfortunately, equations such as (10.2) cannot be (easily) extended beyond the largest exponent due to the difficulty of defining a meaningful scalar product.

10.2 Chronotopic approach and entropy potential

Lyapunov exponents have been introduced with the goal of characterising the time evolution of perturbations of lumped dynamical systems. However, in spatially extended systems, as we have just seen in the previous section, it is important to characterise the spatial evolution as well.

Before proceeding ahead let us note that it would be more appropriate to speak of space-like and time-like variables, as they do not need to coincide with the true space and time axes. In practice the time-like variable is the one which parametrises the evolution while starting from some given initial condition, while the space-like variable is the one used to fix the boundary conditions. For example, in optics, where the stationary (in the true time) propagation of light beams is considered, the spatial direction can be treated as an effective time. Even the exchange of space and time variables may have to be considered for waves propagating in one direction, when driven from a boundary (Pikovsky, 1989).

The *chronotopic* approach (Lepri et al., 1996, 1997) extends the concept of linear instability to the spatial dimension. It is essentially composed of two complementary methods, the first of which deals with the temporal evolution of perturbations with a given exponential spatial profile, while the second is a sort of transfer matrix method, which deals with the spatial evolution. For ease of presentation we refer directly to the tangent space, illustrating the method with reference to a simple setup, where the spatial coupling is of diffusive type

$$u_x(t+1) = F_U[u_x(t) + \varepsilon \mathcal{D} u_x(t)] + b u_x(t-1), \tag{10.3}$$

where $\mathcal{D} u_x \equiv u_{x-1} - 2u_x + u_{x+1}$ is the discrete Laplacian operator. This structure is borrowed from that of a chain of Hénon maps (A.11), where $F_U = -2U_x$ is obtained by iterating the model in real space. One can also interpret F_U as a stochastic variable, as we will do in the following, to facilitate the corresponding numerical studies. Given this structure, let us assume in (10.3) that the perturbation has an exponential profile $u_x(t) = e^{\mu x} v_x(t)$, where μ is a free parameter. It follows that v_x evolves according to

$$v_x(t+1) = F_U\left[\varepsilon e^{\mu} v_{x+1}(t) + (1-2\varepsilon)v_x(t) + \varepsilon e^{-\mu} v_{x-1}(t)\right] + b v_x(t-1). \tag{10.4}$$

Now, assuming a finite length L with periodic boundary conditions, we can determine the Lyapunov density spectrum in the thermodynamic limit. For $\mu = 0$, the standard definition is recovered; otherwise one obtains the *generalised* temporal Lyapunov density spectrum $\lambda(\rho, \mu)$, which tells us how fast an exponentially shaped perturbation grows. A graphical representation of the accessible λ values is presented in Fig. 10.3.

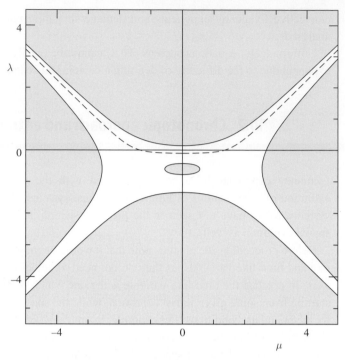

Fig. 10.3 Chronotopic representation of the tangent-space dynamics (10.4) for $b = 0.3, \varepsilon = 0.15$ and F_U uniformly distributed in the interval $[-1.5, 1.5]$. The white region corresponds to the allowed (λ, μ) pairs.

The upper border of the shaded region corresponds to the maximal exponent for different μ values. We see that $\lambda(0, \mu)$ has a minimum for $\mu = 0$. This is quite a general property; it expresses the fact that exponential perturbations tend to grow faster because of the diffusion from the larger neighbouring sites. The range of values covered by the standard spectrum corresponds to the points along the y axis that belong to the allowed (shaded) region. The presence of a hole means that the Lyapunov spectrum is composed of two separate bands. The left-right symmetry reflects the left-right symmetry of the spatial evolution ($\mu \rightarrow -\mu$), while the vertical symmetry around $\log b/2$ is less obvious. As noticed in Kuptsov and Politi (2011), it follows from the pseudo-symplectic structure of the model. Finally, if one proceeds to larger μ values, the LEs concentrate around the line $\mu = \lambda$, which reflects the nearest neighbour coupling.

The generalised temporal exponents can be used to determine the growth rate whenever the profile of a perturbation is locally exponential. One example is the propagation of initially localised perturbations, which is discussed in the next section. Another example is when a chaotic spot is embedded in an ordered environment: the perturbation eventually grows in time in the stable region, too, but its amplitude decreases exponentially with the distance from the chaotic spot (see also Chapter 11).

The two axes in Fig. 10.3 correspond to the growth rate in two directions, space and time, respectively. One can exchange their role, propagating in space a given "initial" temporal profile. This is tantamount to a transfer matrix approach and leads to the definition of the so-called spatial Lyapunov exponents. In practice, with reference to the same stochastic model, one has to iterate the following set of equations

$$u_{x+1}(t) = -\frac{1-2\varepsilon}{\varepsilon}u_x(t) + \frac{u_x(t+1)}{\varepsilon F_U} - u_{x-1}(t) - b\frac{u_x(t)}{F_U}$$

along the x axis. Analogously to the previous case, one can now assume an exponential profile in time, $u_x(t) = e^{\lambda t}v_x(t)$, where λ is a free parameter. In practice one obtains

$$v_{x+1}(t) = -\frac{1-2\varepsilon}{\varepsilon}v_x(t) + \frac{v_x(t+1)}{\varepsilon F_U}e^{\lambda t} - v_{x-1}(t)e^{-\lambda t} - b\frac{v_x(t)}{F_U}.$$

As long as one is interested in the long time limit, there is no problem of setting boundary conditions; any choice works equally well (periodic boundary conditions are anyhow assumed to reduce finite-size corrections). In practice one ends up determining $\mu(\sigma, \lambda)$, where $\sigma = j/T$ is the integrated density of states. At variance with the previous case, λ is an independent parameter, while μ is the dependent one. By exchanging μ with λ and ρ with σ, one can proceed as before to identify the admissible μ values for any given λ. As a result, one identifies the same regions as with the direct approach. Notice also that the "dynamics" is invariant under space reversal; this means that the spectrum is symmetric as in symplectic dynamics. This property does not hold in systems where there is no left-right symmetry.

Problems of Anderson-type localisation (see Section 12.1) can be studied by invoking this setup. In one-dimensional systems, under the assumption of a temporally constant state, the problem reduces to a two-dimensional map. In such a case μ can be read as the inverse of the localisation length.

This analysis can be summarised as follows: given a generic perturbation, characterised by the average growth rates μ and λ along the spatial and temporal directions, respectively, its linear stability is fully determined by the integrated densities of the two variables: $\rho(\lambda, \mu)$ and $\sigma(\lambda, \mu)$.

It turns out that the characterisation can be inverted by using ρ and σ to uniquely identify each perturbation and thereby attributing to them the rates $\lambda(\rho, \sigma)$ and $\mu(\rho, \sigma)$. In Fig. 10.4, we report the isolines of constant λ (solid curves) and of constant μ (dashed curves) in the σ, ρ plane. There, each point identifies a different perturbation. The upper and lower borders in the previous representation are mapped onto the points $(1,0)$ and $(1,2)$, while the left and right borders are mapped onto $(2,1)$ and $(0,1)$, respectively. Finally, the border of the inner circle is mapped onto the point $(1,1)$. Conversely, the four sides of the square correspond to the points ad infinitum in the four horns. This representation preserves the original symmetries (left-right and time reversal).

In practice, Fig. 10.4 reveals that ρ and σ can be equivalently used as independent variables. In the following, we argue that other pairs provide a more compact description by interpreting them as the real and the imaginary part of an analytic complex function. We anticipate that the following considerations are just heuristic but are fully supported by numerical simulations. In practice, we introduce the complex expansion rates $\tilde{\lambda} = \lambda + i\sigma$ and $\tilde{\mu} = \mu + i\rho$. This way of thinking is analogous to the node-counting approach for the Schrödinger equation, where the number of spatial oscillations (i.e. the density of states) is the imaginary component of the spatial eigenvalues. Accordingly one can combine the

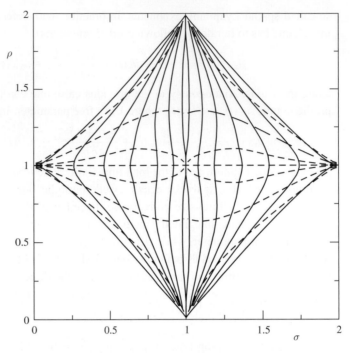

Fig. 10.4   Chronotopic representation in the σ, ρ plane of the same system characterised in Fig. 10.3. Dashed (solid) curves are the isolines with constant μ (λ).

information on the linear stability into the complex function $\tilde{\lambda}(\tilde{\mu})$. As long as $\tilde{\lambda}(\tilde{\mu})$ is analytic, one can invoke the Cauchy-Riemann conditions, which require the existence of a real function Ψ such that

$$\lambda = \frac{\partial \Psi}{\partial \rho}, \qquad \sigma = -\frac{\partial \Psi}{\partial \mu}.$$

The very existence of an analytical complex function can easily be proven in the case of of homogeneous periodic solutions, but in more general contexts it is confirmed only by a large body of numerical simulations.

These equations suggest that λ and ρ can be considered as conjugate variables in a Legendre transform, the conjugate function being the so-called entropy potential,

$$\Phi = \lambda \rho - \Psi.$$

In fact, it is easily seen that

$$\rho = \frac{\partial \Phi}{\partial \lambda}, \qquad \sigma = -\frac{\partial \Phi}{\partial \mu},$$

i.e. the partial derivatives of Φ yield the two integrated densities ρ and σ. Note that adding a linear combination of λ and μ ($\Phi' = \Phi + a_1\lambda + a_2\mu$) corresponds simply to shifting the two densities ρ and σ by some suitable amount. We find it convenient to set $a_1 = 1$, $a_2 = 0$ since, with this choice, Φ is constant above the maximum temporal Lyapunov exponent $\tilde{\lambda}(0, \mu)$. Exploiting also the fact that a potential is known up to a constant, we

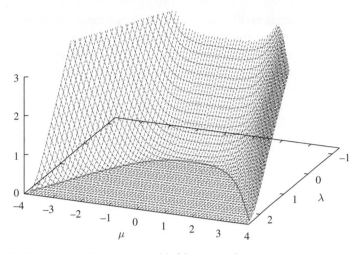

Fig. 10.5 Entropy potential in the μ, λ plane for the same model of the previous figures.

can set it equal to zero. The resulting shape for the previous model is plotted in Fig. 10.5, where the solid line corresponds to the border of the area characterised by $\Phi = 0$.

Φ is called the *entropy potential*, since it generalises the Kolmogorov-Sinai entropy (Section 10.1). Let us first notice that, since h_{KS} is the area below the curve $\lambda(\rho)$ until it crosses 0, it can be defined by exchanging the role of ρ and λ, to obtain h_{KS} as

$$\tilde{h}_{KS} = \int_0^{\lambda_0} d\lambda \rho(\lambda).$$

Having defined Φ so as to be equal to zero along the line $\lambda(0, \mu)$, this formula implies that $\Phi(0,0) = 0$.

Let us now consider a yet more general setup, where the time and space axes are rotated by some angle θ. Let $\vec{u} = (\sin\theta, \cos\theta)$ be a unit vector which defines the orientation of the corresponding time axis and determine, for instance, the exponents λ_θ. This requires to move along the line $\lambda = \mu \tan\theta$ in the (μ, λ) plane. The very existence of the entropy potential implies that the corresponding density ρ_θ is obtained as

$$\rho_\theta(\lambda_\eta) = \vec{u} \cdot \nabla \Phi,$$

where $\nabla = (\partial\mu, \partial\lambda)$. Moreover, one can extend the definition of the entropy, obtaining

$$\tilde{h}_{KS}(\theta) = \int_0^{\eta_{max}} d\lambda \rho_\theta(\lambda),$$

where the integral is now to be performed along the tilted line $\mathfrak{L}$ defined as $\lambda = \mu \tan\theta$. The existence of the entropy potential implies that as long as the line $\mathfrak{L}$ does not intercept the upper horns of the density (i.e. as long as $\theta < \pi/4$), the integral is independent of the orientation. Thus, h_{KS} is a "super" invariant: this is because it corresponds to the entropy density of the two-dimensional space-time pattern. Above $\pi/4$ the line intersects the right boundary of the spectral region; one can say that we are no longer entitled to call it a time-like variable, but rather a space-like one. In such a case, we know that a side-wise iteration typically leads to divergences, meaning that the invariant set is not an attractor but rather a repeller of the corresponding "dynamics". In repellers, one knows that the escape rate has to be subtracted from the sum of the positive exponents to obtain the correct value of the Kolmogorov-Sinai entropy (cf. Eq. (6.14)). In spatially continuous models such as the complex Ginzburg-Landau equation, the upper border of the spectral region has a parabolic shape, i.e. no matter how small θ is, the line $\mathfrak{L}$ will always intersect the upper part of the spectrum. This practically means that one has to face the instability of high-frequency modes. This instability, however, contributes not to the dynamical entropy but just to a non-vanishing escape rate.

Finally, notice that the chronotopic approach allows the definition of an imaginary component for the Lyapunov exponents in one-dimensional spatially extended systems, which can be interpreted as the density of exponents themselves.

10.3 Convective exponents and propagation phenomena

An alternative way of looking at spatio-temporal properties of the perturbation dynamics is offered by the *convective* exponent, which describes the growth of an initially localised perturbation (Deissler and Kaneko, 1987) within an open infinite environment. In Fig. 10.6 we present a typical example of the chaotic evolution of a δ-like perturbation placed in $x = 0$ at time $t = 0$. The idea behind the convective exponent is to measure the local expansion rate $\mathscr{L}(v)$ in a frame that moves away from the perturbed region with the velocity v. In a more formal way, assuming that the initial perturbation $u(x, 0)$ is strictly different from zero only in a finite (possibly small) interval around the origin, one expects that

$$\|\mathbf{u}(x, t)\|_x \propto e^{\mathscr{L}(v = x/t)t}\|\mathbf{u}(0, 0)\|, \tag{10.5}$$

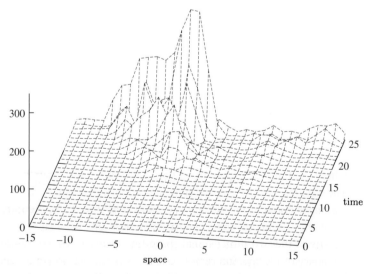

Fig. 10.6   Example of a perturbation propagation in a chain of maps.

where $\|\mathbf{u}(x,t)\|_x$ denotes the norm of the perturbation in the site x at time t (if the local dynamics is multi-dimensional, it may correspond to the local Euclidean norm). For finite times, it is very likely for the growth rate to depend on both the initial condition in real and tangent space. As for the usual Lyapunov exponent, the infinite-time limit removes such uncertainties (at least when the fluctuations are sufficiently well behaved),

$$\mathcal{L}(v) = \lim_{t \to \infty} \frac{1}{t} \ln \frac{\|\mathbf{u}(vt,t)\|_x}{\|\mathbf{u}(0,0)\|}. \tag{10.6}$$

Computationally speaking, it is not convenient to kill the statistical fluctuations by considering (10.6) for extremely long times (as normally done for the Lyapunov exponents). In fact, the perturbation covers increasingly wider areas, with a continuous growth of the memory demand and a slowing down of the calculations.

In practice, it is more convenient to set a time t_θ and determine $L(v)$ as

$$\mathcal{L}(v) = \frac{1}{t - t_\theta} \left\langle \ln \frac{\|\mathbf{u}(vt,t)\|_x}{\|\mathbf{u}(vt_\theta, t_\theta)\|_x} \right\rangle. \tag{10.7}$$

In this way, transient phenomena are automatically discarded, while the ensemble average $\langle \cdot \rangle$ (performed by seeding the initial perturbation in different space-time points) allows us to get rid of the statistical fluctuations.

An example of the expected quality of the various approaches is presented in Fig. 10.7a, which reports the results for a one-dimensional chain of coupled logistic maps (see Eq. (A.10)) with $a = 2$ and $\varepsilon = 1/3$. Dashed and dotted lines have been obtained by averaging the direct definition (10.5) for $t = 10$ and 50, respectively. One can see that the asymptotic shape has not yet been reached. The solid curve instead corresponds to the second method (Eq. (10.7)) by choosing $t_\theta = 10$ and $t = 50$. This curve

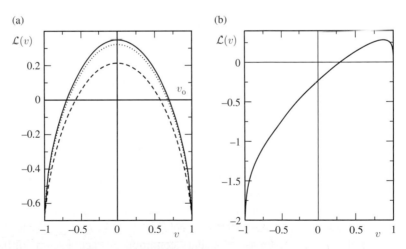

Fig. 10.7 (a) Spectra of convective Lyapunov exponents for a chain of diffusively coupled logistic maps (A.10) with $a = 2$ and $\varepsilon = 1/3$. Dashed and dotted curves have been obtained from Eq. (10.5) with $t = 10$ and 50, respectively, after averaging over 10^5 initial conditions. The solid line has been obtained by implementing Eq. (10.7) with $t_\theta = 10$ and $t = 50$. (b) Spectrum of convective LEs for the asymmetric chain (10.8), with $\varepsilon_- = 0.8$ and $\varepsilon_0 = \varepsilon_+ = 0.1$.

is practically indistinguishable from the asymptotic one, thus showing the superiority of the latter approach.

Some comments are in order: first of all, in this example the function $\mathscr{L}(v)$ is symmetric around $v = 0$. This reflects the left-right symmetry of the evolution on the lattice. Moreover, the maximum $\mathscr{L}(0)$ coincides with the maximum $\lambda(\rho = 0)$ of the Lyapunov density spectrum. This equality tells us that an initially localised perturbation when observed in the same place where it has been generated grows with the maximal rate in spite of the fact that it is not constrained to a finite volume. We are not aware of any proof of this property. Finally, notice that upon increasing $|v|$, $\mathscr{L}(v)$ decreases until it becomes negative for $v > v_0$. The velocity v_0 is the maximal propagation velocity of (infinitesimal) perturbations and is often referred to as the *linear* velocity.

The negative exponents detected for $|v| > v_0$ mean that although the perturbation is able to reach sites lying at a distance vt in a time t, it does so with an exponentially decreasing amplitude. This is somehow similar to evanescent waves.

Whenever left-right symmetry is broken, the maximum of $\mathscr{L}(v)$ is attained for a non-zero velocity. One such example is the asymmetric lattice of coupled maps

$$U_x(t+1) = F[\varepsilon_- U_{x-1}(t) + \varepsilon_0 U_x(t) + \varepsilon_+ U_{x+1}(t)], \tag{10.8}$$

where $F[U]$ is some one-dimensional map and $\varepsilon_- + \varepsilon_0 + \varepsilon_+ = 1$. If $\mathscr{L}(0) < 0$, while the maximum is positive, the system is said to be *convectively unstable*. This is, for instance, the case of logistic maps (i.e. $F(x) = 2 - x^2$) depicted in Fig. 10.7b. If a convectively unstable system is open, localised perturbations travel away and the system, locally, relaxes back to the previous state, so that no instability is observed. In the case of closed systems, when perturbations are reflected from the boundary or travel because of periodic boundary conditions, convective instability turns into an absolute instability, so that the perturbation grows also in the initially excited sites.

An example of convectively unstable dynamics

Chains of unidirectionally coupled elements represent an interesting example of a convectively unstable dynamics, where spatial and temporal Lyapunov exponents are somehow "exchanged". Let us consider the following chain of ordinary differential equations

$$\frac{dU_k}{dt} = -U_k + f(U_{k-1}), \tag{10.9}$$

which can be physically interpreted as a unidirectional chain of nonlinear amplifiers. Additionally, we assume that $f(x) = ax(1 - x)$.

Equation (10.9) should be supplemented with appropriate boundary conditions. We will consider an open system setup, where the spatial coordinate k runs from 1 to ∞, while the field $U_0(t)$ is given as a boundary condition. In such a setup, all the temporal Lyapunov exponents are equal to -1, because the system (10.9) is skewed. Nonetheless, the model generates some chaotic dynamics, as can be seen by exchanging the roles of space and time. Indeed, Eq. (10.9) can be rewritten as

$$U_k(t) = \hat{C}f(U_{k-1}(t)), \tag{10.10}$$

where $\hat{C}$ is the linear operator

$$\hat{C}V(t) = \int_0^\infty V(t - t')e^{-t'} dt'.$$

The recursive relation (10.10) describes the evolution along the discrete "time" axis k of the field $U(t)$ parametrised by the continuous "space variable" t. The evolution consists of the nonlinear local transformation f combined with a linear "coupling" induced by the action of the operator $\hat{C}$. In this representation $U_0(t)$ plays the role of an initial condition, and the system (10.10) can be altogether considered as a spatially continuous version of a coupled map lattice. If the field $U_0(t)$ is periodic with period T, then model (10.10) can be solved with periodic boundary conditions; otherwise, the entire semi-infinite interval $[0, \infty)$ should be considered. In this formulation, the evolution in k may be very well be chaotic, with a standard Lyapunov density spectrum as described in Section 10.1. The chaotic structure is due to the nontrivial arrangements of the various k-layers.

10.3.1 Mean-field approach

In coupled map lattices, approximate expressions for the convective Lyapunov exponents can be derived by following the approach introduced in Section 8.1.3 with reference to sparse matrices. In the asymmetric model (10.8), the evolution in tangent space is

$$u_x(t+1) = F'(x,t)[\varepsilon_- u_{x-1}(t) + \varepsilon_0 u_x(t) + \varepsilon_+ u_{x+1}(t)].$$

Accordingly, $u_x(t)$ can be seen as the sum of the contributions arising from all possible paths which connect the origin $(0,0)$, where the perturbation has been seeded, to (x,t). As shown in Fig. 10.8, each path is composed of three types of links (minus, zero and plus), which are classified according to the spatial shift when moving backwards in time.

In practice, one can write

$$u_x(t) = \sum_j e^{\Lambda_j t} \varepsilon_-^{m_-} \varepsilon_0^{m_0} \varepsilon_+^{m_+},$$

where m_-, m_0 and m_+ are the numbers of minus-, zero- and plus-links. Moreover $e^{\Lambda_j t} = \prod_{t' < t} F'(x, t')$ and we have implicitly assumed that the multipliers are positive. It is easily seen that $m_- = (t + x - m_0)/2$ and $m_+ = (t - x - m_0)/2$. The number of paths connecting the origin to (x,t) with a given number m_0 of zero-links is

$$N(m_0, x, t) = \frac{t!}{m_0! m_-! m_+!} \approx e^{n(y,v)t},$$

where $v = x/t$ and $y = m_0/t$. By invoking the Stirling approximation,

$$n(y, v) = -y \ln y - \frac{1 - y + v}{2} \ln \frac{1 - y + v}{2} - \frac{1 - y - v}{2} \ln \frac{1 - y - v}{2}.$$

On the other hand, the coupling contribution can be written as

$$\varepsilon_-^{m_-} \varepsilon_0^{m_0} \varepsilon_+^{m_+} = e^{-g(y,v)t},$$

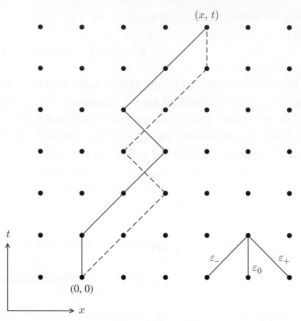

Fig. 10.8 Schematic representation of perturbation propagation in a coupled-map lattice. Two of the many paths connecting the origin $(0, 0)$ with (x, t) are depicted.

where

$$g(y, v) = -y \ln \varepsilon_0 - \frac{1 - y + v}{2} \ln \varepsilon_- - \frac{1 - y - v}{2} \ln \varepsilon_+ .$$

By now, assuming that all paths are mutually uncorrelated, one can write

$$u(x, t) \approx \int dy d\Lambda e^{(n(y,v) - g(y,v) + \Lambda - S(\Lambda))t},$$

where we have introduced the large-deviation function $S(\Lambda)$.

By recalling the definition of the generalised Lyapunov exponent $\mathcal{L}(1)$ and estimating the integral over y with the usual saddle-point method, one finds that

$$\mathscr{L}(v) = \mathcal{L}(1) + n(y_0, v) - g(y_0, v), \tag{10.11}$$

where y_0 is determined by the implicit equation

$$\frac{\partial n}{\partial y}(y_0, v) = \frac{\partial g}{\partial y}(y_0, v),$$

which yields

$$y_0 = \frac{1 - v^2}{1 + \sqrt{v^2 + 4(1 - v^2)\frac{\varepsilon_- \varepsilon_+}{\varepsilon_0^2}}}. \tag{10.12}$$

$\mathcal{L}(1)$ is a proper value to be selected only if $n(y_0, v) - S(\mathcal{L}(1)) > 0$, which ensures that this growth rate is generically observable in one of the possible paths. When this is not the case, a smaller Λ-value should be selected, such that $n(y_0, v) - S(\Lambda) = 0$ (see Livi et al. (1992) for a more detailed discussion). For $v = 0$ and a symmetric coupling ($\varepsilon_- = \varepsilon_+ = \varepsilon$), it follows that $y_0 = 1 - 2\varepsilon$, so that the two contributions $n(y_0, 0)$ and $g(v_0, 0)$ cancel each other out, yielding $\lambda = \mathscr{L}(0) = L(1)$.

These results are exact only in the absence of fluctuations, when $\mathcal{L}(0) = \mathcal{L}(1)$, as the assumption of *independent paths* is clearly false, since many paths share the same links. As long as fluctuations are negligible, Eqs. (10.11, 10.12) nevertheless provide a good approximation. An important type of fluctuations are those induced by the presence of positive and negative signs. In this case, as discussed by Pikovsky (1993), two phases may exist: (i) one where one sign prevails and the LE is determined by the growth rate of the mean amplitude and (ii) one where the LE is determined by the variance of the perturbations.

10.3.2 Relationship between convective exponents and chronotopic analysis

It can be easily shown that $\mathscr{L}(v)$ is connected to the largest temporal Lyapunov exponent $\lambda(0, \mu)$ via a Legendre-type transform. Eq. (10.5) implies that the perturbaton has a locally exponential profile with a spatial rate

$$\mu = \frac{d\mathscr{L}(v)}{dv} \tag{10.13}$$

in the point $x = vt$. From the chronotopic analysis, we know that such a perturbation evolves as

$$u(vt, t) \simeq \exp\left[(\lambda(0, \mu) + \mu v)t\right]. \tag{10.14}$$

Notice that with these conventions $\mu < 0$ in the positive x region. By combining Eqs. (10.5) and (10.14), one obtains

$$\mathscr{L}(v) = \lambda(0, \mu) + \mu \frac{d\lambda(0, \mu)}{d\mu},$$

which, together with Eq. (10.13), can be interpreted as a Legendre transform from the pair $(\mathscr{L}, v)$ to the pair (λ, μ). The inverse transform reveals the further constraint

$$v = -\frac{d\lambda(0, \mu)}{d\mu}.$$

The corresponding geometrical construction is presented in Fig. 10.9.

It is important to remark that, numerically, it is more convenient to determine $\mathscr{L}(v)$ from the Legendre transform, as it is not necessary to deal with continously expanding lattices.

In order to determine the velocity corresponding to a given μ-value, it is necessary to compute the derivative of $\lambda(\mu)$. Since the numerical computation of derivatives is always affected by large numerical errors, it is convenient to perform a few more analytical steps. By following Politi and Torcini (1992), we introduce

$$v_x(t, \mu + d\mu) = v_x(t, i) + z_x(t, \mu)d\mu$$

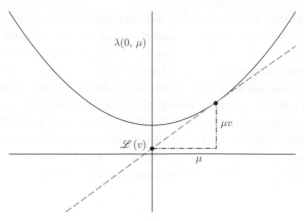

Fig. 10.9   Illustration of a relationship between the convective exponent $\mathscr{L}(v)$ and the chronotopic spectrum.

in the recursive relation (10.4), obtaining an equation for the deviation $z_x(t)$ (for the sake of simplicity, we now drop the dependence on μ):

$$z_x(t+1) = F_U \left\{ \varepsilon e^{-\mu x}[z_{x-1}(t) - v_{x-1}(t)](1 - 2\varepsilon)z_x(t, \mu) + \right.$$
$$\left. + \varepsilon e^{\mu x}[z_{x+1}(t) + v_{x+1}(t)] \right\} + b z_x(t-1). \tag{10.15}$$

The knowledge of z_x and of u_x allows us to determine $\lambda'(\mu)$. In fact, by computing the μ derivative of the definition of the chronotopic Lyapunov exponent,

$$\lambda(\mu) = \frac{1}{2} \lim_{t \to \infty} \frac{\log \|\mathbf{v}(t)\|^2}{t},$$

one obtains

$$\lambda'(\mu) = \lim_{t \to \infty} \frac{\mathbf{u}(t) \cdot \mathbf{z}(t)}{t\|\mathbf{u}(t)\|^2},$$

where $\cdot$ stands for the scalar product.

In order to better understand the selection process of the propagation velocity, it is convenient to go back to the evolution of a single exponential profile $w_x(t) = \exp[\lambda(\mu)t - \mu x]$. Its velocity is obviously $V(\mu) = -\lambda/\mu$. From the Legendre transform it follows that

$$\frac{dV}{d\mu} = \frac{1}{\mu}\left(\frac{d\lambda}{d\mu} - \frac{\lambda}{\mu}\right) = -\frac{\mathscr{L}(v)}{\mu^2}.$$

Since the perturbation velocity v_0 is identified by the condition $\mathscr{L}(v_0) = 0$, we see that v_0 corresponds also to the minimum of $V(\mu_0)$. In other words, as long as the evolution is controlled by linear mechanisms, it coincides with the velocity of the slowest front. By comparing with wave propagation, the velocity of the single front corresponds to the phase velocity, while v_0 is equivalent to the group velocity.

This analysis has basically shown that the maximal convective expansion rate can be determined by Legendre transforming $\lambda(0, \mu)$. It is tempting to extend the concept to a generic spectral component $\lambda(\rho, \mu)$ with $\rho > 0$, to obtain the propagation velocity of the other exponents. It is, however, not yet clear whether this concept is purely formal or not.

Kenfack Jiotsa et al. (2013) have shown that the resulting convective exponent is consistent with a more direct determination of $\mathscr{L}(\rho, v)$, obtained by following the perturbation amplitude within a finite window and imposing suitable boundary conditions. Unfortunately, no direct definition has yet been given of such generalised convective exponents, and the question of whether they are meaningful observables is still unanswered. There is, nevertheless, an interesting point to notice: as can also be seen from Fig. 10.3, there exists a critical value of ρ_c above which $\lambda(\rho, \mu)$ has a local maximum, rather than a minimum, in $\mu = 0$. This implies that the Legendre transform would yield a $\mathscr{L}(\rho, 0)$ that is strictly smaller than $\lambda(\rho, 0)$. In principle this phenomenon might occur for the maximum exponent, as well, but it is not clear whether some general properties forbid a different concavity for $\lambda(0, \rho)$.

10.3.3 Damage spreading

In many systems, the linear velocity v_0 coincides with the propagation velocity of directly measurable physical observables such as the sound velocity in the Fermi-Pasta-Ulam chain (A.15) or the propagation of correlations in the complex Ginzburg-Landau equation (A.18) (Giacomelli et al., 2000). This is, however, not always the case; for instance, in the so-called hard-point gas (a chain of elastically colliding free particles), the sound velocity is larger than the value one could infer from a linear stability analysis (Delfini et al., 2007). It is, therefore, instructive to look at the propagation of finite-amplitude perturbations, or, as otherwise called, *damage spreading*, interpreting the perturbation as a sort of damage to a given configuration.

Even the front of a finite propagating perturbation must be preceded by a leading edge, where the perturbation is infinitesimal. Therefore, the velocity v_f of the front must be equal to the velocity $v(\mu^*)$ of some exponential profile $e^{-\mu^* x}$. Were the propagation controlled by linear mechanisms, the spatial profile of the front would be characterised by a rate $\mu_0 = (d\mathscr{L}/dv)(v_0)$. It is hard to imagine a mechanism by which μ^* is smaller than μ_0; it is more natural to expect $\mu^* > \mu_0$ and, therefore, $v_f > v_0$ (steeper fronts propagate faster).

A numerical study of the coupled maps (A.13, A.14) indeed confirms such expectations. The solid and dashed curves in Fig. 10.10 correspond to v_f and v_0, respectively. There, we see that v_f is strictly larger than v_L for $\eta < \eta_c \approx 1.2 \times 10^{-3}$, while above η_c the two velocities coincide within numerical accuracy. One can also notice that the linear velocity is not defined for $\eta < \eta^*$, where the system is linearly stable and no propagation of infinitesimal perturbations can occur at all.

As discussed by Torcini et al. (1995), the mechanism responsible for the finite difference between v_0 and v_f is that perturbations of increasing amplitude (which are naturally present in the bulk of the perturbation) propagate faster and thereby tend to push the corresponding front. The occurrence of velocities larger than the linear one is indeed fairly general and not just restricted to the aforementioned abstract model (see, e.g., Cencini and Torcini (2001) and Delfini et al. (2007)). This phenomenology is conceptually equivalent to what is observed in the context of front propagation (see, e.g., fronts connecting steady states in reaction-diffusion systems discussed by van Saarloos (1988, 1989)), which is effectively described by the famous Fisher-Kolmogorov-Petrovsky-Piskunov equation (Fischer, 1937; Kolmogorov et al., 1937).

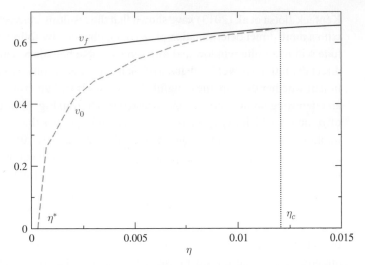

Fig. 10.10   Linear (v_0, solid curve) and front (v_f, dashed curve) velocity versus η for the model (A.13, A.14) with $\varepsilon = 1/3$, versus the width η of one of the three intervals defining the local map F. Standard deterministic chaos exists only for $\eta > \eta^*$. Beyond η_c, $v_0 = v_f$.

Altogether, the evidence of a front-velocity v_f strictly larger than the linear velocity v_0 can be considered as an extension of the concept of stable chaos to a regime where local instabilities are present but not strong enough to control spreading phenomena (Politi and Torcini, 1994).

Binary variables

As already discussed in Section 10.1.1, the usual Lyapunov exponents cannot be defined at all in discrete-state systems due to the impossibility of introducing infinitesimal perturbations. In such a case, one is bounded to use the velocity v_f of finite-amplitude perturbations as a tool to quantify the degree of instability.

If the local variables are binary, another approach has been proposed that is based on Boolean algebra. In this case, the difference between two states can be expressed through the XOR operation, which gives 0 if the states are equal and 1 if they differ. The smallest possible perturbation (a so-called damage) is just a change of state U_x in a single site, which also can be represented by the XOR operation: $U'_x = U_x$ XOR 1. Let us denote with $\mathbf{U}(0)$ and $\mathbf{U}'(0)$ the two initial configurations and ask how the difference $\delta \mathbf{U}(t) = \mathbf{U}(t)$ XOR $\mathbf{U}'(t)$ evolves. There are two limit cases: (i) the perturbation eventually dies out so that after some time $\mathbf{U} = \mathbf{U}'$, this is the stable scenario; (ii) the perturbation grows in time, implying an unstable dynamics. So long as the updating rule is local, this growth looks like spreading, therefore the name "damage spreading" (Kauffman, 1969; Herrmann, 1990).

The distance $\delta \mathbf{U}(t)$ can, at most, grow linearly in time, with a velocity $2v_f$ (as it grows in two directions). A more Lyapunov-like indicator has been introduced, by referring to a pool of perturbations, by Bagnoli et al. (1992). Whenever the evolution of a given pattern

$\mathbf{U}'(t)$ gives rise to $m > 1$ defects, $m - 1$ replicas are added to the pool, with a possible exponential growth of the cardinality of the pool itself. In practice, there is no need to effectively handle a growing number of configurations; it is sufficient to introduce the Boolean-Jacobian matrix $\mathbf{J}$, whose entries,

$$\mathbf{J}_{xy} = \frac{\partial U_x(t+1)}{\partial U_y(t)} = U_x(t+1) \text{ XOR } U'_x(t+1), \quad U'_x(t) = U_x(t) \text{ XOR } \delta_{xy},$$

are either equal to 0s or 1s. Next, the *perturbation vector* $\mathbf{N}$ is introduced, whose component N_x counts the number of replicas with a defect in x. $\mathbf{N}(t)$ evolves according to

$$\mathbf{N}(t+1) = \mathbf{J}\mathbf{N}(t).$$

A Boolean Lyapunov exponent λ_B can be finally defined in a usual way, as an asymptotic growth rate

$$\lambda_B = \lim_{T \to \infty} \frac{1}{T} \frac{\|\mathbf{N}(T)\|}{\|\mathbf{N}(0)\|}.$$

λ_B is more an entropy-like rather than Lyapunov-like observable, as it arises from the counting of possible future patterns, although it is not exactly equivalent to the dynamical entropy since the procedure does not take into account the correlations among different neighbouring sites. Some applications of this approach can be found in the work of Bagnoli et al. (1992); a generalisation has been discussed by Martin (2007).

10.4 Examples of high-dimensional systems

Many different dynamical systems are characterised by a large number of variables. In this section we discuss the most relevant model classes.

10.4.1 Hamiltonian systems

Obtaining analytical estimates for the LEs in multidimensional systems is highly difficult. In Hamiltonian models there is at least the advantage that the invariant measure is known from the Liouville theorem. In this section we show that some encouraging results can be obtained with the help of formulas derived for random systems. Here, we illustrate the potentiality of two approaches to estimate the largest LE and the entire spectrum for the Fermi-Pasta-Ulam chain. From the evolution equation (A.16) in real space, it is found that, in tangent space,

$$\ddot{q}_x = q_{x-1} - 2q_x + q_{x+1}$$
$$+ 3(Q_{x-1} - Q_x)^2(q_{x-1} - q_x) + 3(Q_{x+1} - Q_x)^2(q_{x+1} - q_x). \tag{10.16}$$

Largest Lyapunov exponent

Casetti et al. (1995) developed a geometric approach to estimate the largest Lyapunov exponent by deriving an equation for the so-called sectional curvature. Here, we briefly illustrate the key steps, by following a substantially equivalent, but purely dynamical, derivation. Given Eq. (10.16), it is known that the largest Lyapunov vector of Hamiltonian systems is strongly localised (see Chapter 11 for a detailed discussion of the Lyapunov vectors). A good approximation of the tangent-space dynamics is thus obtained by assuming that the maximum of the Lyapunov vector sits in x and neglecting the $q_{x\pm1}$ terms:

$$\ddot{q}_x + [2 + 3(Q_{x-1} - Q_x)^2 + 3(Q_{x+1} - Q_x)^2]q_x = 0. \tag{10.17}$$

This equation represents a big step forwards, since it reduces the problem to that of a noisy oscillator of the type investigated in Section 8.2.2. Let us now define

$$E = 2 + 6\langle(Q_{x-1} - Q_x)^2\rangle, \quad \Delta^2 = 4\mathrm{Var}(1 + 3(Q_{x-1} - Q_x)^2).$$

Both E and Δ^2 can be computed by making use of the invariant microcanonical measure. Eq. (10.17) can be now identified with Eq. (8.26), under the additional assumption that the diffusion coefficient of ξ is given by

$$\sigma^2 = \Delta^2 \tau/2,$$

where the time constant τ is still to be determined.

Casetti et al. (1995) use some geometrical considerations to propose

$$\frac{1}{\tau} = \sqrt{\frac{2\Delta^2}{E}} + \sqrt{\frac{E}{2}}.$$

Now E and σ^2 are known and one can make use of Eq. (8.30) to determine the generalised Lyapunov exponent $\mathcal{L}(2)$. The detailed expressions of $\mathcal{L}(2)$ can be found in Pettini (2007). They are remarkably close to the true maximum LE for an extremely wide range of energy densities. It should be noted, however, that this approach is not always so effective, the weakness being presumably the estimation of the time constant τ.

The Lyapunov density spectrum

Determining the entire spectrum of LEs is a yet more difficult task. Some results have been derived by Eckmann and Wayne (1988) for the Fermi-Pasta-Ulam chain of oscillators, making use of the theory developed by Newman for a product of random matrices (see Chapter 8).

In the case of N such oscillators, the Jacobian is composed of 4 $(N \times N)$ blocks

$$K = \begin{pmatrix} 0 & \Omega \\ 1 & 0 \end{pmatrix},$$

where Ω is a tridiagonal matrix of the type

$$\Omega = \begin{pmatrix} \omega_1 & -\omega_1 & 0 & \cdots & 0 \\ -\omega_1 & \omega_1 + \omega_2 & -\omega_2 & \cdots & 0 \\ 0 & -\omega_2 & \omega_2 + \omega_3 & \cdots & 0 \\ \vdots & \vdots & \vdots & \ddots & \vdots \\ 0 & 0 & 0 & \cdots & \omega_{N-1} \end{pmatrix},$$

where $\omega_i = 1 + 3(Q_{i+1} - Q_i)^2$. Free boundary conditions have been assumed, but as long as one is interested in the thermodynamic limit, this is not a relevant point.

In order to clarify the various approximations involved in the whole procedure, it is instructive to split it into four steps.

Step 1: Short correlation time

If the correlation time τ of the matrix entries is so short that the nearest-neighbour coupling has not yet propagated to the next neighbours, one can write the finite-time evolution in tangent space approximately as

$$H_\tau = \exp(K\tau). \tag{10.18}$$

The Lyapunov exponents can be then determined by multiplying infinitely many H_τ matrices, which can be assumed to be mutually uncorrelated. This assumption is expected to be reasonably accurate at high energies, when the dynamics is highly chaotic.

Step 2: Thermodynamic equilibrium

Upon assuming that the Q_i variables are distributed according to the equilibrium measure, one can analytically determine the distribution $F_\Omega(\omega)$ of the random variables ω_i. This is analogous to the previous approach for the computation of the largest LE.

Step 3: Spectral properties of random matrices

Given F_Ω, it is possible to determine the spectral density $\tilde{F}_J(\nu)$ of the eigenvalues ν of K and thereby the density $F_J(\mu)$ of the eigenvalues of $(H_\tau^\mathsf{T} H_\tau)^{1/2}$. The corresponding equations are quite cumbersome and the determination of the final expression technically involved. The interested reader is invited to consult the original paper by Eckmann and Wayne (1988).

Step 4: From the single matrices to their product

By finally assuming that the matrices $H_\tau^\mathsf{T} H_\tau$ are rotation invariant, one can invoke Eq. (8.16), with $F_J(\mu)$ playing the role of $S(\mu)$. In view of the symplectic structure of H_τ, it is easily seen that $\lambda(\rho)$ is symmetric around $1/2$, as expected. Moreover,

$$\lambda_{max} = \frac{1}{2} \ln \int F_J(\mu)\mu^2.$$

Testing the accuracy of the various steps is a difficult task. It is at least possible to check the last step, the conceptually most important one, since it establishes a connection between the spectrum of single matrices with that of an infinite product. Here, we show some numerical results obtained for the Fermi-Pasta-Ulam chain with

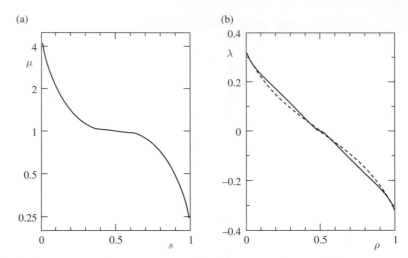

Fig. 10.11 (a) Average spectrum of the short time Jacobian for an FPU chain with energy density 10, 32 oscillators and $\tau = 0.062$ and (b) Lyapunov density spectrum for the same model (solid curve) and the one estimated from Eq. (8.16) (dashed curve).

periodic boundary conditions. H_τ has been determined by numerically integrating the equations in the tangent space for a time τ and averaging over a large number of trajectories.

The results are plotted in Fig. 10.11a for an FPU chain with $N = 32$ oscillators and a high energy density ($\mathcal{H}/N = 10$). In this condition, a strongly chaotic dynamics is expected. The Jacobians H_τ have been determined by directly integrating the equations over a time interval $\tau = 0.062$; it is sufficiently short to expect that the approximation (10.18) is valid. In order to facilitate the comparison with the final Lyapunov density spectrum (see panel (b)), the μ value is plotted versus the integrated density s (i.e. $s(\mu) = \int F_J d\mu$). The vertical logarithmic scale allows recognising the symmetry $\mu(s) = \mu(1 - s)$, typical of symplectic matrices. In panel (b), the outcome of direct simulations (solid curve) is superposed to the expression deriving from the application of Eq. (8.16) (dashed curve). Both spectra are approximately, though not exactly, linear. Quite remarkably, a quasi-linear shape is a general feature that is found also in products of random symplectic matrices of the type (10.18) for τ of order 1, where the arguments developed in step 4 do not apply (see Paladin and Vulpiani (1986) and Livi et al. (1987)).

As for the relatively good quantitative agreement, one should not be too excited. In fact, while the shape of the spectrum obtained from Eq. (8.16) is substantially independent of selection of the τ value, the same is not true for its scale. If τ is selected by following the previously discussed geometrical approach, it is found that $\tau = 0.124$ and thereby $\lambda_{max} \approx 0.69$ – twice as high, compared with the expected value $\lambda_{max} = 0.32$. One can thus conclude that this theory provides a reasonable description of the shape of the spectrum, although we cannot expect it to be exact; direct studies of the Lyapunov vectors indeed show that they are localised, i.e. isotropy is not satisfied. The most delicate point is, however, the need for an appropriate protocol for the selection of the optimal τ value.

10.4.2 Differential-delay models

As soon as a delayed interaction term is introduced in an otherwise one-dimensional, ordinary differential equation, the phase-space dimensionality becomes infinite. Let us consider, for instance, the equation

$$\dot{U} = F(U(t), U(t - \tau)),$$

where τ is a given delay. The initial condition can be assigned by defining the function $U(t)$ over a time interval of length τ (e.g. $t \in [-\tau, 0]$). The dynamical complexity of such a model is equivalent to that of a spatially extended system. This can be appreciated by decomposing the time variable as $t = n\tau + x$, where $x \in [0, \tau]$ is a continuous space-like variable, while the integer n plays the role of time (see Arecchi et al. (1992), Kuznetsov and Pikovsky (1986) and Fig. 10.12).

The only (important) difference with respect to spatially continuous, time-discrete systems is in the boundary conditions, since here they involve two different times: $U(n, 0) = U(n-1, \tau)$. Since for large values of τ one expects the boundary conditions to be substantially irrelevant, one can safely conclude that long delays are equivalent to spatially extended systems in the thermodynamic limit. In order to illustrate the scaling of the Lyapunov spectra, we consider the Ikeda equation (A.20) for different delays. The spectra reported in Fig. 10.13 confirm that the delay plays the role of the system size in space-time chaos ($\rho_\tau = (i - 1/2)/\tau$). Moreover, analogously to spatially continuous systems, the density ρ_τ is unbounded from above. However, at variance with diffusively coupled spatial units, where the Lyapunov spectrum is proportional to $-\rho^2$, here the decrease is typically

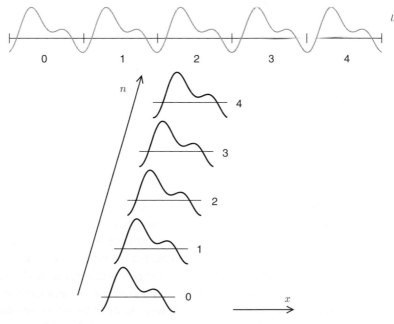

Fig. 10.12   Spatial interpretation of a delayed system.

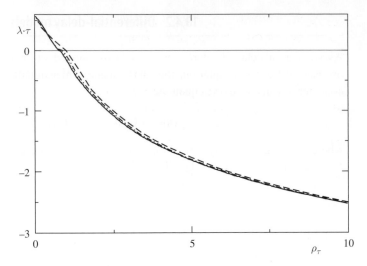

Fig. 10.13 Rescaled Lyapunov spectra for the Ikeda delay equation (with $a = 5$ and $U_0 = 0$) and different delays: the dashed, dotted-dashed, dotted and solid curves correspond to $\tau = 2.5, 5, 10$ and 20, respectively.

logarithmic (Farmer, 1982). An important difference with spatial models is that here, the Lyapunov exponents scale with the "system size": they indeed decrease as $1/\tau$ (see the vertical axis). This is due to the fact that the time-like variable n is related to the true time through the relation $n = \lfloor t/\tau \rfloor$. As a consequence of both scalings, the Kolmogorov-Sinai entropy does not increase with the value of delay time, but rather it stays constant: the same amount of instability is simply distributed over an increasing number of variables. On the other hand, the fractal dimension is extensive, as in spatially extended systems, since it depends on the number of the positive Lyapunov exponents, but not on their actual values.

So far, we have referred to the case of a scalar variable. If $\mathbf{U}$ is a vector, such as, for instance, in Rössler oscillators (A.8) with delayed feedback, the scenario may differ in the long-delay limit. In fact, if the local dynamics is chaotic by itself, one or more positive Lyapunov exponents may remain finite even for $\tau \to \infty$. This can be understood with the help of a self-consistency argument. In full generality, the tangent space dynamics is ruled by the following equation,

$$\dot{\mathbf{u}} = \frac{\partial \mathbf{F}}{\partial \mathbf{U}}\mathbf{u} + \frac{\partial \mathbf{F}}{\partial \mathbf{U}_\tau}\mathbf{u}_\tau,$$

where the subscript τ means that the corresponding variable is estimated at a delayed time. Now, if there exists a finite and positive Lyapunov exponent λ, the corresponding $\mathbf{u}_\tau$ is exponentially negligible with respect to $\mathbf{u}$ (being smaller by a factor of $\exp(-\lambda\tau)$), so that the last term in this equation can be neglected. As a result, the value of λ is obtained by simply linearising the original equation, as if the delayed term were absent (notice that this is not true for the evolution in real space). In other words, if the simplified linearised equation admits a positive exponent, this is also a true exponent of the delayed dynamics. It is well known that no such solution can be obtained in one-dimensional differential-delay equations. The occurrence of this short-term instability was initially defined as an "anomalous" exponent by Lepri et al. (1994); in more recent studies, the term *strong chaos* has been introduced to distinguish the regime where a finite LE survives for long delays, from the *weak chaos*, when this does not happen (Heiligenthal et al., 2011). Transitions from weak to strong chaos are naturally encountered in networks of oscillators with delayed coupling upon tuning some control parameter (Heiligenthal et al., 2011). An intermittent presence of short-term instabilities seems to be responsible for the low-frequency fluctuations observed in semiconductor lasers with optical feedback (Yanchuk and Wolfrum, 2010).

In some systems the delay may be state dependent. The corresponding mathematical model is by far more difficult to treat; the solutions resulting from an initial value problem have less smoothness properties than in usual problems. Very little is known about the stability properies of such systems. Hartung et al. (2006) give a fairly complete review of both physical contexts where such equations may arise and of their dynamical properties.

10.4.3 Long-range coupling

When coupling involves interactions at arbitrary distances, the way the Lyapunov spectrum scales in the thermodynamic limit is no longer obvious. In fact, one cannot view the whole system as the juxtaposition of many weakly interacting subsystems. In this section we nevertheless show that the extensivity hypothesis proves to be essentially correct, although some notable exceptions are present.

Global coupling: chaotic oscillators

We first consider an ensemble of globally coupled maps of the type (A.23). The evolution in tangent space is ruled by the recursive equation

$$u_j(t+1) = F'_j(t)\left[(1-\varepsilon)u_j(t) + \varepsilon\bar{u}(t)\right],$$

where $\bar{u}(t) = (1/N)\sum_i u_i(t)$.

In Fig. 10.14, we report the Lyapunov spectra obtained by simulating ensembles of tent maps ($F(V) = a(1/2 - |1/2 - V|)$) of different sizes. One can see that the spectra become flatter and flatter in the thermodynamic limit. This can be understood in the following way. Since the Lyapunov vectors are typically localised in a chaotic environment (see Chapter 11), the contribution $\varepsilon\bar{u}(t)$ of the coupling to the tangent dynamics becomes increasingly negligible in the thermodynamic limit. The mutual coupling contributes only to the self-consistent determination of the mean field $E = (\sum_i U_i)/N$, which, in the absence of collective chaos, is independent of time. As a consequence, the dynamics of each map is described by the effective equation (see Appendix A)

$$U_j(t+1) = F((1-\varepsilon)U_j(t) + \varepsilon E),$$

and all Lyapunov exponents are equal to one another and equal to $\lambda_e = \langle \ln|F'((1-\varepsilon) U_j(t) + \varepsilon E)|\rangle$. In the case of tent maps, since the slope is everywhere the same, $\lambda_e = (1-\varepsilon)\ln a$.

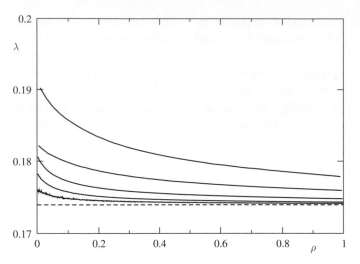

Fig. 10.14 Lyapunov spectrum for an ensemble of globally coupled tent maps with $a = 1.7$ and $\varepsilon = 0.3$. The solid curves refer, from top to bottom, to $N = 50, 100, 200, 400$ and 800. The dashed line corresponds to the theoretical expectation.

The correctness of this analysis is confirmed by the nice agreement between λ_e and the Lyapunov spectra obtained for the larger system sizes (see Fig. 10.14, where one can see that even the maximum LE which is affected by the largest deviations is clearly converging towards the expected asymptotic value).

In general, however, it is natural to conjecture that the Lyapunov density spectrum $\lambda(\rho)$ of an ensemble of generic d-dimensional chaotic oscillators is composed of d perfectly flat steps, which correspond to the d LEs of a single oscillator forced by a suitable mean field.

Although this scenario is substantially confirmed by numerical simulations, two important exceptions are found, as well. The first one is induced by collective chaos, i.e. by a chaotic dynamics of the mean field. In fact, in the presence of collective chaos, some Lyapunov vectors may be extended, rather than localised, thus implying that the coupling term is no longer negligible, as assumed here (see Chapter 11 for a more detailed analysis of this point).

The second exception is induced by coupling sensitivity. This effect, also discussed in Chapter 11, leads to a non-extensive number of exponents which differ from the expected plateau at λ_e. The reason why neither of these effects is present in the aforementioned tent maps is that no collective chaos is present and the finite-time LEs do not fluctuate.

So far we have discussed ensembles of identical oscillators. If the oscillators differ from each other, one can keep arguing as before and conclude that the diversity should remove the degeneracy typical of identical systems. This is indeed what happens when each map is characterised by a different parameter a.

The spectra reported in Fig. 10.15 indeed show a convergence to the curve $\lambda = \ln[(2 - \rho/2)(1 - \varepsilon)]$ that is obtained assuming that the maps are eventually uncoupled and the a-parameter uniformly distributed within the interval $[1.5, 2]$. In practice we see that the extensivity of globally coupled systems is quite trivial: it follows from the effectively

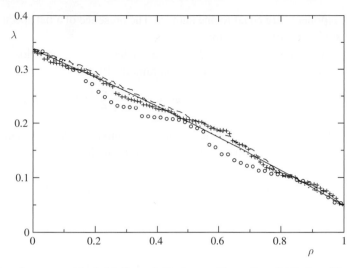

Fig. 10.15 Lyapunov spectrum for an ensemble of globally coupled tent maps with a uniform distribuiton of the parameter a in the interval $[1.5, 2]$ and $\varepsilon = 0.3$. Circles and plusses correspond to $N = 50$ and 100, respectively, while the dashed and dotted lines correspond to $N = 100$ and 200, respectively. Finally, the solid curve is the theoretical expectation.

vanishing coupling among the various elements. The effect of the interactions is only the self-consistent determination of a parameter, which in the case of tent maps does not even affect the Lyapunov exponent.

The only degeneracy that is not resolved by adding a diversity of the oscillators is the degeneracy of the zero Lyapunov exponent, which is always present in autonomous continuous-time dynamical systems. In the next section we analyse such exponents.

Global coupling: phase oscillators

The splay state offers the chance of studying the effect of global coupling on the zero LEs in a simple context. The splay state is a microscopically periodic regime where the phases ϕ_i of N identical oscillators are equispaced, i.e. $\phi_i - \phi_{i-1} = 2\pi/N$. If the Lyapunov exponents (or, more appropriately, the Floquet exponents) are arranged from the largest to the smallest, according to their modulus, analytical and accurate numerical studies have revealed that $|\lambda_k| \approx \exp(-\beta k)$ in pulse-coupled oscillators, when the velocity field $\dot{\phi}$ is sufficiently smooth (Olmi et al., 2012). The main effect of increasing N is indeed the inclusion of increasingly smaller exponents (allowing for larger k values). Altogether, this implies that $\lambda(\rho)$ is flat (and equal to zero) with a singularity on either its left or right, depending on whether the splay state is stable or unstable. More specifically, the number $N(\delta)$ of exponents whose magnitude is larger than δ is of the order of $|\ln \delta|$. In this case, the reason why the coupling is negligible for most but not all eigenvalues is not the localisation of the Lyapunov vectors but the fact their average nearly vanishes.

Let us recall that the fully synchronous solution $U_j = U$ (see the discussion in Section 9.2.2) is characterised by one zero (longitudinal) exponent, while all the other (transverse) exponents are equal to one another. The presence of a flat spectrum there is due to the high symmetry of the solution.

The presence of disorder induces a less trivial scenario, as it can be understood, by studying the Kuramoto model (Kuramoto, 1975)

$$\dot{\phi}_i = \omega_i + \frac{g}{N} \sum_i \sin(\phi_j - \phi_i), \tag{10.19}$$

where the bare frequencies ω_i are distributed according to some distribution $P(\omega)$. In the thermodynamic limit, if $P(\omega)$ is unimodal with the peak located in $\omega = 0$,[1] for $g < g_c = 2/(\pi P(0))$ the oscillators behave as if they were uncoupled, while for $g > g_c$ a finite fraction of oscillators are mutually locked (Acebrón et al., 2005). The transition is revealed by the order parameter

$$Re^{i\psi} = \frac{1}{N} \sum_i^N e^{i\phi}.$$

For $g < g_c$, $R = 0$, while for $g > g_c$, the value of R is determined by the implicit condition (Acebrón et al., 2005)

$$g \int_{-\pi/2}^{+\pi/2} d\phi P(gR \sin \phi) \cos^2 \phi = 1.$$

Since the equations of motion can be rewritten as

$$\dot{\phi}_i = \omega_i + gR \sin(\psi - \phi_i),$$

all oscillators with a bare frequency $|\omega| < gR$ are locked and their phase is given by $\sin \phi = \omega/(gR)$ (the origin can be set in such a way that $\psi = 0$). Accordingly, their stability is determined by the exponent

$$\lambda = -\sqrt{g^2 R^2 - \omega^2}.$$

One can now determine the Lyapunov density spectrum $\lambda(\rho)$ by linking ρ with ω. This is ensured by the relationship

$$\rho = 1 - \int_{-\omega}^{+\omega} d\tilde{\omega} P(\tilde{\omega}),$$

which expresses that the most stable oscillator is the one with $\omega = 0$, while the least stable one is characterised by a frequency $\omega(\rho_c) = gR$. For $\rho < \rho_c$, where $\omega(\rho_c) = gR$, $\lambda(\rho) = 0$, as the oscillators are unlocked (see Radons (2005)).

The solid curves in Fig. 10.16 are two prototypical spectra. Below transition, as for $g = 0.8g_c$, the Lyapunov spectrum is flat and equal to zero, since no locking is present. Above the transition (e.g. for $g = 1.2g_c$), two components are present: a flat branch, which corresponds to the unlocked oscillators, followed by a negative branch, which corresponds to the entrained oscillators. The two dashed curves correspond to the spectra obtained for a specific realisation of the disorder and $N = 400$. The most important finite-size effect

---

[1] This is not a severe restriction, as the frequency can be arbitrarily shifted without altering the equation structure.

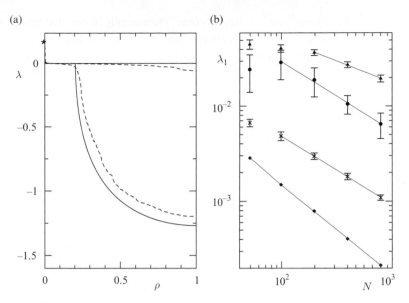

Fig. 10.16 (a) Lyapunov density spectrum of the Kuramoto model (10.19) for a Gaussian distribution of frequencies with unit variance and $g = 0.8g_c$ (see the two upper curves) and $g = 1.2g_c$ (lower curves). The solid lines correspond to the asymptotic spectra; the dashed lines have been obtained for $N = 400$. The asterisk denotes the maximum exponent for $g = 1.2g_c$ and $N = 400$. (b) Average maximum LE versus the system size N. Triangles ($g = 0.8g_c$) and circles ($g = 1.2g_c$) correspond to a Gaussian distribution of frequencies as in panel (a); crosses and diamonds correspond to a uniform distribution, respectively, perfectly equispaced frequencies within the interval $[-1, 1]$ and $g = \pi g_c/8$.

is the presence of positive LEs, which culminates in a rather large maximum exponent $\lambda_1(N)$ (see, for instance, the asterisk in Fig. 10.16a which refers to $g = 1.2g_c$). This reveals that, although no instabilities are present in the thermodynamic limit, finite ensembles are weakly chaotic. No theory has so far been developed which is able to predict the scaling behaviour of $\lambda_1(N)$. Popovych et al. (2005) found that $\lambda_1(N) \approx N^{-1}$ when the frequencies are equispaced. This is substantially confirmed by the data reported in Fig. 10.16b (see the crosses), which are well fitted by a decay $N^{-0.93}$. For truly random distributions, it appears that the decay is of the type $N^{-\alpha}$ with $\alpha < 1$. For a Gaussian distribution and $g = 0.8g_c$, it is found that $\alpha \approx 0.47$, while for $g = 1.2g_c$, $\alpha \approx 0.73$ and, finally, for a uniform distribution and $g = \pi g_c/8$, $\alpha \approx 0.71$. Altogether, it is not clear whether there exist a few universality classes; the simulations at least convincingly show that the sample-to-sample fluctuations of the Lyapuov exponents do not vanish when the LE itself decreases to zero.

Networks

So far, we have discussed models where each variable is coupled either to a finite subset of nearby variables (this is the case of many spatially extended systems) or "democratically" to all other variables (this is the case of the mean-field models).

In recent years it has become increasingly clear that many setups lie somewhere in between such two extreme cases. These are the network structures, where each variable interacts with a subset of variables that are not necessarily located nearby. As a representative example, we consider an ensemble of one-dimensional maps,

$$U_i(t+1) = (1-\varepsilon)F(U_i) + \frac{\varepsilon}{K}\sum_j S_{ij}F(U_j), \tag{10.20}$$

where the network structure is determined by the matrix entries S_{ij}, which are equal either to 0 or to 1, while K denotes the average connectivity of the single elements. There exist many structures, most of which have been poorly explored from the point of view of their dynamical properties. Here, we limit ourselves to discuss the Erdös-Renyi networks, where the links are randomly selected. Roughly speaking, there are two large subfamilies: that of *massive* and of *sparse* networks. In the former case, K is assumed to be proportional to the network size (i.e. to the total number N of elements), while in the latter one, K is independent of N. The random matrices illustrated in Section 8.1.3 belong to the latter class.

From the point of view of Lyapunov exponents, massive networks are not much different from globally coupled networks. As the coupling term is the sum of an increasing number of terms (when the thermodynamic limit is attained), its fluctuations can be neglected, and such networks are substantially equivalent to globally coupled systems, with a suitably rescaled coupling strength.

More intriguing is the case of sparse networks, since they are characterised by a nontrivial asymptotic spectral density $\lambda(\rho)$. An example is reported in Fig. 10.17, where the spectra obtained for a sparse network of logistic maps are reported for different network sizes.

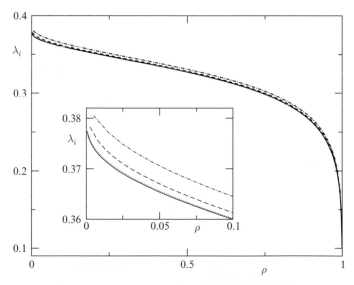

Fig. 10.17 Lyapunov spectra for the model (10.20) ($\rho = (i-1.2)/N$), for $F(U) = aU(1-U)$, $a = 3.9$, $\varepsilon = 0.1$ and $K = 10$. The dot-dashed, dashed, dotted and solid curves correspond to $N = 100, 200, 500$ and 1000, respectively. The inset contains an enlargement of the initial part of the spectrum.

The various curves reveal a clear convergence towards an asymptotic shape for increasing N. As shown in Luccioli et al. (2012), the existence of a well-defined spectral density is not specific to the logistic maps but arises also in other contexts. The result is remarkable, since, at variance with standard spatially extended models, here the presence of "long-range" links prevents viewing the network dynamics as the evolution of several almost uncoupled subunits. In other words, here, chaos is extensive without being *additive*, at least in the standard phase space. One should, in fact, notice that a chain of nonlinear oscillators with nearest-neighbour coupling, which is clearly additive when seen in real space, no longer seems to be so, when viewed in Fourier space, where all modes are mutually coupled. In other words, one cannot exclude that the extensivity revealed by the sparse networks is a consequence of additivity in some yet-to-be-discovered representation.

11

High-dimensional systems: Lyapunov vectors and finite-size effects

In this chapter we discuss some scaling properties of Lyapunov exponents and Lyapunov vectors. In the case of spatially extended systems, a fruitful analogy with roughening phenomena (Pikovsky and Politi, 1998) is used to characterise the spatial structure of the Lyapunov vectors and to establish the convergence of the LEs towards their asymptotic value.

The localisation of the (largest) Lyapunov vector allows simplifying the evolution equation for the largest LE and thereby deriving analytic expressions. A first instance of this strategy was discussed in Section 10.4.1 in chains of coupled oscillators. It also helps to revisit coupling sensitivity in globally coupled ensembles of dynamical systems. In Section 11.2, we show that the effect is so strong as to sustain a finite increase of the LE even in the thermodynamic limit.

Globally coupled systems offer also the opportunity to discuss the chaotic properties of collective dynamics. In Section 11.3, we introduce the "macroscopic" LEs and briefly discuss the possible relationship between microscopic and macroscopic instabilities.

Section 11.4 is devoted to a quantitative analysis of the scaling properties of LE fluctuations and of their relationship with problems such as dimension variability in spatially extended systems.

Finally, in the last section yet another class of Lyapunov spectral densities is introduced that is appropriate to establish the scaling property of the invariant measure in infinitely extended dyanamical systems.

11.1 Lyapunov dynamics as a roughening process

Numerical simulations of a wide class of spatially extended dynamical systems have revealed that the Lyapunov vectors are often localised in space. An example of the typical structure of the first Lyapunov vector is presented in Fig. 11.1 with reference to a chain of coupled Hénon maps (see also Eq. (A.11))

$$U_\ell(t+1) = a - [U_\ell(t) + \varepsilon \mathcal{D} U_\ell(t)]^2 + b U_\ell(t-1). \tag{11.1}$$

The local amplitude $\|\mathbf{u}_x\| = \sqrt{u_x^2(t) + u_x^2(t-1)}$ of the perturbation is plotted in different ways. The snapshot plotted in panel (a) clearly reveals the presence of a localised structure; the same pattern plotted in logarithimic scale (see panel (b)) reveals a rough profile, while a stochastic-like structure emerges in the space-time plot of panel (c), where only those

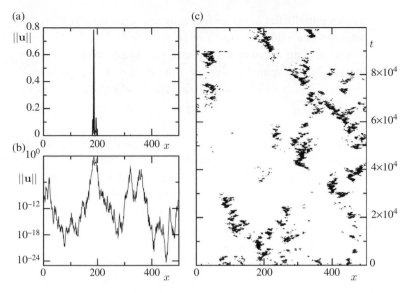

Fig. 11.1 Amplitude of the first Lyapunov vector in the chain of Hénon maps (11.1) with $a = 1.4, b = 0.3$ and $\varepsilon = 0.025$. The local amplitude of the perturbation is plotted in linear and logarithmic scales in panels (a) and (b), respectively. In panel (c) a space-time representation is plotted; a dot is plotted when the local amplitude is larger than 0.06.

points where the local amplitude of the Lyapunov vector is larger than some fixed threshold are plotted. Altogether, it is convenient to interpret

$$h(t, x) = \ln\|\mathbf{u}(t, x)\|, \tag{11.2}$$

as the height of a moving and roughening interface. In fact, the exponential growth of $\|\mathbf{u}(t, x)\|$ is equivalent to a linear growth of $h(t, x)$, i.e. to a finite average velocity of an interface. It is therefore natural to implement the arsenal of tools developed in the study of rough interfaces, starting from the selection of the interface square-width W^2

$$W^2(t, L) = \left\langle \overline{h^2(\mathbf{x}, t)} \right\rangle - \left\langle \overline{h(\mathbf{x}, t)}^2 \right\rangle, \tag{11.3}$$

as the most appropriate observable. The angular brackets denote an average over an ensemble of different trajectories, while the overline denotes a spatial average. Under the action of random forces (here due to the fluctuating degree of instability), the zero width of an initially flat interface is expected to grow according to the Family-Viczek scaling ansatz (Family and Vicsek, 1985)

$$W^2(t, L) = L^{2\alpha} \mathcal{F}(t/L^z), \tag{11.4}$$

where $\lim_{v \to \infty} \mathcal{F}(v) = \mathcal{O}(1)$. In other words, this equation implies that the asymptotic value of the square-width is $W^2(\infty, L) \approx L^{2\alpha}$, so that α measures the asymptotic roughness of the interface. The scaling ansatz (11.4) is completed by the assumption that $\mathcal{F}(v) \approx v^\beta$, for $v \ll 1$, where β measures the initial growth rate. Finally, since $t \approx L^z$ is the crossover time from the growing to the saturation regime, one finds that $z = 2\alpha/\beta$. This implies that an initially flat interface first roughens according to $W^2(t, L) \approx t^\beta$ and then eventually saturates at a value $W^2(\infty, L) \approx L^{2\alpha}$.

The scaling behaviour can be better understood in Fourier space. Let us start by introducing the spatial power spectrum of the interface profile:

$$S(t, k) = \left\langle \left| \int_0^L dx \, h(t, x) e^{2\pi i k x} \right|^2 \right\rangle.$$

Since one is eventually interested in long spatial scales, i.e. small k values, it is legitimate to assume that the spatial variable is continuous and thereby represent the Fourier transform as an integral, even if the underlying dynamics is defined on a discrete lattice. During the roughening process, $S(t, k)$ develops a divergence $k^{-2\alpha-1}$ which progressively extends to increasingly small wave numbers (with a lower cutoff $1/L_c(t)$ that corresponds to the inverse of a time-dependent correlation length).

As a result, W^2 can be approximated as

$$W^2(t, k) \approx \int_{1/L_c}^{1/L_m} dk S(k) \approx \frac{1}{k_c^{2\alpha}}, \tag{11.5}$$

where the wavelengths above $1/L_m$ can be neglected either because the system is discrete or because they are naturally cutoff (the interface is smooth below some scale), while those below $1/L_c$ can be neglected, since they have not yet been generated during the roughening process. One thus recovers Eq. (11.4) by simply assuming

$$L_c(t) \approx L\mathcal{F}(t/L^z),$$

so that we can interpret the saturation of $\mathcal{F}$ as a constraint on L_c, which cannot become larger than the actual system size L.

11.1.1 Relationship with the KPZ equation

Dimension one

Some light on the evolution of the first Lyapunov vector can be shed with the help of a simple model. It is reasonable to assume that the perturbation field $w(t, x) = \|\mathbf{u}(t, x)\|$ satisfies the stochastic differential equation

$$\frac{\partial w}{\partial t} = (a + \xi(t, x))w + \nu \frac{\partial^2 w}{\partial x^2}, \tag{11.6}$$

where ν gauges the strength of the spatial diffusion, a represents the average local exponential rate, and the white Gaussian multiplicative noise $\xi(t, x)$ ($\langle \xi(t', x')\xi(t, x) \rangle = 2D\delta(t - t')\delta(x - x')$) mimics a fluctuating (in space and time) local rate of instability. This simple equation contains the minimal ingredients that are expected to be present in a generic space-time chaotic dynamics; for a lump dynamics it reduces to the elementary stochastic model (8.24).

Upon introducing the change of variables (11.2) (which, in this context, corresponds to the so-called Hopf-Cole transformation (Hopf, 1950; Cole, 1951)), this stochastic equation is mapped onto the one-dimensional Kardar-Parisi-Zhang (KPZ) equation (Kardar et al.,

1986; Barabási and Stanley, 1995; Halpin-Healy and Zhang, 1995), also known as the noisy Burgers equation,

$$\frac{\partial h}{\partial t} = \nu h_{xx} + \nu h_x^2 + a + \xi(t, x), \tag{11.7}$$

widely used to describe roughening phenomena.

Notice that, here, the coefficients of the quadratic and diffusive terms are equal to one another. This is by no means a restriction, since all parameters of the KPZ equation can be scaled away in one spatial dimension.[1]

One does not expect Eq. (11.6) to reproduce exactly the behaviour of a generic spatially extended system; the effective noise is not typically white, additional derivatives may be present and the coupling might not even be diffusive. In spite of such relevant differences, numerical simulations suggest that the long-time, large-scale behaviour of many models (including the complex Ginzburg-Landau equation, chains of oscillators and symplectic maps and even differential-delay equations) is described by the KPZ equation (Pikovsky and Politi, 1998). The reason is that all such systems belong to the same broad *universality* class. The KPZ equation is the minimal stochastic model able to reproduce a scaling behaviour characterised by $\alpha = 1/2$ and $\beta = 2/3$ (i.e. $z = 3/2$) (exact renormalisation group calculations provide such results (Halpin-Healy and Zhang, 1995)). In all dynamical systems so far investigated, the sign of the quadratic term is positive; this means that the spatial coupling *increases* the degree of instability. Mathematically, this is presumably because the coefficients of both the quadratic and the diffusive terms originate from the same multiplicative noise term in the equation ruling the evolution of the perturbation (11.6) (see the derivation of Eq. (11.7)). Physically, this effect is the spatial version of the coupling sensitivity observed in weakly coupled systems (Section 9.1): if two neighbouring regions tend to grow with a different rate, the faster region tends to increase the growth rate of the slower one, rather than vice versa. From the point of view of the scaling behaviour, this is not, however, a restriction, as the sign of the quadratic term can be changed by mapping $h \rightarrow -h$.

The emergence of roughening is illustrated in Fig. 11.2, where the results for a chain of logistic maps (see Eq. (A.10)) are reported. The rescaled square-width $W^2(t)$ is plotted for different system sizes in panel (a). There we see that the convergence to the predicted behaviour is rather slow. Relevant deviations are still present for $N = 6400$. In panel (b), one can instead appreciate how the roughening process progressively affects long wavelengths: the low-frequency cutoff decreases until it reaches the minimal value, which corresponds to the maximal wavelength that can be supported by the chain length. Notice also that, since $\alpha = 1/2$, the power-law divergence is $1/k^2$, i.e. the profile of the interface corresponds to a spatial Brownian motion (because of the periodic boundary conditions, it is called Brownian bridge; see also Fig. 11.1).

Although the KPZ universality class is rather broad (it basically includes all cases where interactions are short-range and the space-time correlations of noise decay exponentially), there are a few exceptions where different exponents can be observed. One example is that

---

[1] In three dimensions and above, the story is more complicated – see the next sub-section.

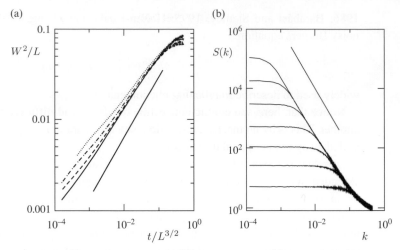

Fig. 11.2 (a) Growth of the square width for a chain of logistic maps (A.10) with $\varepsilon = 1/3$ and $a = 2$ and different lengths L: $800, 1600, 3200$ and 6400 (dotted, dotted-dashed, dashed and solid lines, respectively). (b) Interface growth in Fourier space: the power spectrum for $L = 8192$ is plotted at times $2, 8, 32, 128, 512, 2048$ and 8192 (from bottom to top).

of Hamiltonian models such as the FPU and Φ^4 chains, where the "interfaces" are rougher than predicted for the KPZ equation (Pikovsky and Politi, 2001), and it is not even clear whether the same dynamical exponent z as for KPZ holds. Since a reshuffling (in both time and space) of the local multipliers reinstates the standard KPZ exponents, it is natural to conjecture that the origin of the different scaling behaviour can be traced back to the presence of long-range correlations of the effective noise. In the FPU case, the presence of anomalous correlations is consistent with its anomalous (diverging) thermal conductivity, which can in turn be attributed to the spontaneous onset of a Levy-walk-type superdiffusive behaviour (Lepri et al., 2003). In the Φ^4 model, however, the origin of the long-range correlations is less clear, since this system is characterised by normal transport properties.

Higher dimensions

In higher dimensions, the correspondence between the largest Lyapunov vector and KPZ dynamics still holds. Simulations performed with a two-dimensional lattice of logistic maps (Pazó et al., 2013) are consistent with the best estimates of the KPZ critical exponents, i.e. $\alpha \approx 0.393(3)$, $z = 1.606(7)$ (Marinari et al., 2000). The Galilean invariance implies that $z + \alpha = 2$ in any dimension (Barabási and Stanley, 1995). In three spatial dimensions, the scenario is more complex: there exists a critical value of the noise amplitude (coupling strength) below which W^2 remains finite for $L \to \infty$, while above a "standard" roughening is observed with $\alpha = 0.313(5)$. It is important to notice that in a three-dimensional space, the KPZ equation must be complemented by an ultraviolet cutoff, to remove the unphysical contribution of very large wave numbers. No numerical studies of dynamical systems are available in three dimensions.

Finite-size corrections to the Lyapunov exponent

The relationship between the Lyapunov vector and the KPZ universality class allows the calculation of the finite-size corrections which affect the computation of the largest LE. As already mentioned, the LE is nothing but the average velocity of the corresponding "interface". The convergence of the velocity to its asymptotic value is controlled by the quadratic term in the KPZ equation. In fact, the interface velocity v is obtained by averaging Eq. (11.7),

$$v = a + v \left\langle h_x^2 \right\rangle.$$

The quadratic term expresses the fact that the velocity of a tilted interface is different (either larger or smaller, depending on the sign of v) from that of a flat one. In our context, this is equivalent to saying that the Lyapunov exponent of a perturbation with an exponential profile $e^{\mu x}$ is larger than the standard Lyapunov exponent (see the chronotopic approach in Chapter 10).

Before proceeding further, it is instructive to discuss the role of the norm adopted to compute the LE. In the context of spatially extended systems it is natural to introduce the generalised q-norm ($0 \leq q < \infty$),

$$M_q(t) = \left[\frac{1}{L} \int_0^L w^q(t, x) \, dx \right]^{1/q}.$$

In numerical simulations, the Euclidean norm is typically used, as it follows from a scalar product; it corresponds to $q = 2$. Sometimes, the maximum norm is also used; it corresponds to $q = \infty$. Within the interface interpretation, it is natural to define the Lyapunov exponent as the average velocity of the interface height (i.e. from the logarithm of the local amplitude w); it corresponds to $q = 0$. In any case, since all points along the interface move with the same average velocity (the interface width saturates), any norm yields the same result (as it should) in the infinite-time limit.

The fluctuations of the finite-time Lyapunov exponents Λ (see Chapter 7) instead depend substantially on q, as can be seen in Fig. 11.3, where the variance $\Delta \Lambda_q = \langle \Lambda_q^2 \rangle - \langle \Lambda_q \rangle^2$ (the sub-index q is used here to avoid confusion with the q-dependence of the generalised exponents) is plotted for different chain lengths. The behaviour of $\Delta \Lambda_q$ can be understood as follows. Since the dynamics of different regions are weakly correlated, so are the instantaneous growth rates (velocities). One thus expects that $\Delta \Lambda_q$ decreases with the system size (roughly as $1/L$). On the other hand, no decrease is expected when the maximum norm is used (i.e. for $q \to \infty$), as it involves the value in one spatial position (the maximum). The absence of fluctuations for $q = 0$ does not imply, however, that the limit $T \to \infty$ is not necessary. In fact, the contribution to the Lyapunov exponent arising from the quadratic term in the KPZ equation depends on the size of the system, as shown in Eq. (11.8).

Given a system of size L and a uniform initial infinitesimal perturbation, let us define the time-t Lyapunov exponent with reference to the 0-norm (from now on we skip the subindex) as

$$\Lambda(t, L) = \frac{\langle \log M_0(t) - \log M_0(0) \rangle}{t}.$$

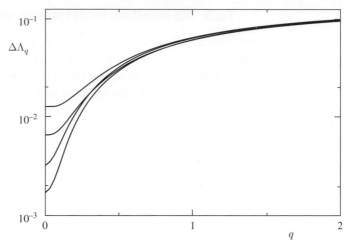

Variance of the instantaneous Lyapunov exponent for the same chain of logistic maps as in the previous figures. The various curves correspond to different lengths ($L = 256, 512, 1024$ and 2048, from top to bottom).

According to the theory of KPZ dynamics (Krug and Meakin, 1990; Halpin-Healy and Zhang, 1995), the following scaling relation holds:

$$\Lambda(t, L) - \lambda_\infty = L^{-1} f(tL^{-3/2}), \tag{11.8}$$

where $\lambda_\infty = \Lambda(\infty, \infty)$ is the largest Lyapunov exponent for an infinite system and $f(y)$ is a scaling function that approaches a finite value for large y and scales as $y^{-2/3}$ for small y. In practice, the r.h.s. is the inverse of the square root of the r.h.s. in Eq. (11.5). This can be understood by going in Fourier space. In fact,

$$\lambda_\infty - \Lambda(t, L) \approx \int dx[(h_x)_\infty^2 - (h_x)_L^2] \approx \int_0^{1/L_c} dk S(t, k)k^2 \approx 1/L_c.$$

The relation (11.8) has several consequences. First, it gives the leading correction to the largest Lyapunov exponent,

$$\lambda(\infty, L) - \lambda_\infty \sim L^{-1}.$$

This formula can be used to improve the estimate of the asympototic Lyapunov exponent from finite-size calculations. Moreover, it reveals that the transient length T_{tr} needed to reach the statistically stationary state of the Lyapunov vector scales with the system size is $T_{tr} \sim L^{3/2}$. Finally, since for "short" times

$$\lambda(t, \infty) - \lambda_\infty \sim t^{-2/3},$$

one could estimate the asymptotic Lyapunov exponent by linearly extrapolating $\lambda(t, L)$ vs. $t^{-2/3}$ for large L, as can be appreciated from the bottom curve in Fig. 11.4.

We finally compare the finite-size corrections emerging for different values of q. In Fig. 11.4 we have reported the finite-time Lyapunov exponent obtained for $q = 0, 0.5, 1$ and 2 (bottom to top). There, we see that while the convergence for $q = 0$ is from below, it

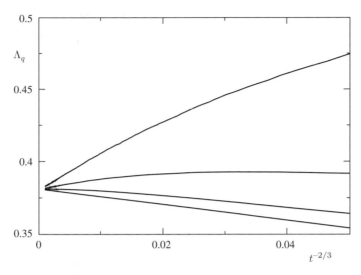

Fig. 11.4 Finite-time Lyapunov exponent for a chain of logistic maps of length 8192. From bottom to top the curves refer to $q = 0, 0.5, 1$ and 2.

is from above for both $q = 1$ and 2. The value of $q = 0.5$ is nearly optimal. There is, however, no theory stating whether the optimal q value is independent of the model.

11.1.2 The bulk of the spectrum

In this section the roughening interface interpretation is extended to the whole Lyapunov spectrum. Given the covariant Lyapunov vector $\mathbf{u}_\rho(t, x)$ for a generic density ρ ($\rho = 0$ corresponds to the maximum LE), the starting point is again the definition of its square-width from the interfacial representation (11.2). This straightforward definition (11.3), however, poses a problem: one cannot start from a constant initial condition since flat vectors converge towards the first Lyapunov vector. The problem can be circumvented by subtracting the initial shape of the vector, i.e. by introducing the following alternative definition:

$$W_\rho^2(t_f - t_i, L) = \left\langle \overline{((h(\mathbf{x}, t_f) - h(\mathbf{x}, t_i))^2)} \right\rangle - \left\langle \overline{(h(\mathbf{x}, t_f) - h(\mathbf{x}, t_i))}^2 \right\rangle.$$

$W_\rho^2(t)$ is expected to behave as W^2 (the reader can easily check the statement in the case of the first vector, when both definitions can be implemented), the only difference being that W_0^2 saturates, at large times, to a value that is twice the saturation value of W^2 (whenever the latter can be defined).[2]

Since no stochastic differential equation has yet been identified that is expected to reproduce the observed phenomenology, one can rely only on numerical simulations, which are unfortunately quite computationally demanding. For a chain of coupled logistic maps, the scaling behaviour of the interface corresponding to $\rho = 0.195$ is shown in Fig. 11.5.

---

[2] Practically speaking, since the covariant Lyapunov vectors are generated by iterating backwards in the tangent space, it is more convenient to increase $t = t_f - t_i$ by keeping the final time t_f fixed and letting the initial time t_i decrease.

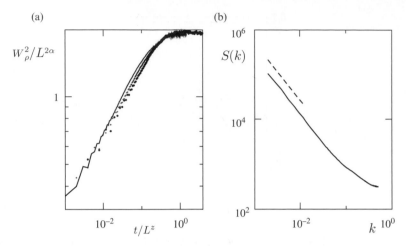

Fig. 11.5 (a) Growth of the square-width of a Lyapunov vector in the bulk for $\rho = 0.195$ and different system sizes for a chain of logistic maps as in the previous section. Circles, plusses and the solid curve correspond to $L = 256, 512$ and 1024, respectively ($z = 1, 2\alpha = 0.28$). (b) Interface structure in Fourier space: the asymptotic power spectrum for $L = 512$ (same ρ value as in panel (a)). The slope of the dashed line is -1.28.

From both the asymptotic value of the square width W_ρ^2 in panel (a) and its power spectrum in panel (b), it is evident that the interface is much smoother than in the KPZ case. The numerical data suggest that $2\alpha \approx 0.28$, but the data collapse is not very convincing. More refined simulations of various models (Pazó et al., 2013) indeed suggest that $z = 1$ and $2\alpha = \beta = 0.15$. The lack of a reference stochastic model (such as the KPZ equation for the first vector), however, prevents drawing definite conclusions. In particular, the small value of α is rather puzzling; one cannot entirely exclude the hypothesis that it is a finite-size effect and that the asymptotic spectrum is of $1/k$ ($1/f$) type (see also Section 11.3.2). It appears anyway that the scaling behaviour in the bulk of the spectrum is independent of the density ρ, at least in a finite range of densities.

A different behaviour is expected around the zero crossing of the Lyapunov spectrum, where symmetries and conservation laws come into play (see also the last section of this chapter) and close to band edges such as in a chain of Hénon maps (Kuptsov and Politi, 2011).

A more generic property is instead the appearance of *straight* (non-rough) Lyapunov vectors deeply in the region of negative exponents of spatially continuous models. Here, we illustrate the phenomenon with reference to the Kuramoto-Sivashinsky equation (A.19), simulated with a length $L = 96$ and periodic boundary conditions (Takeuchi et al., 2011b). In this case, there are ten positive Lyapunov exponents. As can be seen in Fig. 11.6, the Lyapunov vectors that correspond to sufficiently negative Lyapunov exponents have an approximately sinusoidal shape (their Fourier spectra are highly peaked around increasingly large wave numbers k), i.e. their width does not increase with the system size. This behaviour can be understood by analysing Eq. (A.19) in the absence of nonlinearities. The resulting equation can be easily solved in Fourier space,

$$\dot{u}_k = (k^2 - k^4)u_k.$$

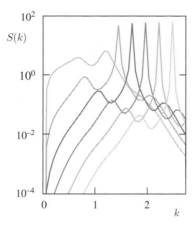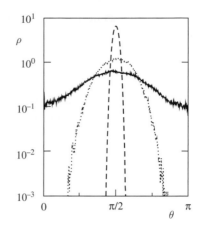

Fig. 11.6   (a) Spatial Fourier spectrum of the Lyapunov vectors, 32, 44, 52, 60, 68 and 76 (from left to right) in the Kuramoto-Sivashinsky equation. (b) Distribution of the angles between three pairs of Lyapunov vectors (39, 40) (solid line), (43, 44) (dashed), (20, 60) (dotted). (Data: courtesy of K. A. Takeuchi et al.)

For large enough k the quartic term dominates, ensuring the stability of the corresponding Fourier mode. The numerical study of the full Kuramoto-Sivashinsky equation is thus telling us that high-order covariant Lyapunov vectors are not significantly affected by the nonlinear fluctuations. This could be expected on the basis of inertial manifold theory, which predicts that the chaotic dynamics of a finite system is embedded into a finite-dimensional smooth manifold (the inertial manifold), all other degrees of freedom being slaved. The theory is, however, still unable to predict the dimensionality of the manifold and its structure.

The covariant Lyapunov vectors allow identification of the dynamically irrelevant directions as those that are everywhere transverse to the attractor. In fact, by measuring the angles between pairs of consecutive vectors, it is found that there exists a critical index i_c ($i_c = 40$ in the Kuramoto-Sivashinsky equation), below which they can assume arbitrarily small values, while above they are bounded away from zero. Furthermore any vector below i_c is transversal to any vector above i_c (Takeuchi et al., 2011b)[3].

It would be interesting to know if, in the thermodynamic limit, the onset of transversality configures itself as a phase transition from rough to straight Lyapunov vectors.

11.2 Localisation of the Lyapunov vectors and coupling sensitivity

Another way of analysing the structure of the Lyapunov vectors is by investigating their localisation properties. This can be done by transforming the local perturbation amplitude $\mathbf{u}_x$ into a pseudo-probability p_x (typically, defining $p_x \equiv \|\mathbf{u}_x\|^2$ and using the fact that the

---

[3] For the sake of correctness, one should take into account that above i_c, LEs come in pairs, as if they were sine and cosine modes. Such modes cannot be mutually discriminated.

Euclidean norm of $\mathbf{u}$ is equal to 1) and thereby defining an effective occupation number N_q as

$$N_q = \exp\left\{\frac{-\log \sum_x |\mathbf{u}_x|^{2q}}{q-1}\right\},$$

where the expression in curly brackets is nothing else but the order-q Renyi entropy. N_q, which, by definition, ranges in the interval $[1, N]$, expresses the number of sites where the perturbation is effectively different from zero. In general, it is sufficient to consider a single q value (different q values correspond to giving different weights to the tail of the distribution of the pseudo-probabilities). Two choices are typically encountered in the literature: (i) $q = 1$, in which case N_1 corresponds to exponential of the Shannon entropy, and (ii) $q = 2$, much used in the context of Anderson localisation, where $Y_2 = 1/N_2$ is also named inverse participation ratio.

The localisation of a vector is assessed by looking at the scaling behaviour of N_q with the system size; a vector is said to be localised if N_q remains finite for $L \to \infty$, while it is extended if $N_q \propto L$. Finally, a vector is considered to be weakly localised (or critical) if $N_q \approx L^\gamma$ with $\gamma < 1$. In the case of one-dimensional linear systems with quenched disorder, all eigenfunctions (the equivalent of the covariant Lyapunov vectors) are typically exponentially localised (see, e.g., the Anderson problem discussed in Chapter 12). In the case of the Lyapunov vectors discussed in the previous sections, $\|\mathbf{u}_x\| \approx \exp(-x^\delta)$, where x is the distance from the maximum and $\delta = 1/2$ for the first Lyapunov vector, while $\delta \approx 0.15$ in the bulk of the spectrum. This implies that such vectors are also localised. On the other hand, the sinusoidal "spurious" Lyapunov vectors that seem to characterise the stable modes which do not contribute in any way to the asymptotic dynamics are extended.

Another context where localisation emerges (as a result of a different mechanism) is that of globally coupled systems, where it contributes to determining the value of the largest Lyapunov exponent in a highly nontrivial way. Here, we discuss the phenomenon with reference to the simple setup of one-dimensional maps (see Eqs. (A.21, A.22, A.23)) in the absence of a collective dynamics (i.e. when the mean field $\overline{U}$ is constant). In such a case, if one assumes that $N = \infty$, each map evolves independently according to the evolution rule $F[(1 - \varepsilon)U + \varepsilon\overline{U}]$, where $\overline{U}$ has to be determined self-consistently. Thus, if one first takes the limit $N \to \infty$, all Lyapunov exponents are expected to be equal to one another and equal to the exponent λ_0 of such a single map (see also the discussion in Chapter 10). Simulations of large ensembles indeed confirm the existence of an almost flat spectrum but reveal also that the first exponents are larger than expected.

The tangent-space dynamics of the globally coupled maps is written as

$$u_j(t+1) = \mu_j(t)\left[u_j(t) + \frac{\varepsilon}{(1-\varepsilon)N}\sum_i u_i(t)\right], \tag{11.9}$$

where $\mu_j(t) = (1 - \varepsilon)F'(V_j(t))$. For the sake of simplicity, we assume that the terms $\ln|\mu_j(t)|$ denote independent, identical random processes with average $\langle\ln|\mu_j(t)|\rangle = \lambda_0$ and diffusion coefficient D.

Under the additional assumption (to be verified a posteriori) that the vector $u_j(t)$ is localised, the coupling term can be approximated by its maximum addendum $\varepsilon u_M(t)/N$.

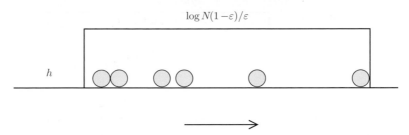

$$\log N(1-\varepsilon)/\varepsilon$$

h

Fig. 11.7 Schematic representation of the tangent space dynamics in a globally coupled ensemble of identical units in a logarithmic scale.

In spite of the prefactor $1/N$, which is increasingly small in the thermodynamic limit, the effect of coupling cannot be neglected. The problem is better understood by representing the tangent-space dynamics in a logarithmic scale, i.e. by introducing $h_i = \ln |u_i|$. As long as the positions h_i do not significantly differ from each other, the system can be viewed as a set of particles drifting with an average velocity λ_0 and freely diffusing. As soon as the distance of the ith particle from the rightmost one becomes larger than $\ln[N(1-\varepsilon)/\varepsilon]$ (i.e. $u_M/u_i > (1-\varepsilon)N/\varepsilon$), the coupling term in Eq. (11.9) becomes suddenly predominant. Its action can be schematised as a sudden force, which prevents the distance from becoming larger than the maximum allowed value. In other words, as depicted in Fig. 11.7, the various particles are confined within a box of width $\log[N(1-\varepsilon)/\varepsilon]$, whose right edge corresponds to the position of the rightmost particle. The left edge of the box acts as a barrier, which prevents the particles from diffusing away from the box itself. As a result of the action of the (left) barrier, the average velocity of the box (i.e. the Lyapunov exponent of the entire system) becomes larger than λ_0. This phenomenon is an extreme form of the coupling sensitivity discussed in Section 9.1.

A quantitative analysis can be carried out from the evolution equation for the distribution $P(s, t)$ of particles in a frame moving with a velocity λ' ($s = h - \lambda' t$). Under the previous assumptions, $P(s, t)$ satisfies the Fokker-Planck equation

$$\frac{\partial P}{\partial t} = \frac{\partial}{\partial s}\Delta\lambda P + \frac{D}{2}\frac{\partial^2 P}{\partial s^2},$$

where $\Delta\lambda = \lambda' - \lambda_0$. If λ' coincides with the (yet unknown) velocity of the box, we can set the left edge in $s = 0$ and thereby assume that $P(s, t) = 0$ for $s < 0$. As shown by Takeuchi et al. (2011a), this equation can be solved with a self-consistent argument. The stationary solution is $P(s) = (2\Delta\lambda/D)\exp(-(2\Delta\lambda s/D))$. Therefore, given N particles distributed according to $P(s)$, the typical position s_{max} of the rightmost particle is given by the condition $P_>(s_{max}) = 1$, where $P_>(s)$ is the probability to find at least one out of N particles above s,

$$P_>(s) = \int_s^\infty NP(s)ds = N\exp(-(2\Delta\lambda s/D)).$$

By then imposing that s_{max} is equal to the box size, $s_{max} = \ln[N(1-\varepsilon)/\varepsilon]$, one obtains an equation for $\Delta\lambda$, whose solution is

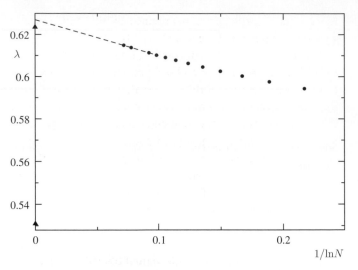

Fig. 11.8 Largest Lyapunov exponent in an ensemble of globally coupled maps of the type (A.21, A.23, A.22), where $F(V) = bV [F(V) = b(1 - V)/(b - 1)]$ if $V \leq 1/b$ [$1/b < V \leq 1$], with $\varepsilon = 0.02$ and $b = 4$. The two triangles along the vertical axis correspond to the Lyapunov exponent of a single map (the lower point) and the extrapolated largest Lyapunov exponent according to Eq. (11.10).

$$\Delta\lambda \approx \frac{D}{2}\left[1 - \frac{\log(1 - \varepsilon)/\varepsilon}{\log N}\right]. \qquad (11.10)$$

As a result, the increase of the Lyapunov exponent remains finite even in the thermodynamic limit, although the effective coupling strength eventually vanishes. In other words, the limits $N \to \infty$ and $t \to \infty$ do not commute. By replacing the value of $\Delta\lambda$ into the expression of the probability distribution, we find that $P(s) = \exp(-s)$, which implies that the distribution $Q(u)$ of the amplitudes $u = \exp(s)$ scales as $Q(u) \approx 1/u^2$. This means that the Lyapunov vector is weakly localised (the integral of $Q(u)u$ converges because of the corrections to the decay exponent of $Q(u)$).

In Fig. 11.8 the largest Lyapunov exponent is plotted for an ensemble of skew tent maps for different values of N.[4] The lower triangle corresponds to the Lyapunov exponent $\lambda_0 = 0.530(2)$ of the single map $F[(1 - \varepsilon)U + \varepsilon\overline{U}]$, for the mean field $\overline{U} = 0.511(0)$, while the upper triangle corresponds to the the asymptotic value as from Eq. (11.10). The theoretical prediction is close to the value extrapolated from the numerical simulations. The only point of disagreement with the theory is the coefficient of the $(1/\ln N)$ correction, which is smaller than predicted. The difference is possibly due to: (i) the weak localisation of the perturbation vector, so that the coupling strength is stronger than assumed by the theory, and (ii) the fluctuating velocity of the "box": no matter how large N, is the position of the box depends on the highly fluctuating position of the rightmost particle.

In spite of the underlying approximations, the prediction Eq. (11.10) appears to be rather accurate: simulations performed with various models confirm that the increase of

---

[4]   Skewness is a necessary condition for the observation of coupling sensitivity since it ensures the presence of fluctuating multipliers and, thereby, of a finite variance D.

the Lyapunov exponent is determined by the diffusion coefficient of the single-system Lyapunov exponent (Takeuchi et al., 2011a).

The most extreme form of coupling sensitivity can be found in the so-called Hamiltonian mean-field model (A.29). Although the dynamics reduces, in the thermodynamic limit, to a collection of periodic orbits (with different periods), a strictly positive largest Lyapunov exponent is found. The origin can be traced back to the existence, for any N, of sufficiently many oscillators that stay long enough and close enough to the unstable direction of the saddle point of the self-generated poential to be able to pull the "box" (see Ginelli et al. (2011) for a detailed discussion of the problem). Coupling sensitivity also plays a major role in determining the scaling behaviour of the largest LE in the Kuramoto model that we encountered in Section 10.4.3. A quantitative theory is, however, still lacking.

11.3 Macroscopic dynamics

Besides microscopic chaos, high-dimensional systems may exhibit a nontrivial collective dynamics, where not only the single variables fluctuate irregularly in time, but also the coarse-grained ones do so. The simplest setup where this phenomenon manifests itself is that of mean-field models, where each element interacts with all the others in an equal manner, independently of the physical distance (the limit case of long-range interactions). Two examples are briefly illustrated in Fig. 11.9. In panel (a), the recursive evolution of the mean field is plotted for an ensemble of globally coupled maps (see Eq. (A.21) for a definition of the model) with $F(V) = a(1/2 - |1/2 - V|)$, $a = 1.7$ and $\varepsilon = 0.3$. From the figure, one can appreciate the existence of a two-band seemingly chaotic dynamics (see also the inset where the attractor has been suitably tilted and enlarged to emphasise the existence of a non-trivial fine-grain structure). Fig. 11.9b instead portrays the temporal evolution of the mean field for an ensemble of Stuart-Landau oscillators (see Eq. (A.27)) with $c_1 = -2$, $c_2 = 3$ and $K = 0.47$; the presence of an irregular dynamics is transparent.

In the thermodynamic limit, the problem of characterising the collective motion can be addressed by introducing the instantaneous density $\rho(\mathbf{U}, t)$ of dynamical units whose variable lies in the infinitesimal region spanned by $d\mathbf{U}$. In the case of maps of the interval, the variable $\mathbf{U}$ is a scalar quantity and the density ρ satisfies the self-consistent Perron-Frobenius equation

$$\rho(U, t+1) = \sum_j \frac{\rho(F_j^{-1}((1-\varepsilon)U + \overline{U}, t), t)}{|F_j'|}. \tag{11.11}$$

This equation is nonlinear, since the mean field $\overline{U}(t)$ is a function of $\rho(t)$; in the $N \to \infty$ limit it is written as

$$\overline{U} = \int U\rho(U, t)dU. \tag{11.12}$$

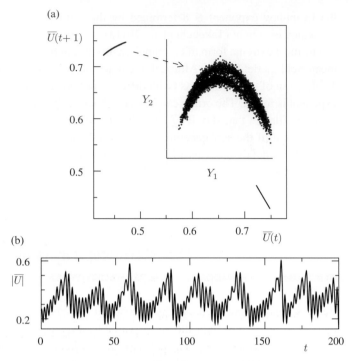

(a) Recursive plot of the mean-field dynamics for 10^5 globally coupled tent maps with $a = 1.7$ and $\varepsilon = 0.3$, after a sufficiently long transient. In the inset the variables $Y_1 = \overline{U}(t) + 0.6\overline{U}(t+1)$, $Y_2 = \overline{U}(t+1) - 0.6\overline{U}(t)$ are used to better resolve the chaotic structure of the upper band (obtained for an ensemble of 10^9 maps). (b) Behaviour of the mean field of an ensemble of 10^7 Stuart-Landau oscillators for the parameter values reported in the text. (Data: courtesy of Takeuchi et al.).

In the case of the Stuart-Landau oscillators, the collective motion is described by the nonlinear Liouville equation

$$\frac{\partial \rho}{\partial t} = -\frac{\partial}{\partial U}\left[(U - (1 + ic_2)|U|^2 U + K(1 + ic_1)(\overline{U} - U)\right]\rho, \qquad (11.13)$$

where the definition (11.12) of $\overline{U}$ still holds with the proviso that U is now a complex variable.

The computation of the leading Lyapunov exponents of the collective motion is far from trivial: Eqs. (11.11, 11.13) are functional equations, i.e. the phase space is infinite-dimensional. Moreover, additional difficulties may further complicate the task, such as the presence of singularities in the distribution and even the fractal structure of the support itself.

Whenever the probability distribution is sufficiently smooth, it can be projected onto a suitable functional basis and thereby identify a finite (although possibly large) number of effective modes. If the variable of the single map is phase-like (i.e. $U + 2\pi \equiv U$), it is convenient to expand ρ into Fourier modes, $\rho(U, t) = \sum_j \psi_k(t)e^{ikU}$. The corresponding

Perron-Frobenius equation is

$$\psi_k(t+1) = \sum_m R(k,m)\psi_m(t),$$

where

$$R(k,m) \equiv \frac{1}{2\pi}\int_0^{2\pi} dU e^{i(mU-kF(U))}.$$

This approach is quite effective in the case of the model (A.24), where the mean field $\overline{U}$ is defined by Eq. (A.25). More specifically,

$$R(k,m) \equiv \frac{1}{2\pi}\int_0^{2\pi} dU e^{i[(m-2k)U-2k\varepsilon(1-2a^2\overline{U}^2)\sin U]} = J_{m-2k}[2k\varepsilon(1-2a^2\overline{U}^2)],$$

where J_α denotes the Bessel function of order α. In the present case, $\overline{U} = \psi(1)$ and all ψ_k are real. As a result, the Perron-Frobenius operator is

$$\psi_k(t+1) = 2k\varepsilon \sin(\psi_1(t))\sum_m J_{m-2k}\psi_m(t).$$

Upon iterating the aforementioned equations together with their linearisations, one can easily compute the Lyapunov spectrum: the first 10 exponents are plotted in Fig. 11.10. From the figure we infer that the collective motion is low-dimensional: only the first mode is positive and the expansion is already compensated by the second exponent ($\lambda_1 + \lambda_2 < 0$). Moreover, the Lyapunov spectrum is discrete; upon increasing the number of Fourier modes, additional negative exponents add up, which correspond to increasingly stable, high-frequency modes. In practice, the scenario is reminiscent of the continuum-limit in spatially extended chaos. This is not, however, always the case. Coupled logistic maps appear to be characterised by an infinite-dimensional collective dynamics; this has been conjectured by adding noise to the microscopic dynamics (in order to smooth out

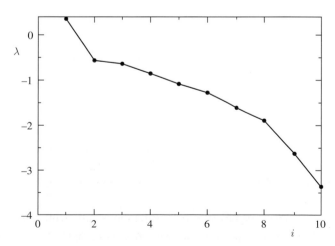

Fig. 11.10 Collective Lyapunov spectrum for the model of globally coupled maps (A.24, A.25) for $a = 2.2$ and $\varepsilon = -1.15$.

the singularities of ρ) and thereby decreasing the noise amplitude (Shibata et al., 1999; Takeuchi and Chaté, 2013). As a result, it has been found that the dimension increases logarithmically with the decreasing of the noise amplitude.

11.3.1 From micro to macro

Since any collective dynamics is ultimately the consequence of microscopic dynamical rules, it is legitimate to ask whether the "microscopic" Lyapunov exponents are of any usefulness for its characterisation.

One can approach the problem with the help of finite-amplitude exponents. We illustrate the idea by referring again to the ensemble of tent maps shown in Fig. 11.9. This model has been studied by Cencini et al. (1999). Given a generic macroscopic configuration $\{U_j\}$ (selected according to the invariant measure), a second initial condition has been generated, by adding the same perturbation δ to each component and thereby determining the finite amplitude exponent $\ell(\Delta)$ as described in Chapter 7. The resulting behaviour is plotted in Fig. 11.11 for four different system sizes after rescaling the perturbation amplitude by $\sqrt{N}$. There, one can notice the presence of two plateaus: the first one, which occurs on small scales, corresponds to the microscopic maximal Lyapunov exponent: over such scales, the evolution of finite perturbations is correctly described by the linearised equations. The lower plateau corresponds to the macroscopic Lyapunov exponent (the one obtained by iterating the Perron-Frobenius operator); the size of the plateau increases with N progressively extending to smaller scales. In fact, the data collapse reveals that it starts at a resolution of order $1/\sqrt{N}$, the amplitude of statistical fluctuations. In practice, the microscopic dynamics seems to contribute to what, macroscopically, appears as internal noise.

One is therefore tempted to conclude that microscopic and macroscopic worlds are mutually exclusive. There are, however, examples where the spectral properties of

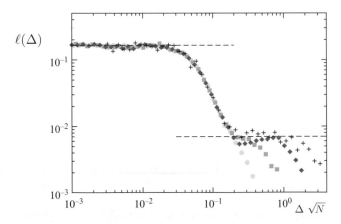

Fig. 11.11 (a) Finite amplitude Lyapunov exponent for coupled tent maps ($f(U) = a(1 - |1/2 - x|)$) with $a = 1.7$ and $\varepsilon = 0.3$. Circles, squares, diamonds and plusses refer to $N = 10^4, 10^5, 10^6$ and 10^7, respectively. (Data: courtesy of M. Cencini et al.)

the macroscopic dynamics extend down to the infinitesimal scales that are typical of microscopic dynamics. One example is that of the splay states, which emerge in pulse-coupled oscillators (see Chapter 10). In fact, in some models, it has been proven that the leading eigenvalues of the corresponding Liouville-type operators coincide with the largest exponents as determined by following the microscopic dynamics (Olmi et al., 2012): there exists a crossover, beyond which the microsocpic Lyapunov exponents perceive the finiteness of the system and are therefore different in the two setups. Such a striking, almost one-to-one, correspondence is not too surprising, as it holds in the absence of microscopic chaos, which creates a sort of "barrier" of statistical fluctuations, below which the macroscopic analysis is no longer valid.

It is therefore legitimate to ask when and whether the validity of the microscopic linear equations can extend up to those scales, where a perturbative analysis of the macroscopic deviations can be carried out, and vice-versa. A crucial prerequisite for a microscopic Lyapunov exponent to be part of the macroscopic spectrum is that the "direction" of the corresponding covariant vector be oriented as expected for the "modes" of the Perron-Frobenius operator. In particular, this implies that the microscopic vector should be extended rather than localised.

Preliminary studies of Stuart-Landau oscillators (Takeuchi and Chaté, 2013) suggest that the microscopic Lyapunov spectra contain a signature of collective properties, in the vicinity of the largest Lyapunov vector. From the data in Fig. 11.12, one can in fact see that the distribution of the inverse participation ratio Y_2 shrinks to zero upon increasing the system size, and, from the inset, we can even see that it asymptotically scales as $1/N$, suggesting that the corresponding Lyapunov vector is eventually extended.

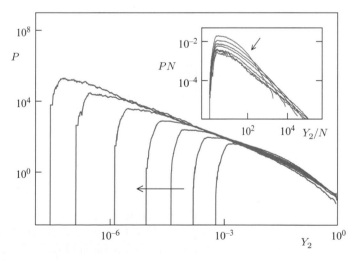

Fig. 11.12   Distribution of the inverse participation ratio for the first Lyapunov vector in the globally coupled ensemble of Stuart-Landau oscillators. From right to left, the curves refer to $N = 2048, 8192, 32{,}768, 131{,}072, 10^6$, 10^7 and 5×10^7. In the inset, the distribution is rescaled to emphasise its scaling with N. (Data: courtesy of K. A. Takeuchi et al.)

11.3.2　Hydrodynamic Lyapunov modes

The use of linear stability properties of the microscopic equations to infer macroscopic properties is particularly challenging in spatially extended systems, where collective dynamics manifests itself as a long-wavelength motion. In such setups, at variance with mean-field models, the study of Liouville-type functional equations is a prohibitive task, so that it is highly desirable to extract as much information as possible directly from the microscopic equations.

As in the previous section, the main working hypothesis is the interpretation of the extended nature of the covariant Lyapunov vectors as the signature of the collective dynamics, i.e. the identification of degrees of freedom that correspond to a motion of the system as a whole. The study of space-time chaos (e.g. in the Kuramoto-Sivashinsky equation) has revealed that the relaxation dynamics can be partly interpreted as a collective process. As mentioned in Section 11.1, beyond a critical index (which depends on the system size), the Lyapunov vectors are close to Fourier modes and, thereby, extended (Takeuchi et al., 2011b). The extensivity of such vectors is, however, not so relevant, since it refers to degrees of freedom that are basically inactive (they characterise the relaxation towards the manifold which contains the attractor, rather than the asymptotic dynamics itself).

A more interesting possibility is suggested by the mapping between the first Lyapunov vector and the KPZ dynamics of rough interfaces. It is, in fact, known that in more than two dimensions interfaces need not be rough. Would the potential extensivity of the logarithm of a Lyapunov vector be a sufficient condition implying the existence of some collective motion? This is an interesting question that has not yet been explored.

In general, the possibility to extract information on large-scale behaviour from the dynamics of Lyapunov vectors was first explored in Hamiltonian models. While studying a fluid of hard disks, Posch and Hirschl (2000) noticed that the Lyapunov vectors corresponding to the smallest (in the absolute sense) Lyapunov exponents – notably those obtained with the orthogonalisation procedure – correspond to collective perturbations. McNamara and Mareschal (2001) went further, by developing a kinetic theory approach to characterise such vectors that were named *hydrodynamic Lyapunov modes*.

At variance with the interface representation discussed in Section 11.1, where the logarithm of the amplitude is considered, hydrodynamic Lyapunov modes are better characterised by looking directly at the vector amplitude, i.e. by introducing the structure function

$$\mathcal{S}(t,k) = \left\langle \left| \int_0^L dx u(t,x) e^{2\pi i k x} \right|^2 \right\rangle.$$

$\mathcal{S}(t,k)$ undoubtedly reveals the existence of long-wavelength modes for the vectors associated with nearly vanishing LEs. In (Ginelli et al., 2007) it has even been noted that in three symplectic lattice systems such covariant Lyapunov vectors exhibit a $1/f$ (spatial) power spectrum; Yang and Radons (2013) have found that hydrodynamic Lyapunov modes emerge in dissipative systems as well, provided that some continuous symmmetry is present. A clear connection between the tangent-space dynamics and physical properties such as transport coefficients is, however, still lacking.

11.4 Fluctuations of the Lyapunov exponents in space-time chaos

In Chapter 5 we have seen that finite-time Lyapunov exponents do fluctuate, and their fluctuations can be quantified by introducing a suitable large-deviation function. In this section we analyse the dependence of such fluctuations on the system size in spatially extended systems. This is done by determining the diffusion matrix D (see Eq. (5.12)). This approach proves to be an alternative method to understand the extensivity of the underlying dynamics. In the following we shall always refer to the $q = 0$ norm that is naturally associated with the interface-interpretation of the Lyapunov vectors. This makes no difference for the identification of the elements of D, while different scaling functions (see later in this section) are expected for different q-values (Pazó et al., 2013).

For the sake of simplicity, we refer to a chain of L Hénon maps (A.11) with $a = 1.4$, $b = 0.3$, $\varepsilon = 0.025$ and periodic boundary conditions. For such parameter values, the Lyapunov spectrum is composed of two bands of positive and negative exponents, respectively. Since the Jacobian matrix J satisfies the symplectic-like condition $\mathsf{JAJ}^{\mathsf{T}} = -b\mathsf{A}$, with A being a generic antisymmetric matrix, one can conclude that Lyapunov exponents come in pairs, such that $\lambda_i + \lambda_{N+1-i} = \ln b$. The numerical results are shown in Fig. 11.13. In panel (a) we report the self-diffusion coefficients D_{ii}. The clean overlap of the scaled curves indicates that $D_{ii}(\rho) \approx 1/L^\gamma$ with $\gamma \approx 0.85$. This means that the LE fluctuations vanish in the thermodynamic limit, i.e. the Lyapunov exponents self-average.

In Fig. 11.13b, D_{ij} is plotted along the column $j = 2L/5$ (i.e. for $\rho = 2/5$). The off-diagonal terms decrease as $1/L$, so that the matrix D becomes increasingly diagonal

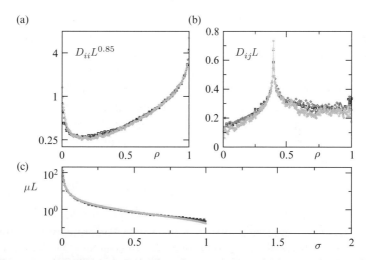

Diffusion coefficients in a chain of Hénon maps. In all panels, squares, diamonds, plusses and circles refer to $L = 40, 80, 160$ and 320, respectively. The results have been obtained by iterating the chain over 5×10^6 time steps. Panel (a) contains the diagonal elements $-\rho = (i - 1/2)/L$; panel (b) refers to the column $j = 2L/5$, $\rho = i/L$ and panel (c) refers to the eigenvalues of D, ordered from the largest to the smallest one $-\sigma = k/L$.

in the thermodynamic limit. Nevertheless, the contribution of the off-diagonal terms is not negligible. As a consequence of the aforementioned symplectic-like structure of J, it turns out that half of the eigenvalue spectrum of D is equal to zero. Moreover, the amplitude of the non-zero eigenvalues decreases as $1/L$ with the system size rather than as $L^{-\gamma}$, thus implying that the eigenvalues of $\mathsf{Q} = \mathsf{D}^{-1}$ (see also Eq. (5.11)) are proportional to L (see Fig. 11.13c), i.e. that the large deviation function S is an extensive observable. This behaviour is confirmed by the study of other models, such as symplectic maps and chains of Stuart-Landau oscillators (Kuptsov and Politi, 2011). Actually, the extensivity of S has a deeper meaning than usually expected for spatially extended systems. Consider, for instance, a chain of uncoupled chaotic dynamical systems. The system is trivially extensive in the sense that its fractal dimension and dynamical entropies are proportional to the system size. Because of the lack of coupling, however, the diffusion matrix D is (block) diagonal so that all its eigenvalues remain finite in the thermodynamic limit.

As soon as coupling is switched on, only the leading eigenvalue μ_1 remains finite, while all the others scale as $1/L$. The different scaling exhibited by μ_1 manifests itself as the $1/\sigma$ singularity that can be seen in Fig. 11.13c. Some implications of this phenomenon are discussed at the end of this section, in the context of dimension variability.

The scaling behaviour of the diagonal elements of D can be connected with the roughening of the corresponding Lyapunov vectors. It is convenient to define the diagonal elements of the diffusion matrix using the interface language. We start by introducing the variance

$$\chi^2 = \langle (\bar{h}(t))^2 \rangle - \langle \bar{h} \rangle^2, \tag{11.14}$$

where $h(t)$ is the logarithm of the amplitude of the ith covariant Lyapunov vectors, and, for the sake of simplicity, we drop the Lyapunov index i; finally $\bar{h}(t)$ is nothing but λt. The diffusion coefficient is

$$D = \lim_{t \to \infty} \frac{\chi^2}{t}.$$

By comparing Eq. (11.14) with the definition (11.3) of the square width, one can see a rather similar structure, although a different asymptotic behaviour is expected for the two variances: χ^2 diverges linearly in time, while W^2 saturates. In any case, the two quantities being dimensionally equivalent, it is reasonable to expect that they exhibit the same scaling behaviour, namely

$$\chi^2(t, L) = L^{2\alpha - z} \mathcal{G}(t/L^z) t,$$

where the factor t is made explicit to emphasise the asymptotic linear growth of χ^2 (i.e. $\mathcal{G}(u)$ converges to a constant for $u \to \infty$). Accordingly, the diffusion coefficient is

$$D(L) = L^{2\alpha - z} \mathcal{G}(\infty),$$

so that the aforementioned exponent γ can be expressed in terms of the interface roughening exponent,

$$\gamma = z - 2\alpha.$$

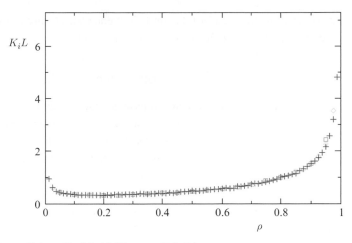

Fig. 11.14 Rescaled diffusion coefficients K_i of the LE differences $\delta\lambda$ in Hénon maps.

This relationship accounts for the anomalous scaling of the diffusion constant exhibited by the first Lyapunov vector. Since this vector is characterised by the KPZ dynamics, it follows that $\gamma_1 = 1/2$, as indeed observed by Kuptsov and Politi (2011). In the bulk of the spectrum γ is close to 1 since $z = 1$, and the smallness of 2α reflects itself in small deviations of γ from 1 (for a more detailed discussion see Pazó et al. (2013)).

We conclude this section by discussing how this analysis can be used to assess the hyperbolicity of the underlying dynamics. In Section 5.7.1, we saw that the diffusion coefficient K_i of $\delta_i = \Lambda_i - \Lambda_{i+1}$ is a useful observable to determine whether the Oseledets splitting is dominated (Bochi and Viana, 2005). Eq. (5.32) allows us to express K_i in terms of the diffusion matrix D.

The results of numerical simulations for a chain of Hénon maps are plotted in Fig. 11.14, where one can see that $K_i \propto 1/L$ in the bulk of the spectrum. Therefore the standard deviation of δ_i is much larger ($1/\sqrt{L}$) than δ_i itself, which is of order $1/L$ (this follows from the very existence of a limit Lyapunov density spectrum), and one must thereby conclude that order exchanges of the LEs may occur with a finite probability. In order to establish whether the Oseledets splitting is dominated, one should, however, focus on the i value which corresponds to the dimension of the unstable manifold. In the case of the Hénon maps (for the parameter values considered in this chapter), this happens for $\rho = 1$, where there is a gap in the Lyapunov spectrum. There, even though K_L decreases slower (namely, as $1/\sqrt{L}$), since δ_L is finite, the probability of order exchanges goes to zero, revealing that stable and unstable manifolds are mutually transversal and the system is effectively hyperbolic. The existence of a gap is, however, not a general property of dynamical systems, and it is therefore reasonable to conjecture that generic dynamical systems are not typically hyperbolic in the thermodynamic limit.

Another interesting property has been detected, however, which makes the scenario more intriguing. In a chain of Stuart-Landau oscillators, it has been found that somewhere in the negative part of the Lyapunov spectrum K_i vanishes. It has been conjectured that this is associated with the separation between the physical modes (which contribute to the attractor dynamics) and irrelevant or slaved modes (which govern the relaxation towards the inertial manifold) (Kuptsov and Politi, 2011).

Dimension variability

High-dimensional chaotic systems may be so complex that their local dimension exhibits fluctuations that are larger than one unit, and it may even happen that the periodic orbits embedded in the attractor have a different number of unstable directions (Kostelich et al., 1997). Such features are manifestations of a non-hyperbolic dynamics. A statistical analysis of Lyapunov exponents allows us to quantify dimension variability.

In Chapter 6, we saw that the dimension m of an attractor[5] corresponds to the dimension of a box V_m, whose volume expansion rate $\mathcal{S}_m = \Lambda_1 + \Lambda_2 + \cdots + \Lambda_m$ vanishes. Because of LE fluctuations, m fluctuates as well, implying a dimension variability across the invariant measure. A semi-quantitative analysis can be carried out by arguing as follows. Within a Gaussian approximation, the distribution of $\mathcal{S}_m$ values over a finite time τ is approximately given by

$$P(\mathcal{S}_m) \approx \exp\left(-\frac{\mathcal{S}_m^2 \tau}{2D^V(m)}\right), \tag{11.15}$$

where $D^V(m)$ can be expressed in terms of the entries of the diffusion matrix D,

$$D^V(m) = \sum_{i,j \leq m} D_{ij}.$$

The quantity $D^V(m)$ can be determined only numerically. The results of simulations made with chains of Hénon and logistic maps are reported in Fig. 11.15, where the scaled diffusion coefficients are plotted versus $\rho = m/L$, together with the integrated Lyapunov spectrum $\mathcal{S}$, which enables the location of the Kaplan-Yorke dimension density as the point where $\mathcal{S} = 0$. In the case of Hénon maps D^V is symmetric around $\rho = 1$; this is a consequence of the pairing rule of the Lyapunov exponents. For $\rho > 1$, negative Lyapunov exponents progressively contribute to the total volume expansion rate, but their fluctuations cancel those of their positive "companions", as they are perfectly anticorrelated. One expects to observe the same symmetry in any symplectic dynamical system. No symmetry is instead present in the chain of logistic maps, but the good data collapse reveals that D^V is proportional to L in both models. This is because D^V is the sum of a number of elements of order L^2, all being positive (at least as long as positive Lyapunov exponents are considered) and of order $1/L$. This in turn implies that the covariance along the direction $\mathbf{u}_m \equiv (1,1,1,\ldots,0,\ldots)/\sqrt{m}$ (the different entries correspond to the different Lyapunov exponents ordered from the largest to the smallest one, while m is the number of "1"s) is of order $\mathcal{O}(1)$ (recalling that m is of order L). Accordingly, one understands that the previously discussed divergence in the spectrum of D is due the large LE fluctuations observed along directions that can be expressed as a linear (positive) combination of many components in the original basis. A physical justification of this behaviour is still lacking.

---

[5] Here, for the sake of simplicity, we neglect fractional corrections and assume m to be an integer.

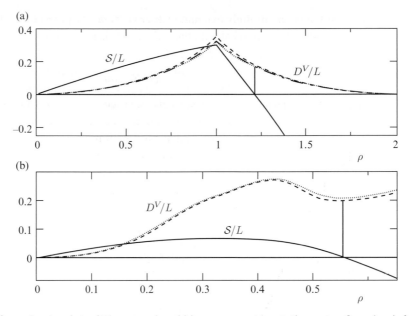

Fig. 11.15   Volume fluctuations in a chain of Hénon maps (panel (a): same parameters as in the previous figures) and of logistic maps (panel (b): $\varepsilon = 1/3, a = 2$). The system sizes are: 20, 40, and 80 (dotted, dashed and dotted-dashed lines) in the former case; 100 and 200 (dotted and dashed) in the latter case. The solid curves correspond to the rescaled volume expansion rate $\mathcal{S}/N$.

Within a linear approximation, $\mathcal{S}_\mathcal{D} = \mathcal{S}_m + (\mathcal{D} - m)\lambda_m$. By identifying $\mathcal{D}$ with the value where $\mathcal{S}_\mathcal{D} = 0$, it follows that $\mathcal{D} = m - \mathcal{S}_m/\lambda_m$. By combining this change of variables with Eq. (6.6), which allows transforming the dependence on τ into a dependence on ε, Eq. (11.15) can be rewritten as

$$P(\mathcal{D}) \approx \varepsilon^{-\frac{(\mathcal{D}-m)^2 |\lambda_m|}{2D^V(m)}} \, ,$$

so that $\sqrt{D^V/|\lambda_m|}$ quantifies the amplitude of the dimension variability. Since D^V is proportional to L, this implies that the dimension fluctuations grow with the system size, but they are increasingly negligible if expressed as a dimension density.

Similar arguments can be developed to estimate the variability of the unstable manifold (whenever the Lypaunov spectrum crosses the zero axis in a continuous way) to conclude that it is again a relatively negligible effect. Anyway, it should be noted that all of these arguments are exclusively based on numerical observations; further theoretical progress is needed to determine the true scaling behaviour and the possible universality of the scenario.

11.5 Open system approach

In this section we study the invariant measure in spatially extended systems by adopting an "open" system point of view, i.e. we consider the projection of the thermodynamic measure (defined over an infinitely extended lattice) onto a subspace of length L. This is equivalent to comparing the canonical with the microcanonical ensemble in statistical mechanics. The main difference is that in the case of dissipative systems, the invariant measure does not fill the entire phase space, and this is a major source of technical problems.

We will make use of the resolution-dependent, or ε-entropy $H(\varepsilon)$, introduced by Kolmogorov and Tikhomirov (1961) to characterise a given measure with a finite resolution. It is defined as[6] $H(\varepsilon) = -\sum p_i \ln p_i$, where p_i is the probability of the ith box (of size ε).[7] Given $H(\varepsilon)$, one can thereby introduce the "effective" dimension as

$$\mathcal{D}_c(\varepsilon) = -\frac{dH}{d\ln\varepsilon}.$$

In principle, an ε-dependent dimension is an ill-defined concept since it is not invariant upon space re-parametrisation. For this reason, it is necessary to consider the limit $\varepsilon \to 0$. In fact, in this limit, not only does $\mathcal{D}_c(0)$ become a dynamical invariant that can be related to other dynamical invariants such as Lyapunov exponents, but also the leading correction terms turn out to be universal, as discussed next. This is precisely one of the reasons why ε-entropies have been, for instance, used to quantify the cardinality of various classes of functions, e.g. the entire functions (see Kolmogorov and Tikhomirov (1961) and Collet and Eckmann (1999)).

In the study of spatially extended systems, one is faced with the difficulty of combining the dependence on ε and on the system size L. Several studies of closed systems have revealed that, for sufficiently large L, the coarse-grained dimension is an extensive quantity,

$$D_c(0, L) \propto dL,$$

where d can be interpreted as the dimension density (i.e. the contribution to the dimension per lattice site) that can be determined from the Kaplan-Yorke formula.

In an open system $D_c(0, L) = L$, since the infinitely many degrees of freedom ruling the outer part of the chain act as a sort of external noise. In order to understand how the two results can be reconciled, it is therefore necessary to investigate the simultaneous dependence of $H(\varepsilon, L)$ on both ε and L. Cipriani and Politi (2004) conjectured that

$$D_c(\varepsilon, L) \approx dL - \eta \ln \varepsilon, \tag{11.16}$$

where d is again the dimension density, while η is a suitable parameter. Notice that Eq. (11.16) allows reconciling the apparently contradictory expectations for closed and open systems. In fact, if the limit $\varepsilon \to 0$ is taken before the limit $L \to \infty$, D_c/L diverges (though, in reality, it cannot become larger than 1 – this inconsistency is due to the perturbative character of this formula). If the order of the limits is reversed, then D_c/L converges to the expected (closed system) value d. Before developing a general method which justifies this conjecture, it is first useful to summarise some special results.

---

[6] For the sake of simplicity, we consider only the standard Shannon entropy, rather than the whole set of Renyi entropies.

[7] Rigorously speaking, one should refer to an optimal covering. We implicitly assume to have made such a choice.

If the invariant measure fills a linear subspace, it can be characterised with global methods such as principal component analysis (Politi and Witt, 1999). In such a case, it has been found that

$$\ln \varepsilon = -LF(D_c/L), \qquad (11.17)$$

where the function $F(x)$ is identically zero for $x < d$, while it increases monotonously for $x > d$, starting with a finite slope. This equation can be solved for $x = D_c/L$. By then expanding the resulting inverse function, $F^{-1}(\ln \varepsilon)$, for small values of $\ln \varepsilon$, one finds the perturbative expression

$$D_c(\varepsilon, L) = dL - \frac{\ln \varepsilon}{\beta_1},$$

which has the same structure as Eq. (11.16). Thus, Eq. (11.17) appears as a natural extension to arbitrarily small scales.

In the case of the complex Ginzburg-Landau equation, a rigorous upper bound has been obtained by Collet and Eckmann (1999):

$$D_c(\varepsilon, L) < B_0 L + B_1/\varepsilon^2.$$

The approach that has led to such a formula is the same discussed in Chapter 6 to derive the Kaplan-Yorke formula for standard finite-dimensional attractors: given a set of boxes that cover the attractor, a finer covering is obtained by simply letting each box evolve in time. As a result, time evolution can be turned into an increase of the resolution. We anticipate that the main difference between closed and open systems is that in the latter case, the Lyapunov exponents exhibit a slow convergence that is absent in closed systems.

Let us now proceed by introducing some notation: let the vectors $\mathbf{U}_\parallel$ and $\mathbf{U}_\perp$ correspond to the state variables within and, respectively, outside the window W_L of size L we are interested in. The aim is to infer the fractal properties of the underlying attractor from the time evolution of the probability $P(t, \mathbf{U}_\parallel, \mathbf{U}_\perp)$. Let the initial condition $P(0, \mathbf{U}_\parallel, \mathbf{U}_\perp)$ denote a homogeneous distribution confined to a unit hypercube which contains the attractor and is contained in its basin of attraction.[8] The probability density can be then represented as

$$P(t, \mathbf{U}_\parallel, \mathbf{U}_\perp) = \int d\mathbf{U}_\perp^0 Q(t, \mathbf{U}_\parallel, \mathbf{U}_\perp | \mathbf{U}_\perp^0) P_\perp(\mathbf{U}_\perp^0), \qquad (11.18)$$

where $Q(t, \mathbf{U}_\parallel, \mathbf{U}_\perp | \mathbf{U}_\perp^0)$ denotes the probability density at time t conditioned to the initial configuration $\mathbf{U}_\perp^0$ of the external "hidden" variables, while $P_\perp(\mathbf{U}_\perp^0)$ represents their distribution. The integral over the variables $\mathbf{U}_\perp^0$ represents the first relevant difference with respect to closed systems: the ignorance about their current values contributes to dressing the probability density. We briefly discuss this problem in Section 11.5.2.

There exists, in fact, a second difference: pairs of initial conditions that differ only inside the window W_L evolve in a way that their separation propagates in the outer region as well. This suggests the need of defining a different, more appropriate Lyapunov spectrum.

---

[8] If this is not possible, one should decompose the following reasoning into various steps at the expense of requiring additional technicalities.

11.5.1 Lyapunov spectra of open systems

For the sake of simplicity we restrict our considerations to a one-dimensional lattice and we assume that each site is characterised by a scalar variable. Let $\{\mathbf{u}^{(n)}(t)\}$, $(n = 1, \ldots, L)$ denote L independent perturbation vectors, all restricted to a window of length L, i.e. $u_x^{(n)}(0) = 0$ for $x \leq 0$ and $x > L$ and $\forall n = 1, \ldots, L$, where $u_x^{(n)}$ stands for the amplitude of the nth vector at site x. Such perturbations are allowed to freely evolve within a formally infinite environment for a time $T = gL$, to determine the volumes spanned by the projections $\mathsf{P}_L \mathbf{u}^{(n)}(t)$, where P_L is the projection operator,

$$[\mathsf{P}_L \mathbf{u}^{(n)}(t)]_x = \begin{cases} u_x^{(n)}(t) & \text{if } 0 < x \leq L, \\ 0 & \text{otherwise .} \end{cases}$$

In practice, one proceeds as in the standard computation of Lyapunov exponents, except for two differences: (i) the environment is unbounded (as for the convective exponents), and (ii) volumes are determined only with reference to a fixed finite window. The open-system Lyapunov exponents Λ_i^o are finally defined by taking the $L \to \infty$ limit for a fixed g value, where g is a free, independent parameter. For $g \to \infty$, one expects that all open-system Lyapunov exponents converge to the maximum Lyapunov exponent, since the space covered by the perturbations grows and this procedure becomes similar to the computation of an increasingly tiny fraction of a Lyapunov spectrum. On the other hand, for $g \to 0$, we expect to recover the usual Lyapunov spectrum, since propagation effects are negligible.

Here, we illustrate the results with reference to a chain of logistic maps (see Eq. (A.10)). In Fig. 11.16a we report the spectra obtained for $\varepsilon = 1/6$, $a = 2$, $L = 150$ and different values of g (from 0.1 to 0.5). The nice data collapse reveals that the spectral change essentially amounts to an expansion along the horizontal axis: the Lyapunov label i has indeed been scaled by the effective length $L_e = f(g)L$, where the factor f has been determined so as to maximise the overlap. This result can be defined as evidence of hyperscaling (scale invariance of the spectrum under both changes in the system size and propagation times).

It is, a priori, reasonable to expect that the effective length increases linearly with time, namely that $L_e = L + 2v_o T$, where the factor 2 accounts for the fact that the growth occurs on both sides of the window, while v_o is some unknown velocity. By then recalling that $g = L/T$, the aforementioned equation implies $L_e/L = 1 + 2v_o g$. The data plotted in Fig. 11.16b indeed confirm a linear growth and suggest that $v_o \approx 0.085$ (see the dashed line). This is a rather small velocity compared with the propagation speed measured from the convective exponents (see Section 10.3); in fact it measures a different quantity: how an ensemble of perturbations is able to fill the available phase space, rather than the amplitude along a specific direction.

11.5.2 Scaling behaviour of the invariant measure

Now we can return to the problem of extending the Kaplan-Yorke formula to open systems. We start by neglecting the integration over the external degrees of freedom (see

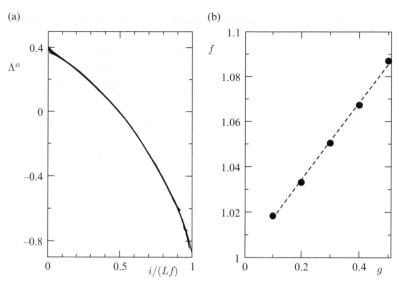

Fig. 11.16 (a) Open-system Lyapunov spectra for a chain of logistic maps: the factor f is empirically determined so as to ensure the best overlap. The various curves correspond to $g = 0, 0.1, 0.2, 0.3, 0.4$ and 0.5. In panel (b), the scaling factor f is plotted versus g. The dashed line has been obtained with a linear fit; its slope is 0.16.

Eq. (11.18)). In this approximation, we are entitled to use Eqs. (6.6) and (6.3), with the warning that now the Lyapunov exponents do depend on time. Under the assumption that the hyperscaling holds, one can still invoke the proportionality of the dimension to the system size, with the length L replaced by its effective value after a time T:

$$\tilde{D}_{KY}(T, L) = d(1 + 2v_o g)L = dL + 2dv_o T. \qquad (11.19)$$

We can now use Eq. (6.6) to transform the dependence of $\tilde{D}_{KY}$ on time T into a dependence of D_{KY} on the resolution ε. Indeed, inverting Eq. (6.6), we have

$$T = -\frac{\ln \varepsilon}{|\lambda_d|}, \qquad (11.20)$$

where $\lambda_d \doteq \lambda(\rho = d)$. Upon inserting Eq. (11.20) into Eq. (11.19), we obtain

$$D_{KY}(\varepsilon, L) = dL - \frac{2vd}{|\lambda_d|} \ln \varepsilon, \qquad (11.21)$$

which reveals a logarithmic dependence on the resolution of the type heuristically conjectured in (11.16).

This formula represents, however, a lower bound on the dimension D_{KY}, since we have neglected the effect of the integral over $\mathbf{x}_\perp^0$ in Eq. (11.18) that accounts for the uncertainty on the inner variables induced by the initial lack of knowledge on the outer degrees of freedom. An accurate quantification of the corresponding propagation of information is far from trivial, since it may be controlled by nonlinear mechanisms, as it happens for instance for stable chaos (Politi and Torcini, 2010), where nonlinear mechanisms are responsible for the self-sustainment of irregular behavior even in the presence of a negative

Lyapunov spectrum. In the following, we limit ourselves to presenting a heuristic argument that leads to an upper bound. From the convective Lyapunov exponents, we know that a perturbation of amplitude 1 originated at the boundary of the window at time 0 is "amplified" by a factor $\exp[T\mathscr{L}(x/T)]$ after a time T at a distance x (this is the point where the assumption that nonlinear effects are negligible is crucial). From Eq. (11.20), we need to probe the perturbation at the time $T\,|\ln\varepsilon/\lambda_d|$. The maximum distance from the boundary where the perturbation is larger than the resolution ε can be then determined by imposing $\exp T\mathscr{L}(x/T) = \varepsilon$, i.e.

$$\lambda_d = \mathscr{L}\left(\frac{x|\lambda_d|}{|\ln\varepsilon|}\right). \tag{11.22}$$

Now let v' denote the velocity such that $\mathscr{L}(v') = \lambda_d$ (if the whole convective spectrum is larger than λ_d, v' must be assumed equal to the maximal possible velocity, which, in the case of nearest-neighbour coupling, is equal to 1). Eq. (11.22) implies that

$$x = \frac{v'}{|\lambda_d|}|\ln\varepsilon|.$$

In other words, x is the maximal distance where a perturbation arisen from the boundary is larger than ε at the time implicitly selected to partition the phase space with boxes of size ε. If we now assume that the distribution over the x sites has the maximal possible dimension, we find that the correction to the dimension due to the external degrees of freedom is xd_v, where d_v is the number of variables per lattice sites in each of the two window edges. Altogether, adding this contribution to Eq. (11.21) one obtains the upper bound

$$D_c(\varepsilon,L) \leq dL - \frac{2(dv + d_v v')}{|\lambda_d|}\ln\varepsilon.$$

This result suggests that the rigorous bound obtained by Collet and Eckmann (1999) with reference to the complex Ginzburg-Landau equation can be improved. For a more detailed discussion of the limits of this derivation (in particular the condition $|\ln\varepsilon| < \alpha L$, where α is a suitable factor) the reader is invited to consult Cipriani and Politi (2004).

We expect the open system approach to be of some relevance in the characterisation of nonequilibrium measures in Hamiltonian systems, since the phase space is filled in a singular way, like in the setup considered in this section.

Applications

12.1 Anderson localisation

Anderson localisation is a general phenomenon occurring in the linear propagation of waves in disordered media. The name comes from the seminal paper by Anderson (1958), where he argued that in a disordered lattice, a wave packet remains localised rather than spreading out. Nearly simultaneously, Gertsenshtein and Vasiljev (1959) found that the almost full reflection of a wave by a disordered waveguide is also a manifestation of localisation. Anderson localisation lies at the heart of several properties of disordered media. It has many applications in solid state physics (see the reviews by Kramer and MacKinnon (1993), Beenakker (1997) and Evers and Mirlin (2008) and the collection edited by Abrahams (2010)) and in optics and acoustics, as well as in general disordered environments, either continuous or discrete.

Two questions are usually addressed: (i) characterisation of the eigenstate structure and (ii) determination of the transport properties (e.g. scattering) in the presence of a disordered layer. Remarkably, in one dimension, a small disorder is sufficient to induce a pure point spectrum of the eigenvalues with eigenmodes that are exponentially localised in space, and an almost perfect reflection with an exponentially small (in terms of the layer length) transmission coefficient.

In both contexts, once a harmonic time-dependence of the field has been assumed, the problem can be reduced to a stationary equation, where the field depends only on the spatial variable(s). In one dimension, this is expressed either as a recursive equation (in lattice systems) or as a differential equation (in spatially continuous formulations).

To be more concrete, let us consider two prototypical formulations: the stationary Schrödinger equation in continuous space,

$$E\psi(x) = -\frac{d^2\psi}{dx^2} + V(x)\psi(x), \tag{12.1}$$

and its lattice analog (called the Anderson model),

$$E\psi_n = V_n\psi_n - \psi_{n-1} - \psi_{n+1}. \tag{12.2}$$

Here, the potential $V(x)$ is a random function (respectively, V_n is a sequence of random numbers).

If Eqs. (12.1, 12.2) are treated as an *initial value problem*, e.g. fixing $\psi(0)$ and $\psi'(0)$ in (12.1) to determine $\psi(x)$ for $x > 0$ (respectively, fixing ψ_0 and ψ_{-1} in (12.2) to determine ψ_n for $n > 0$), a clear analogy with the computation of LEs in stochastic systems arises: the field generally grows as $\psi(x) \sim \exp[\lambda x]$, where λ is the largest Lyapunov exponent. In the continuous setup the computation of the Lyapunov exponents is equivalent to studying the stability of a noise-driven oscillator (see Eq. (8.26)). In a discrete lattice context, the problem is equivalent to the analysis of a product of random (transfer) matrices,

$$\begin{pmatrix} \psi_{n+1} \\ \psi_n \end{pmatrix} = \begin{pmatrix} V_n - E & -1 \\ 1 & 0 \end{pmatrix} \begin{pmatrix} \psi_n \\ \psi_{n-1} \end{pmatrix}$$

(see Section 8.1).

In both cases, however, one is interested not in the initial value problem but in either the *eigenvalue* or the *boundary* problem. Hence, additional care is needed to show that proper solutions of these problems follow the same exponential asymptotics as the initial value problem (see Bougerol and Lacroix (1985); Pastur and Figotin (1992)). Roughly speaking, the LE analysis identifies the correct eigenfunctions under the following assumptions: (i) the asymptotic exponential rates (as defined according to either the Oseledets theorem or, for random matrices, the Furstenberg-Kesten theorem) are valid for *all* perturbations (for a typical realisation of the disorder) and (ii) these rates depend continuously on the control parameters (the energy, in the context of eigenvalue problems).

The fact that the Lyapunov exponent determines the exponential decay rate of the eigenfunction is often called Borland conjecture in the physical literature (Borland, 1963).

Of course, any localised eigenfunction while growing exponentially on one side of the peak decays exponentially on the other side. Since, however, the negative Lyapunov exponent has the same magnitude as the positive one (due to the symmetry $x \to -x$, or $n \to -n$), one can write that away from the localisation area $|\psi| \sim \exp[-\lambda|x|]$. The quantity $\ell = 1/\lambda$ is often referred to as the localisation length of the given eigenfunction. In practice, returning to the initial-value interpretation of the evolution equations, the true eigenfunction is generated by complementing $\psi(0)$ with the special value $\psi'(0)$, such that $\psi(x)$, after having reached its maximum somewhere, eventually decreases back towards the boundary value $\psi(L)$. As for the transmission T through a disordered layer of length L, it is exponentially small, $T \sim \exp[-2\lambda L]$ (the factor 2 is there because the transmission coefficient refers to the "energy" $\sim |\psi|^2$). The Lyapunov exponent depends on E; since it is non-zero for all energy values that belong to the spectrum, there is no room for the propagation of extended modes; i.e the spectrum of the Schrödinger equation in a disordered potential is purely discrete. For the transmission coefficient the situation is slightly more subtle. Here E can be arbitrary (it corresponds to the frequency of the incident wave), while the exponential law for the transmission is valid for typical values of E. There can be exceptional resonant frequencies (corresponding to localised eigenmodes in the bulk of the disordered layer, having close amplitudes at both ends) where the transmission is large. Indeed, if the incident wave is resonant with such a mode, the latter will be excited to very large amplitudes, which will result in a large transmission rate; see examples in Bliokh et al. (2008) and Gredeskul and Freilikher (1990).

In the context of Anderson localisation studies, there is a huge literature on different numerical and analytic approaches to compute the Lyapunov exponents in models of the type (12.1, 12.2). Many of the methods described in Chapter 8 have been, in fact, developed in this field. A detailed account of random potentials with correlations is presented by Izrailev et al. (2012), while problems where the potentials are dynamically generated as quasiperiodic or chaotic functions of a spatial coordinate are discussed by Bourgain (2005). Finally, one should notice that in more complex one-dimensional setups, where the evolution in space involves more than one degree of freedom, several positive and negative exponents are present (e.g. in the presence of next-nearest-neighbour interactions in a lattice model, one deals with 4×4 transfer matrices, so that four exponents are expected). In this case the localisation length is determined by the LE closest to zero, which asymptotically dominates the decay of the eigenfunctions.

The concept of Lyapunov exponent cannot be straightforwardly extended to two- and three-dimensional systems as there is no single time-like coordinate. Also, the stationary problem no longer reduces to an ordinary differential equation but to an elliptic partial differential one. Here, the Lyapunov-exponent approach can be applied to the so-called quasi-one-dimensional setups, i.e. disordered stripes (bars) that are extended only along one dimension, while being bounded in the other dimension(s) (see Kramer and MacKinnon (1993) and Kramer et al. (1987) for details and further references). If there are d transverse dimensions, each of length L, then the vector identifying the given quasi-one-dimensional setup consists of L^d components. Its evolution along the one-dimensional space is governed by a product of $2L^d \times 2L^d$ transfer matrices and is thereby characterised by $2L^d$ Lyapunov exponents (which come in symmetric pairs, as the system is space-reversal). The asymptotic properties of the eigenfunctions are determined by the smallest LE (in absolute value), i.e. the exponent $\lambda_{min}(d, L)$ closest to zero. The localisation properties of a $d + 1$ dimensional disordered medium can be assessed by monitoring the dimensionless parameter $\Gamma = L\lambda_{min}$ for increasing values of the transversal size L (MacKinnon and Kramer, 1981, 1983; Pichard and Sarma, 1981a, b). In the localised phase, Γ increases proportionally to the transversal size L, since λ_{min} remains finite. In the delocalised phase, the minimal Lyapunov exponent tends to zero rapidly enough to make Γ decrease with L. The application of this approach to two-dimensional disordered media ($d = 1$) reveals that all eigenfunctions are typically localised. In the three-dimensional setup ($d = 2$), a so-called mobility edge separates localised from delocalised states. The transition can be observed, e.g., by tuning the strength of the disorder above a critical value (MacKinnon and Kramer, 1981, 1983; Kramer et al., 1987; Kramer and MacKinnon, 1993; Slevin and Ohtsuki, 1999).

12.2 Billiards

Billiards are one of the most studied Hamiltonian setups for testing nonlinear dynamics concepts and investigating connections with statistical mechanics (see Section 12.3). A billiard is a region $\mathcal{R}$, where a pointlike particle is left free to move, colliding elastically with the boundary of $\mathcal{R}$. Two examples are presented in Fig. 12.1: in the triangular billiard drawn in panel (a), collisions occur with an outer boundary; in the Sinai billiard drawn in panel (b), periodic boundary conditions are assumed and collisions occur only with the "inner" disk.

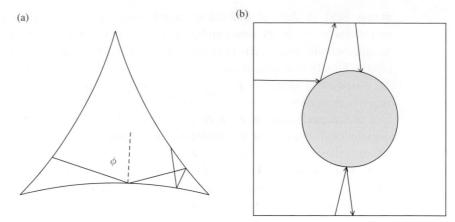

Fig. 12.1 (a) A dispersing triangular billiard. The solid line represents a possible trajectory; ϕ is the angle between the direction of propagation and the normal to the boundary in a given collision. (b) Sinai billiard with periodic boundary conditions.

Depending on whether the concavity of the boundaries points towards inner or outer regions, the billiard is called dispersing or focusing (and can also have a mixed nature). In the former case, the dynamics is undoubtedly chaotic, while the behaviour of focusing billiards is more difficult to assess a priori. The reader interested in a rigorous analysis of such dynamical systems can consult Chernov and Markarian (2006).

The phase state of a plane billiard is determined by three variables: two are needed to identify the position within the billiard region and one to identify the orientation of the velocity (the modulus of the velocity, being constant, can be assumed to be equal to 1 without loss of generality). The evolution in tangent space is conveniently described by referring to a frame that moves with the particle. In such a frame, two variables suffice: the distance u orthogonal to the direction of propagation and the velocity difference v along the same orthogonal direction (the distance along the direction of propagation is irrelevant, as it is neutrally stable and yields a zero Lyapunov exponent).

Let $\{t_n\}$ denote the sequence of collision times of the particle with the boundary ($t_1 < t_2 < \cdots < t_n$). In between two consecutive collisions,

$$\dot{u} = v, \qquad \dot{v} = 0,$$

which integrates to yield

$$u^-_{n+1} = u^+_n + v^+_n \tau_n, \qquad v^-_{n+1} = v^+_n,$$

where $\tau_n = t_{n+1} - t_n$ and the $-/+$ superscript means that the corresponding variable is determined just before/after the collision.

The evolution equations are completed by adding the effect of the collision,

$$u^+_{n+1} = -u^-_{n+1}, \qquad v^+_{n+1} = -v^-_{n+1} + \frac{2u^-_{n+1}}{a \cos \phi_{n+1}},$$

where $-\pi/2 \le \phi_{n+1} \le \pi/2$ is the angle between the direction of propagation and the normal to the boundary (see Fig. 12.1), while a is the local radius of curvature of the boundary. By combining the two steps, the following transformation is obtained:

$$\mathsf{J}_n = - \begin{pmatrix} 1 & \tau_n \\ 2/(a\cos\phi_{n+1}) & 1 + 2\tau_n/(a\cos\phi_{n+1}) \end{pmatrix}.$$

The determinant of J_n is equal to 1, as it should be. The Lyapunov exponent can be determined by multiplying the Jacobians along a generic trajectory. Depending on whether the radius of curvature a is positive or negative, J_n is either hyperbolic or elliptic. In the former case (dispersing billiard) one eigenvalue is larger than 1 and the dynamics is trivially chaotic. It is however, possible to have a positive Lyapunov exponent also in the latter case (focusing billiard), as we have seen in Chapter 8, while discussing random products of elliptic matrices.

The expansion rate between the nth and the $(n+1)$th collision can be expressed as

$$\mu_{n+1} \equiv \frac{u_{n+1}^+}{u_n^+} = -(1 + \kappa_n \tau_n),$$

where $\kappa_n \equiv v_n^+/u_n^+$ denotes the local expanding direction, which can be determined by iterating the recursive equation

$$\kappa_{n+1} = \frac{\kappa_n}{1 + \kappa_n \tau_n} + \frac{2}{a\cos\phi_{n+1}}. \tag{12.3}$$

The positive Lyapunov exponent can be thus expressed as

$$\lambda_+ = \frac{\langle \ln|1 + \kappa_n \tau_n| \rangle}{\langle \tau_n \rangle},$$

where the angular brackets denote a time average. If one is able to determine κ_n in all points of the phase space, the Lyapunov exponent can be equivalently expressed by replacing the time average with an ensemble average.

The quantity κ_n has a simple physical interpretation: it corresponds to the curvature of a "wave front" propagating along the trajectory of reference. In fact, the first term in the r.h.s. of this equation (12.3) corresponds to the linear increase of the radius of curvature $(1/\kappa)$ of the wave front in between two consecutive collisions, while the second term takes into account the reflection with the boundary. In Fig. 12.2, the direction of the Lyapunov vector (i.e. the curvature) is plotted in the phase space of the triangular billiard of Fig. 12.1a.

Going back to the continuous-time representation, it is easily seen that the Lypaunov exponent can be expressed as an average of the instantaneous expansion rate (i.e. the instantaneous value of the curvature),

$$\lambda_+ = \left\langle \frac{v_n^+}{u_n^+ + v_n^+ t} \right\rangle = \left\langle \frac{\kappa_n}{1 + \kappa_n t} \right\rangle.$$

The problem of determining the Lyapunov exponent remains nevertheless quite hard, as no analytic expression is typically available for the curvature. In some limit cases, it is, however, possible to derive approximate formulas. One example is the Lorentz gas, i.e. a collection of circular disks of radius a, in the limit of high dilution, $a \ll 1$. In these

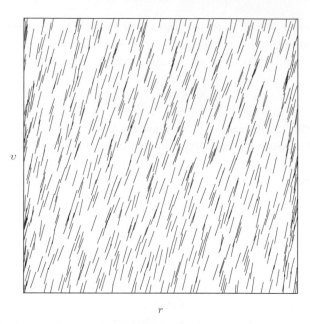

Fig. 12.2 Local phase space for the triangular billiard in the previous figure: r is the position along one of the three arcs composing the boundary, v is the component of the velocity parallel to the boundary. The small segments indicate the local direction of the Lyapunov vector, i.e. the curvature of the corresponding wave front.

conditions, the curvature values in Eq. (12.3) are dominated by the large last term under logarithm on the r.h.s. of this formula. As a result, Eq. (12.3) can be rewritten as

$$\lambda_+ \approx \frac{\ln(2/a) - \langle \ln \cos \phi_n \rangle + \langle \ln \tau_n \rangle}{\langle \tau_n \rangle}.$$

The average time $\langle \tau_n \rangle$ can be determined with the following rough argument: a generic (straight) trajectory hits any disk whose centre is located in a corridor of width $2a$. Accordingly, the typical distance L the particle travels before encountering an obstacle is given by $n2aL = 1$, where n is the density of scatterers, so that $\langle \tau \rangle = 1/(2an)$ – recalling that we have assumed a unit velocity. Therefore,

$$\lambda_+ \approx -4a \ln(an) - 2a \langle \ln \cos \phi \rangle,$$

where we have also assumed that $\langle \ln \tau_n \rangle = \ln \langle \tau_n \rangle$. The leading contribution $-4a \ln a$ is universal, as it does not depend on the details of the scattering process. It was first conjectured by Krylov (1979) and then proved by Friedman et al. (1984). The value of the correction terms depends on the specific physical setup; their estimation requires dealing with the incidence angle ϕ_n. This can be done for instance by focusing attention on the trace of the products of the Jacobians, the knowledge of which suffices for the determination of the Lyapunov exponents in two-dimensional area-preserving maps. By adopting this approach, Dahlqvist (1997) was able to derive an additional constant term in the case of the Sinai billiard (see Fig. 12.1b) in the small scatterer limit.

An alternative approach is based on the estimate of the probability distribution of the curvature in real space (a sort of Boltzmann equation) that is akin to the determination of the density of the direction of Lyapunov vectors that we have encountered in Chapter 8 for products of random matrices. This approach was first developed by van Beijeren and Dorfman (1995) for a planar Lorentz gas with a diluted distribution of scatterers, where the leading correction was determined. Further terms have been later determined by Kruis et al. (2006). Such a method is so powerful that it could be extended to a three-dimensional diluted Lorentz gas as well, where the two positive LEs have been determined to leading orders (Latz et al., 1997) and found to agree with direct numerical simulations (Dellago and Posch, 1997).

12.3 Lyapunov exponents and transport coefficients

Two approaches have been developed that can help to extract information on non-equilibrium properties of statistical systems from the knowledge of dynamical invariants. In both cases, a link is established by expressing the same observable in terms of macroscopic as well as microscopic variables.

12.3.1 Escape rate

This approach starts from the observation that the exponential decay of a suitable probability density can be equivalently expressed in terms of either a "diffusion coefficient" or microscopic dynamical quantities.

Here, we illustrate the idea with reference to the Lorentz gas (see the previous section for a definition). Consider a small particle moving in a region $\mathcal{R}$ of size S, much larger than both the average distance between the scatterers and the mean free path p. Such a particle undergoes a diffusive motion on a length scale that is large compared with p, but small compared with S. Suppose also that $\mathcal{R}$ is surrounded by absorbing boundaries, so that when the particle crosses a boundary, it is removed. It is well known that the probability $P(t)$ that the particle is still inside $\mathcal{R}$ after a long time t decreases exponentially,

$$P(t) \approx e^{-\gamma t}. \tag{12.4}$$

Assuming that we are in the presence of normal diffusion, the rate γ can be obtained by solving a standard diffusion equation with absorbing boundary conditions. In practice γ corresponds to the smallest eigenvalue of this problem, which, for S large enough, has the typical structure $\gamma = Da/S^2$ (Gaspard and Nicolis, 1990), where D is the diffusion coefficient, while a is a constant related to the geometric structure of $\mathcal{R}$ (in a simple one-dimensional setup, $a = \pi^2$).

On the other hand, the particle can be macroscopically seen as following a chaotic trajectory. In this framework, it remains confined forever within the allowed region if and only if it asymptotically converges to some fractal repeller, characterised by a zero Lebesgue measure. The probability for a randomly chosen initial condition to remain inside

$\mathcal{R}$ decays exponentially as described by Eq. (12.4), where γ can now be interpreted as the escape rate from the repeller. Accordingly, from Section 6.3

$$\gamma = \sum_i \lambda_i - h_{KS},$$

where the sum is performed over all positive Lyapunov exponents (in the case of a planar Lorentz gas, only one positive exponent exists), while h_{KS} is the Kolmogorov-Sinai entropy of the repeller.

Upon identifying the two definitions of γ, one obtains the final relationship

$$D = \lim_{S \to \infty} \frac{S^2}{a} \left[\sum \lambda_i - h_{KS} \right]. \tag{12.5}$$

Since the diffusion coefficient D is an intensive quantity that is independent of the size and the form of the region $\mathcal{R}$, this equation implies that the escape rate should vanish quadratically with S and contain an implicit dependence on the geometry, so as to cancel the explicit dependence on a present in the r.h.s. of Eq. (12.5). This appears to be true, but a comprehensive theory is still lacking.

This method can be extended to determine generic transport coefficients in simple fluids. Instead of monitoring the particle position in real space, one needs to consider suitable Helfand moments (Gaspard, 2005). In the case of the shear viscosity η, the appropriate quantity to look at is

$$M_\eta = \frac{1}{\sqrt{V k_B T}} \sum_i x_i p_i^y,$$

where V is the volume, k_B the Boltzmann constant, T the temperature, x_i the x coordinate of the particle and p_i^y the corresponding momentum along the y direction. M_η performs a random walk in phase space. Therefore, one can again follow the previous strategy. On the one hand, the decay rate can be expressed in terms of η as $\gamma = \eta a_\eta / S_\eta^2$, where S_η is the size of the selected phase-space region, while a_η is the corresponding geometrical factor. On the other hand, one can define γ as the escape rate of a suitable repeller, so that a formula of the type (12.5) is again obtained.

The reader interested in a more detailed discussion can consult Gaspard (2005). A weakness of this approach is that properties of repellers in high-dimensional systems are poorly understood, and it is therefore not possible to infer the escape rate directly from the microscopic equations.

12.3.2 Molecular dynamics

A second approach is based on the introduction of deterministic thermostats to steadily maintain a constant temperature even in the presence of an external forcing. Such devices are regularly used in molecular dynamics studies of stationary nonequilibrium states (Hoover, 1991; Evans and Morris, 2008).

We illustrate the approach with reference to a charged particle that moves in a Lorentz gas in the presence of an electric field $\mathbf{E}$. The determination of the corresponding electric conductivity is the object of interest. Between collisions, the particle satisfies the equation

$$\dot{\mathbf{r}} = \mathbf{v}, \qquad \dot{\mathbf{v}} = \frac{q}{m}\mathbf{E} - \alpha(\mathbf{v})\mathbf{v},$$

where m and q are the mass and the charge of the particle, respectively; $-\alpha\mathbf{v}$ is a fictitious force, added to keep the kinetic energy strictly constant. This is ensured by imposing $\dot{\mathbf{v}} \cdot \mathbf{v} = 0$, so that one finds that

$$\alpha(\mathbf{v}) = \frac{q\mathbf{E} \cdot \mathbf{v}}{mv^2};$$

α is a self-generated friction coefficient which determines the electric conductivity. In fact, by averaging over time the definition of α, we obtain

$$\langle \alpha \rangle = \frac{\langle \mathbf{j} \rangle \cdot \mathbf{E}}{k_B T} = \frac{\sigma E^2}{k_B T},$$

where we have made use of the definition of kinetic temperature $\langle v^2 \rangle = k_B T$ and recalled the definition of electric conductivity (j is the electric current).

One should notice that the "frictional" term destroys the Hamiltonian structure of the evolution equation. In particular, the sum of all Lyapunov exponents is no longer equal to zero. The divergence of the velocity field is indeed equal to $-(d-1)\alpha$, where d is the dimension of the space. Since the divergence is also equal to the sum of all Lyapunov exponents, one obtains the final relationship

$$(d-1)\frac{\sigma E^2}{k_B T} = -\sum_i \lambda_i,$$

which is another way of linking microscopic to macroscopic observables. Relationships of this type have been derived by Chernov et al. (1993) and Baranyai et al. (1993), and even earlier with reference to shear viscosity by Evans et al. (1990).

In this case, nontrivial physics is hidden in the fact that the volume contraction rate is expected to scale quadratically with the field E. A weakness of this approach is the ad-hoc character of the additional terms, which have no microscopic justification.

By finally comparing the two approaches, one can notice that the two methods invoke different invariant structures in phase space: the former one assumes the existence of a repeller that is typically fractal along all directions, while the latter assumes a chaotic attractor that is typically fractal only along the stable directions. How and whether the two methods reconcile with one another is still an open problem. A more detailed comparative analysis can be found in Dorfman and van Beijeren (1997).

12.4 Lagrangian coherent structures

Passive tracers are particles which move according to a given velocity field, without affecting it. In an incompressible, turbulent fluid flow, passive tracers mix together quite effectively, so that a uniform distribution is soon established. If the flow is, instead, dominated by relatively stationary large structures (typically, large eddies), a generic initial distribution tends to form rather structured patterns on long but finite time scales. They are called Lagrangian (because they follow the motion of individual fluid elements) coherent structures (LCSs). The theory of these structures, developed by Pierrehumbert and Yang (1993) and Haller (2001), is based on the computation of the finite-time Lyapunov exponents, as described in Chapter 5. Before describing the theory, we give some examples of LCSs. The most spectacular structures are those appearing over large scales as a result of either natural phenomena or man-made disasters. As a result of a volcanic eruption, an ash cloud may be spread by the atmospheric winds and form stripes with large concentration, alternating with basically ash-free regions. In an ocean, such stripes appear as a result of advection of an oil spill or large plankton fields. These and several other examples can be found in a popular article by Peacock and Haller (2013).

Next, we describe the simplest setup of a two-dimensional incompressible, time-dependent flow $(v_x(x,y,t), v_y(x,y,t))$. The particles follow Lagrangian trajectories determined by the dynamical equations

$$\dot{x} = v_x(x,y,t), \qquad \dot{y} = v_y(x,y,t). \tag{12.6}$$

For this dynamical system one can define two Lyapunov exponents, which generally (if the time dependence is random) are both non-zero. Because of incompressibility, their sum is zero, so that they have opposite sign. This means that all trajectories are "saddles" characterised by a stable and an unstable direction. According to the basic interpretation of Lyapunov exponents (see Chapter 2), this means that a small disk around a trajectory evolves into an ellipsoid, extended along the direction of the unstable manifold and compressed along the direction of the stable one. As coherent structures appear over finite times (a uniform mixing eventually settles down), it is natural to consider finite-time Lyapunov exponents.

Let us consider the flow (12.6) over a finite time interval (t_0, t_1). It defines the map

$$x(t_1) = f_x(x(t_0), y(t_0)), \qquad y(t_1) = f_y(x(t_0), y(t_0)).$$

Let us denote the corresponding Jacobian matrix as H. Just as for the definition of the Lyapunov exponents in Chapter 2, the eigenvalues of the symmetric matrix $\mathsf{P}(t_0, t_1) = \mathsf{H}^\mathsf{T}\mathsf{H}$ tell us how an initial small disk is stretched and compressed along different directions (in the context of fluid flow, the matrix P is often called Cauchy-Green strain tensor). The finite-time Lyapunov exponents $\Lambda_{1,2} = 0.5 \log \mu_{1,2}$, where μ_1, μ_2 are the eigenvalues of P, quantify the rate of stretching. Because of the incompressibility of the flow, $\Lambda_1 = -\Lambda_2$. Regions of initial points $x(t_0), y(t_0)$ where the Lyapunov exponent is large are regions where particles will be maximally dispersed (unstable coherent structures). Regions of final points $x(t_1), y(t_1)$ where the Lyapunov exponent is large are regions with a maximal attraction of points for the evolution within time interval (t_0, t_1), thus identifying stable coherent structures. Haller and Yuan (2000) and Haller (2001, 2002) suggest the use of the ridges of the two-dimensional surface $\Lambda_1(x(t_1), y(t_1))$ for the identification of one-dimensional LCSs. In Shadden et al. (2005) it was argued that such ridges are indeed nearly material lines of the Lagrangian flow.

This is illustrated in Fig. 12.3 with reference to the double-gyre flow (Shadden et al., 2005), defined by the stream function

$$\psi = A \sin(\pi(a(t)x^2 + b(t)x)) \sin(\pi y). \tag{12.7}$$

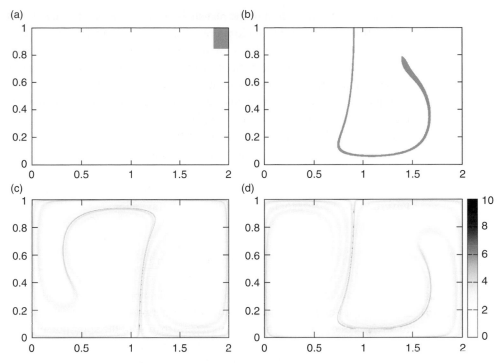

Fig. 12.3 (a) and (b) Initial ($t_0 = 0$) and final ($t_1 = 10$) positions of passive tracers embeddedn in a double-gyre flow (12.7) with $A = 0.1$, $a(t) = 0.2\sin(2\pi t/10)$, $b(t) = 1 - 2a(t)$. (c) and (d) The finite-time Lyapunov exponents as a function of the initial (panel (c)) and the final (panel (d)) coordinates, represented with a grey scale. They define the unstable and stable coherent structures.

Let us recall that the velocity field is defined as $v_x = \partial\psi/\partial y$, $v_y = -\partial\psi/\partial x$. The evolution of particles (see panels (a) and (b)) is compared with the unstable and stable coherent structures (panels (c) and (d)). There one can see that the stable LCS is exactly the region where the tracers are concentrated.

We conclude this section by mentioning that in some works, finite-amplitude exponents (see Chapter 7) have been used, instead of the finite-time ones (d'Ovidio et al., 2004; Bettencourt et al., 2013). As discussed in Section 7.1, in the presence of a well-established chaotic dynamics, the two approaches are substantially equivalent. In the absence of a strong mixing, Lagrangian structures are, instead, more reliably identified through the computation of finite-time Lyapunov exponents (Karrasch and Haller, 2013).

12.5 Celestial mechanics

The solar system is an excellent example of a system where the assessment of the stability is conceptually relevant. It is also an example of how the distinction between order and chaos may be rather fuzzy.

Superficially, the motion of the planets is a manifestation of ordered evolution. Already before the Copernican revolution took place, however, prolonged observations revealed the presence of tiny fluctuations. When the existence of Newtonian gravitational forces became indisputable, it also became clear that such fluctuations originate from the mutual interactions among the various celestial bodies.

Typically, all bodies are treated as point masses (this includes compound bodies, such as the Earth-Moon system, which is assumed to have the mass of the planet plus that of the satellite, and to be located in the barycenter), neglecting tidal friction, solar and planetary oblateness, solar-mass loss, as well as perturbations due to the possible passage of external stars.

In practice, the solar system is modelled as a nonlinear Hamiltonian N-body system, governed by the Newtonian law of universal gravitation,

$$\mathcal{H} = \sum_{j=0}^{N-1} \frac{p_j^2}{2m_j} - G \sum_{j=0}^{N-2} \sum_{k=j+1}^{N-1} \frac{m_j m_k}{|r_j - r_k|},$$

where m_j and r_j denote the mass and the position of the jth body, respectively, p_j is the corresponding momentum, and G is the gravitational constant. Sometimes general-relativity effects are also included, as a perturbation to the solar potential (see, e.g., Laskar (1989)).

The aforementioned Hamiltonian can be expressed as the sum of a Keplerian component, which induces a perfectly integrable dynamics, and an interaction component that is treated perturbatively. Sophisticated techniques have been indeed developed, based on suitable small parameters, such as the relative masses of the celestial bodies, the eccentricity, and the inclination of the orbits. This approach, which proved useful and effective over relatively short astronomical scales, is based on the implicit assumption that the planet motion is quasi-periodic, i.e. that it can be represented as a function $f(\omega_1 t, \omega_2 t, \ldots)$ that is 2π-periodic in each argument, and where ω_i are incommensurate frequencies.

With the progress of nonlinear dynamics, however, it has become clear that even in the presence of extremely small perturbations, the phase space of a Hamiltonian system is filled not only by quasi-periodic orbits (the so-called KAM tori) but also by a web of tiny chaotic layers. The reader interested in a detailed discussion of the various techniques can consult Murray and Dermott (1999).

The main question is whether the current trajectory of the solar system is actually quasiperiodic (in which case the perturbative series is valid over all times), or weakly chaotic, in which case it may eventually lead to the collision of some planets and their ejection from the solar system. Generally speaking, this can be ascertained by integrating the equations of motion and simultaneously computing the corresponding (finite-time) Lyapunov exponent. In practice, the task is very hard, since it is necessary to accurately integrate over a very long time: the gap between the time scale of the fastest motion (a few hours) and that one needed to assess the possible existence of a chaotic dynamics (millions of years) is wider than 10^9.

Because of these difficulties, most of the efforts have been restricted to analysing the motion of the so-called outer solar system (i.e. the planets from Jupiter to Pluto), since the

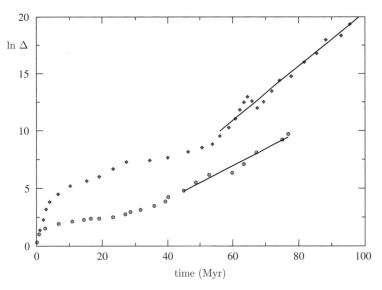

Fig. 12.4 Evolution of a small distance in the solar system (in arbitrary units). Circles and diamonds refer to the simulations by Laskar and Sussman and Wisdom, respectively. Data are freely taken from their respective publications. The slopes, obtained by fitting the last part of the curve, are equal to 0.15 and 0.24 $(\mathrm{Myr})^{-1}$.

inner planets move faster and require a smaller integration step. It was, nevertheless, later understood that the former subsystem is less chaotic than the whole solar system, so that longer simulations are eventually required.

At least two different approaches have been adopted for the integration of the equations. Laskar (1989) made an extensive use of averaging techniques to remove the short periods and to derive effective equations for the secular terms that could be integrated with a much larger time step (about 500 years). In order to have an idea of how complicated the final model is, notice that for eight planets (Pluto was excluded), the model involves about 150,000 polynomial terms. A few pairs of initial conditions have been considered to determine the growth of the relative distance Δ (i.e. in practice, the finite-amplitude exponent has been determined). The result is reproduced in Fig. 12.4 for one such simulation (see the circles): an initially linear increase, which is typical of a quasiperiodic motion, is followed by an exponential growth with an exponent that corresponds to a Lyapunov exponent about 0.16–0.2 $(\mathrm{Myr})^{-1}$. This corresponds to a Lyapunov time scale (the inverse of the Lyapunov exponent) of about 5–6 Myr.

Sussman and Wisdom (1992) have, instead, integrated the equations of motion of the whole solar system (including Pluto), by simulating the effect of the perturbations as that of equispaced delta kicks. This has allowed for an easier integration of the Keplerian component in between consecutive kicks, while keeping a sufficiently high accuracy in the treatment of the perturbations. In practice, the "time step" could be chosen as long as seven days. Their results, obtained again by following a pair of nearby trajectories, correspond to the diamonds in Fig. 12.4. The resulting Lyapunov time is ≈ 4 Myr, a value relatively close to the estimate obtained by Laskar.

In both cases, the exponential growth does not settle from the very beginning; this is to be expected, as the initial perturbation is not aligned along the most unstable direction. Notice that the time needed for the perturbation to reach the proper alignment is larger but comparable to the Lyapunov time (the inverse of the Lyapunov exponent); this suggests that the second Lyapunov exponent is substantially smaller than the maximal one.

Remarkably, the Lyapunov time is much shorter (by a factor 10^3) than the age of the solar system. It should be noted, however, that in nearly integrable Hamiltonian systems, chaos is confined to tiny layers, where only the phase variables (i.e. the position of the planets along their "normal" trajectories) are affected by the chaotic dynamics, while the structure of the orbits is essentially unchanged. The problem of assessing the stability of the solar system is related to the occurrence of substantial orbital changes, which occur on longer time scales. Laskar and Gastineau (2009) have performed careful studies, explicitly including the Moon. Since long simulations are unavoidably affected by the exponential amplification of tiny errors (and by the intrinsic approximations of the model), a statistical approach is more appropriate. Laskar and Gastineau (2009) found significant enhancements in the eccentricity of Mercury's trajectory that are strong enough to trigger collisions with either Venus or the Earth. After simulating 2501 trajectories, they conclude that the probability of a substantial change of Mercury's eccentricity is about 1% over 5 Gyr.

Finally, one should notice that although chaos in the whole solar system is very weak, the motion of small bodies in the gravitational fields generated by the sun and the planets can be more irregular. One example is the chaotic dynamics of Halley's comet, first uncovered by Chirikov and Vecheslavov (1989). They described the motion of the comet as a Kepler orbit, reducing the problem to either a two-dimensional map (if the influence only of Jupiter is taken into account) or a three-dimensional map (if the influence of Saturn is included, too). These maps have regions of stable and chaotic behavior: the current position of the comet was estimated to lie within a chaotic region, rather close to its boundary. The Lyapunov exponent computed by Chirikov and Vecheslavov (1989) was in the range $0.16 \lesssim \lambda \lesssim 0.26$, in units of the comet basic period. Longer calculations revealed a diffusion of the comet's energy due to effectively random kicks from giant planets. As a result they estimated an escape of the comet from the solar system after about 10 Myr, although it is not clear whether the map is still valid on such time scales (e.g. the comet might have evaporated earlier).

12.6 Quantum chaos

The term "quantum chaos" refers to the behaviour of quantum-mechanical systems whose classical counterpart behaves chaotically. This is a broad research field (Stöckmann, 1999; Haake, 2010) which belongs to the even wider class of wave (electromagnetic and acoustic) chaos, where the problem consists in inferring propagation properties whenever the dynamics of rays is chaotic. Here we comment only briefly on those aspects that are related to the concept of Lyapunov exponents.

The definition of LE is intimately related to that of "trajectory", which is not present in quantum mechanics that is instead based on the notion of "quantum" state, whose evolution is ruled by a unitary operator (propagator over a time interval t) $\hat{U}(t)$. Such an evolution is stable in the following sense: the correlation between a given state $|\Psi\rangle$ and a perturbed one $|\Psi'\rangle$, understood as the overlap (scalar product) of these states, is time-independent:

$$\langle\Psi'(t)|\Psi(t)\rangle = \langle\Psi'(0)|\hat{U}^{\dagger}(t)U(t)|\Psi(0)\rangle = \langle\Psi'(0)|\Psi(0)\rangle. \qquad (12.8)$$

By recalling that $\hat{U}^{\dagger}(t) = \hat{U}(-t)$, one can interpret this conservation of the correlation as a nearly perfect invertibility of the quantum evolution: if one lets the forward evolution be followed by the backward one, it is found that a quantum system (nearly) returns to its initial value, even if, at the inversion time, it is slightly perturbed. This is in contrast with the evolution of classical systems, where the memory of the initial condition is lost if the evolution time exceeds the characteristic Lyapunov time.

The classical non-invertibility is often referred to as the Loschmidt paradox: in a discussion about the "arrow of time" in statistical thermodynamics, Loschmidt in 1876 argued that, because of the invariance under time reversal of the classical dynamics, entropy decreases should be also observable. In view of the modern theory of chaos, it is nowadays understood that although individual trajectories are time reversible, the evolution of a typical ensemble of trajectories (a density) leads to an entropy increase in both time directions. In a simple numerical test, the time inversion of a classical chaotic trajectory does not reproduce the initial state because of the unavoidable round-off errors which grow exponentially according to the largest Lyapunov exponent (which, in Hamiltonian systems, has the same value in both time directions), so that the backward trajectory is followed only during the short "predictability time" (? 15). On the other hand, as it follows from Eq. (12.8), the quantum evolution is resilient against the introduction of perturbations (see, e.g., the comparison between the classical and the quantum Chirikov-Taylor standard map by Shepelyansky (1983)).

There is, however, another setup, where the existence of classical chaos manifests itself in a quantum world, inducing an exponential decay of the correlations; this happens if, instead of perturbing the initial state, the Hamiltonian itself is perturbed. Namely, one considers the evolution of the same initial state $|\Psi(0)\rangle$ under two slightly different propagators, $\hat{U}_0(t)$ and $\hat{U}_\varepsilon(t)$. The overlap between unperturbed and perturbed states at time t is called *fidelity*,

$$F_\varepsilon(t) = |\langle\Psi_0(t)|\Psi_\varepsilon(t)\rangle|^2 = |\langle\Psi(0)|\hat{U}_0^{\dagger}(t)U_\varepsilon(t)|\Psi(0)\rangle|^2. \qquad (12.9)$$

One can interpret this relation also as a "Loschmidt echo": fidelity shows how exact the state returns back if the forward and the backward evolution operators differ by a small perturbation $\sim \varepsilon$. As it has been first demonstrated by Jalabert and Pastawski (2001), using a semiclassical approximation, the fidelity (12.9) exhibits a pronounced range of exponential decay in time, with the rate equal (in a certain regime) to the largest classical Lyapunov exponent λ. We do not enter the details of the theory (see Gorin et al. (2006) for a review), but we describe the results according to Goussev et al. (2012). An exponential decay of the fidelity (12.9) according to the Lyapunov exponent λ occurs for perturbations

ε that are neither too small nor too large. For very small ε, the decay rate is smaller than λ, due to the contribution of uncorrelated pairs of classical trajectories which contribute to $F_\varepsilon(t)$. For very large ε, the fidelity decays as $\sim \exp(-t^2)$. Also the time range where an exponential decay of the fidelity is observed is limited. For very small times, the fidelity decays quadratically in time $F \sim 1 - (\eta t/\hbar)^2$, due to the average dispersion η of the perturbation propagator with respect to the initial state. For very large times, the decay of fidelity is saturated by value $F \sim N^{-1}$, where N is the size of the effective Hilbert space (volume of the phase space available for the state during evolution, expressed in units of the Planck's cell). In other words, the exponential decay of $F \sim \exp[-\lambda t]$ is observed only in the time range $\hbar/\eta \lesssim t \lesssim \lambda^{-1} \ln N$.

Reference models

In this appendix we introduce those models that are used in the monograph to illustrate various properties of the LEs. For the sake of simplicity, they are organised in a few categories of increasing complexity.

A.1 Lumped systems: discrete time

The simplest examples of a self-sustained irregular dynamics are given by discrete-time dynamical systems.

The logistic map

The logistic map is a map of an interval

$$U(t + 1) = a - U^2(t). \tag{A.1}$$

This model was originally introduced to describe population dynamics. It expresses the size $U(t+1)$ of the $(t+1)$th generation as a function of the size of the previous generation. It was later acknowledged that the same map approximately describes a wide class of nonlinear phenomena, where all variables but one are very stable (strongly contracting). This map is one-dimensional, i.e. the current state is characterised by a scalar variable $U(t)$. A major limit of the map is that it is not invertible; in fact, given a generic state $U(t + 1)$, there may exist more than one preimage $U(t)$. Quite often other forms of the logistic map are used, obtained from (A.1) by a linear transformation of the variable:

$$U(t + 1) = 1 - bU^2(t), \qquad U(t + 1) = cU(t)(1 - U(t)). \tag{A.2}$$

If there exist exactly two preimages for each point of the interval, the logistic map displays "full chaos". That happens at the so-called Ulam point ($a = 2$, $b = 2$, $c = 4$) where the map is called Ulam map.

The tent map

$$U(t + 1) = \begin{cases} \frac{U(t)}{a} & 0 \le U(t) \le a \\ \frac{1 - U(t)}{1 - a} & a < U(t) \le 1 \end{cases} \tag{A.3}$$

is a piece-wise linear version of the logistic map (A.1) at the Ulam point. The parameter a allows control of the fluctuations of the local multiplier (in the symmetric case, $a = 1/2$; no fluctuations of the absolute value of the multipliers are present).

The Hénon map

$$U(t + 1) = a - U(t)^2 + bU(t - 1) \qquad (A.4)$$

generalises the logistic map to a two-dimensional space; here the state at time t is uniquely identified by the pair of variables $U(t)$ and $U(t - 1)$. An important advantage over the logistic map is that it is invertible: one can easily verify that given the state $U(t + 1), U(t)$, there exists one and only one $U(t - 1)$. Sometimes in the literature one finds the equivalent formulation $U(t + 1) = 1 - aU(t)^2 + bU(t - 1)$ (this latter expression is easily obtained by transforming $U(t) \rightarrow aU(t)$ in (A.4)).

The Lozi map

The Lozi map is basically a piece-wise linear version of the Hénon map,

$$U(t + 1) = 1 - a|U(t)| + bU(t - 1). \qquad (A.5)$$

Like the Hénon map, the Lozi map is invertible.

The Chirikov-Taylor standard map

$$
\begin{aligned}
Q(t + 1) &= Q(t) + P(t) \\
P(t + 1) &= P(t) + K \sin Q(t + 1) = P(t) + K \sin(Q(t) + P(t)),
\end{aligned}
\qquad (A.6)
$$

where both variables $Q(t)$ and $P(t)$ are meant to be taken as modulo 2π. This two-dimensional map is the prototypical model of a chaotic symplectic dynamics, where volumes are conserved and forward and backward dynamics are mutually equivalent. K controls the degree of chaos (the dynamics is typically quasiperiodic for $K \ll 1$.).

A.2 Lumped systems: continuous time

Sets of a few nonlinear ordinary differential equations offer the chance of investigating the continuous-time dynamics.

The FitzHugh-Nagumo model

$$\dot{U} = U - \frac{U^3}{3} - V + I$$

$$\tau \dot{V} = U + a - bV \qquad (A.7)$$

is used as a reference for the onset of an oscillatory dynamics in excitable systems.

The Rössler oscillator

Here $\mathbf{U} = (X, Y, Z)$,

$$\dot{X} = -Y - Z$$
$$\dot{Y} = X + aZ \qquad \qquad (A.8)$$
$$\dot{Z} = b + Z(X - c).$$

This set of equations is called Rössler oscillator. These equations, derived as a simplified model for chemical kinetics, represent one of the prototypical testbed for the study of deterministic chaos. The typical parameter values used in this monograph are $a = 0.1$, $b = 0.1$ and $c = 10$.

The Lorenz system

$$\dot{X} = \sigma(X - Y),$$
$$\dot{Y} = -Y + rX - XZ, \qquad \qquad (A.9)$$
$$\dot{Z} = -bZ + XY,$$

where $U = (X, Y, Z)$ is one of the most popular models for chaos. The "standard" parameter values used by Lorenz are $\sigma = 10$, $r = 28$ and $b = 8/3$.

A.3 Lattice systems: discrete time

Chains of maps (coupled map lattices) provide a natural environment to investigate chaotic properties of spatially extended systems, but they have to be complemented by suitable boundary conditions. Unless otherwise specified, periodic boundary conditions are assumed (the same holds for the other classes of space-time systems).

Logistic maps

In the case of logistic maps, the model is written as

$$U_x(t + 1) = a - [U_x(t) + \varepsilon \mathcal{D} U_x(t)]^2, \qquad \qquad (A.10)$$

where $\mathcal{D} U_x \equiv (U_{x-1} - 2U_x + U_{x+1})$ is the discrete Laplacian operator. This is the simplest example of chaotic spatially extended system. Sometimes in the literature the evolution rule is expressed by referring to the variable $V_x(t) = U_x(t) + \varepsilon \mathcal{D} U_x(t)$. The two formulations are perfectly equivalent.

Hénon maps

Coupled Hénon maps follow the evolution equation

$$U_x(t+1) = a - [U_x(t) + \varepsilon \mathcal{D} U_x(t)]^2 + b U_x(t-1).$$ (A.11)

This is a minimal invertible model for space-time chaos (Politi and Torcini, 1992). It has a symplectic-like structure: Lyapunov exponents come in pairs $\lambda_i, \lambda_{L-i+1}$ (L is the system size) such that $\lambda_i + \lambda_{L-i+1} = \ln b$ (see also Section 2.5.7).

Symplectic maps

A model of coupled symplectic maps can be defined as

$$Z_x(t+1) = Z_x(t) + \mu \big(\sin(U_{x+1}(t) - U_x(t)) - \sin(U_x(t) - U_{x-1}(t)) \big)$$
$$U_x(t+1) = U_x(t) + Z_x(t+1),$$ (A.12)

where both variables are meant to be taken modulo 2π. This is a minimal model often used as a testbed for symplectic dynamics in spatially extended systems in the presence of a conservation law (the sum of the Z variables).

Stable-chaos maps

The following chain of maps provides a paradigmatic stable-chaos model,

$$U_x(t+1) = (1 - 2\varepsilon)F(U_x(t)) + \varepsilon[F(U_{x-1}(t) + F(x_{x+1}(t))],$$ (A.13)

where $F(U)$ is the piecewise-linear function

$$F(U) = \begin{cases} U/a & 0 \le U \le a \\ 1 - (1-b)(U-a)/\eta & a < U < a + \eta \\ b + c(U - a - \eta) & a + \eta < U \le 1. \end{cases}$$ (A.14)

The model is typically studied by fixing $a = 1/2.7$, $b = 0.07$ and $c = 0.1$.

A.4 Lattice systems: continuous time

Allowing for a continuous time in a lattice model offers the possibility of studying more realistic physical systems.

The Fermi-Pasta-Ulam (FPU) model

The most basic model used to test the chaotic properties of one-dimensional Hamiltonian systems is the so-called (β) Fermi-Pasta-Ulam model, which is identified by the Hamiltonian

$$\mathcal{H} = \sum_x \left[\frac{P_x^2}{2} + V(Q_x - Q_{x+1}) \right], \tag{A.15}$$

where the potential is $V(Q) = Q^2/2 + Q^4/4$. The corresponding equations of motion are

$$\ddot{Q}_x = (Q_{x-1} - Q_x) + (Q_{x+1} - Q_x) + (Q_{x-1} - Q_x)^3 + (Q_{x+1} - Q_x)^3. \tag{A.16}$$

Rössler oscillators

A chain of Rössler oscillators

$$\begin{aligned}
\dot{X}_x &= -Y_x - Z_x \\
\dot{Y}_x &= X_x + aZ_x + \varepsilon(Y_{x+1} - 2Y_x + Y_{x-1}) \\
\dot{Z}_x &= b + Z_x(X_x - c),
\end{aligned} \tag{A.17}$$

where $\mathbf{U}_x = (X_x, Y_x, Z_x)$ is a model of dissipative chaotic attractors, often used to investigate phase synchronisation.

A.5 Spatially continuous systems

In many physical setups, the spatial variable is continuous. Partial differential equations provide the natural framework to investigate such systems.

The complex Ginzburg-Landau equation

This is the typical testbed for the study of space-time chaos. It indeed arises as a normal form in the vicinity of a Hopf bifurcation, where the stationary homogeneous solution destabilises. In one-dimensional space it is written as

$$\frac{\partial U}{\partial t} = RU - (1 + i\mu)|U|^2 U + (1 + iv)\frac{\partial^2 U}{\partial x^2}. \tag{A.18}$$

Here, U is a complex variable and the parameter R can, in principle, be scaled out, but here it is kept to emphasize the possibility for the model to operate below the instability threshold (i.e. for $R < 0$).

The Kuramoto-Sivashinsky equation

This partial derivative equation is often used to characterise chemical turbulence, flame-front propagation, and the dynamics of liquid films subject to gravity. In one dimension it is written as

$$\frac{\partial U}{\partial t} + \frac{\partial^2 U}{\partial x^2} + \frac{\partial^4 U}{\partial x^4} + U\frac{\partial U}{\partial x} = 0. \tag{A.19}$$

A.6 Differential-delay systems

Differential-delay equations represent another class of high-(infinite-)dimensional dynamical systems. In this monograph we consider the Ikeda equation,

$$\dot{U} = -U + a\sin[U(t - \tau) - U_0],\tag{A.20}$$

which was introduced to describe the behaviour of an optical bistable resonator.

A.7 Global coupling: discrete time

In many physical systems, the interactions are not restricted to the nearest neighbours but extend to large distances. This happens in plasma physics, neural dynamics and general synchronisation problems. Globally coupled (mean field) models represent the limiting situation, where each variable interacts with all the others, irrespective of their spatial location.

The typical model of globally coupled maps is defined as follows:

$$U_j(t + 1) = F(V_j(t)),\tag{A.21}$$

where

$$V_j = (1 - \varepsilon)U_j(t) + \varepsilon\bar{U}(t),\tag{A.22}$$

while

$$\bar{U}(t) = \frac{1}{N}\sum U_j(t)\tag{A.23}$$

represents the mean field. In this book we also examine a model where

$$V_j = U_j + \varepsilon(1 - 2a^2\bar{U}^2)\sin U_j,\tag{A.24}$$

$$\bar{U}(t) = \frac{1}{N}\sum\cos U_j(t)\tag{A.25}$$

and where the local dynamics is induced by a standard one-dimensional Bernoulli map ($F(V) = 2V \bmod 2\pi$). The variable U_j is, in practice, an angle.

A.8 Global coupling: continuous time

The following equations define a rather general model structure:

$$\dot{\mathbf{U}}_k = \mathbf{F}_k(\mathbf{U}_k, \bar{\mathbf{V}}_k, \bar{\mathbf{Z}}_k)$$

$$\dot{\bar{\mathbf{Z}}}_k = \mathbf{Q}(\bar{\mathbf{V}}_k, \bar{\mathbf{Z}}_k), \qquad \bar{\mathbf{V}}_k = \frac{1}{N}\sum_{j=1}^{N} G_{kj}\mathbf{c}(\mathbf{U}_j).\tag{A.26}$$

Here, $\mathbf{U}_k$ defines the state of the kth unit, while $\overline{\mathbf{V}}_k$ and $\overline{\mathbf{Z}}_k$ account for the global interaction of the kth unit with the rest of the world. The latter variables follow their own dynamics. The Kuramoto model and pulse-coupled oscillators are of this type.

Stuart-Landau oscillators

A widely used model is defined by the equations

$$\dot{U}_j = U_j - (1 + ic_2)|U_j|^2 U_j + K(1 + ic_1)(\overline{U} - U_J), \qquad (A.27)$$

where U_j is a complex variable, while $\overline{U}$ again corresponds to the average value of U. The evolution equation of the single oscillator is the normal form of the Hopf bifurcation.

Hamiltonian mean field

The Hamiltonian mean field (HMF) is the prototypical model for the study of Hamiltonian dynamics in the presence of long-range interactions. It has been derived from a one-dimensional self-gravitating model by truncating the Fourier expansion of the gravitational potential to its first term (Inagaki and Konishi, 1993; Ruffo, 1994). It consists of N unit-mass particles that move on a circle under their mutual attraction. The dynamics of the N particles is ruled by the Hamiltonian

$$H = \sum_{i=1}^{N} \frac{P_i^2}{2} + \frac{1}{2N} \sum_{i,j=1}^{N} \left[1 - \cos(\Theta_i - \Theta_j)\right], \qquad (A.28)$$

where Θ_i and P_i denote particle positions (angles) and velocities. The resulting equations of motion are

$$\dot{\Theta}_i = P_i,$$
$$\dot{P}_i = \frac{1}{N} \sum_j \sin(\Theta_j - \Theta_i) = M \sin(\Phi - \Theta_i), \qquad (A.29)$$

where M is the magnetisation and Φ the associated phase, defined by

$$Me^{i\Phi} = \frac{1}{N} \sum_j e^{i\Theta_j}. \qquad (A.30)$$

The model can also be viewed as the Hamiltonian version of the Kuramoto model (in the case of equal frequencies).

Pseudocodes

In this appendix we report some pseudocodes to be used for the computation of Lyapunov exponents and covariant Lyapunov vectors. The first pseudocode contains a master program for the computation of M Lyapunov exponents of a generic N-dimensional map $\mathbf{F}(\mathbf{U})$. If the QR decomposition is performed after a time τ (as done in continuous-time systems), the LEs are obtained by dividing Λ_j by the total time τT_{av}.

The QR decomposition operates on the input vectors $\{\mathbf{u}_j\}$, returning them after orthonormalisation, together with the diagonal elements $\{\alpha_j\}$ of the triangular matrix R, which are required for the computation of the LEs. The second pseudocode describes the most accurate version of the Gram-Schmidt orthogonalisation (see Section 3.2.1). This routine returns the entire matrix R (the off-diagonal elements are necessary only for the computation of the covariant vectors). When the QR decomposition is performed via a Gram-Schmidt orthogonalisation, the α_j values are automatically positive.

The third pseudocode describes the computation of M covariant Lyapunov vectors. There, $\lceil \cdot \rceil^j$ means that the vector delimited by the brackets is restricted to the first j components (the same holds for $\|\cdot\|^j$). $\lceil \mathbf{v}_j \rceil^j$ denotes the jth covariant Lyapunov vector expressed in the local orthogonal basis (this is the reason why it has j components). Such vectors should be considered as asymptotic only if the time $t < T_{av} - T_{tran}$, where T_{tran} is the transient time needed to ensure convergence. If it is necessary to express the covariant vectors into the physical basis, one has to make use of the vectors $\{\mathbf{w}_j(t)\}$ (see the pseudocode).

The fourth pseudocode describes a QR decomposition performed with Householder reflections. $\lfloor \cdot \rfloor_j$ means that the quantity delimited by the brackets is restricted to the last j components (the same applies to $\|\cdot\|_j$); finally, u_{jj} denotes the jth component of the vector $\mathbf{u}_j$.

Algorithm 1   Computation of M Lyapunov exponents

$\mathbf{U}(0) \leftarrow$ initial conditions for the map .... *[initialisation starts]*
for $j = 1$ to M **do**
   $\mathbf{u}_j(0) \leftarrow$ initial conditions for the linearised variables
   $\Lambda_j \leftarrow 0$
end for .... *[initialisation ends]*

Main loop: T_{trans} is the transient time; T_{av} is the averaging time

---

for $t = 1$ to $T_{trans} + T_{av}$ **do**
   for $j = 1$ to M **do**
      $\mathbf{u}_j(t) = \frac{\partial \mathbf{F}}{\partial \mathbf{U}} \cdot \mathbf{u}_j(t-1)$ .... *[linearised map]*
   end for
   $\mathbf{U}(t) = \mathbf{F}(\mathbf{U}(t-1))$ .... *[nonlinear map]*

   call QR($\{\mathbf{u}_j, \alpha_j\}$) .... *[QR decomposition is performed]*

   if $t > T_{trans}$ **then**
      for $j = 1$ to M **do**
         $\Lambda_j \leftarrow \Lambda_j + \ln |\alpha_j|$ .... *[running average]*
      end for
   end if
end for .... *[end of main loop]*

for $j = 1$ to M **do**
   $\lambda_j \leftarrow \Lambda_j / T_{av}$
end for

Algorithm 2   QR decomposition: Gram-Schmidt orthogonalisation

---

$\alpha_1 \leftarrow \|\mathbf{u}_1\|$
$\mathbf{u}_1 \leftarrow \mathbf{u}_1 / \alpha_1$
$R_{11} \leftarrow \alpha_1$
for $j = 2$ to M **do**
   for $i = j$ to M **do**
      $R_{j-1,i} \leftarrow \mathbf{u}_{j-1} \cdot \mathbf{u}_i$
      $\mathbf{u}_i \leftarrow \mathbf{u}_i - R_{j-1,i}\mathbf{u}_{j-1}$
   end for
   $\alpha_j \leftarrow \|\mathbf{u}_j\|$
   $\mathbf{u}_j \leftarrow \mathbf{u}_j / \alpha_j$
   $R_{jj} \leftarrow \alpha_j$
end for

Algorithm 3	Computation of M covariant Lyapunov vectors

Initialisation

$\mathbf{U}(0) \leftarrow$ initial conditions for the map
for $j = 1$ to M **do**
   $\mathbf{u}_j(0) \leftarrow$ initial conditions for the forward linearised variables
   $\Lambda_j \leftarrow 0$
end for
for $j = 1$ to M **do**
   $\lceil \mathbf{v}_j \rceil^j \leftarrow$ initial conditions for the backward linearised variables
   $s \leftarrow \|\mathbf{v}_j\|^j$
   $\lceil \mathbf{v}_j \rceil^j \leftarrow \lceil \mathbf{v}_j/s \rceil^j$
end for

Forward transient

for $t = 1$ to T_{trans} **do**
   for $j = 1$ to M **do**
      $\mathbf{u}_j(t) = \frac{\partial \mathbf{F}}{\partial \mathbf{U}} \cdot \mathbf{u}_j(t-1)$ .... *[linearised map]*
   end for
   $\mathbf{U}(t) = \mathbf{F}(\mathbf{U}(t-1))$ .... *[nonlinear map]*
   call QR($\{\mathbf{u}_j, \alpha_j\}, \mathsf{R}(t)$) .... *[QR decomposition is performed]*
end for

Forward loop

for $t = 1$ to T_{av} **do**
   $\{\mathbf{w}_j(t)\} \leftarrow \{\mathbf{u}_j\}$ .... *[M $\{\mathbf{u}_j\}$vectors are saved for later use]*
   for $j = 1$ to M **do**
      $\mathbf{u}_j(t) = \frac{\partial \mathbf{F}}{\partial \mathbf{U}} \cdot \mathbf{u}_j(t-1)$ .... *[linearised map]*
   end for
   $\mathbf{U}(t) = \mathbf{F}(\mathbf{U}(t-1))$ .... *[nonlinear map]*
   call QR($\{\mathbf{u}_j, \alpha_j\}, \mathsf{R}(t)$) .... *[QR decomposition is performed]*
end for

Backward loop

for $t = T_{av}$ to $1, -1$ **do**
   for $j = 1$ to M **do**
      for $k = j$ to $2, -1$ **do**
         $v_{jk} \leftarrow v_{jk}/R_{kk}(t)$ .... *[backward iteration of the vector $\mathbf{v}_j$]*
         for $i = 1$ to $k - 1$ **do**
            $v_{ji} \leftarrow v_{ji} - v_{jk}R_{ik}(t)$
         end for
      end for
      $v_{j1} \leftarrow v_{j1}/R_{11}(t)$
      $s \leftarrow \|\mathbf{v}_j\|^j$
      $\lceil \mathbf{v}_j \rceil^j \leftarrow \lceil \mathbf{v}_j/s \rceil^j$
   end for
end for

Algorithm 4	QR decomposition: Householder reflections

for $j = 1, M$ **do**
    $s \leftarrow \|\mathbf{u}_j\|_j$
    $\alpha_j \leftarrow -s\,\mathrm{sgn}(u_{jj})$
    $X \leftarrow \sqrt{s(s + |u_{jj}|)}$
    $u_{jj} \leftarrow u_{jj} - \alpha_j$
    $\lfloor \mathbf{u}_j \rfloor_j \leftarrow \lfloor \mathbf{u}_j / X \rfloor_j$
    for $i = j + 1, M$ **do**
        $\sigma \leftarrow \lfloor \mathbf{u}_j \cdot \mathbf{u}_i \rfloor_j$
        $\lfloor \mathbf{u}_i \rfloor_j \leftarrow \lfloor \mathbf{u}_i - \sigma \mathbf{u}_j \rfloor_j$
    end for
end for

for $i = 1, M$ **do**
    $\mathbf{y}(i) \leftarrow \mathbf{e}_i$
    for $j = M, 1, -1$ **do**
        $\sigma \leftarrow \lfloor \mathbf{u}_j \cdot \mathbf{u}_i \rfloor_j$
        $\lfloor \mathbf{y}_i \rfloor_j \leftarrow \lfloor \mathbf{y}_i - \sigma \mathbf{u}_j \rfloor_j$
    end for
end for
for $i = 1, M$ **do**
    $\mathbf{u}_i \leftarrow \mathbf{y}_i$
end for

In this appendix we prove some results for the products of random matrices that are discussed in Chapter 8. More specifically, the following two sections are devoted to the computation of the Lyapunov spectrum in a discrete-time and continuous-time case, respectively.

C.1 Gaussian matrices: discrete time

An ensemble of $N \times N$ matrices A that are characterised by the symmetry properties mentioned in Section 8.1.2 satisfy Eq. (8.14). Such an equation can be rewritten as

$$\mathcal{S}_k = \langle \ln \| \mathsf{A}\mathbf{e}_1 \wedge \ldots \wedge \mathsf{A}\mathbf{e}_k \| \rangle = \langle \ln \| \mathsf{A}\mathbf{e}_1 \wedge \ldots \wedge \mathsf{A}\mathbf{e}_{k-1} \| \| \mathsf{P}_k \mathsf{A}_k \mathbf{e}_k \| \rangle,$$

where P_k is the projector onto the $(N-k+1)$-dimensional subspace orthogonal to the subspace spanned by $\mathsf{A}\mathbf{e}_1, \mathsf{A}\mathbf{e}_2, \ldots \mathsf{A}\mathbf{e}_{k-1}$. In the Gaussian matrix ensemble, the vectors $\mathsf{A}\mathbf{e}_1, \mathsf{A}\mathbf{e}_2, \ldots$ and $\mathsf{A}\mathbf{e}_{k-1}$ are independent of $\mathsf{A}\mathbf{e}_k$. Moreover, $\mathsf{A}\mathbf{e}_k$ is a random vector that has the same distribution for any rotation of A, so that we can replace P with the projector onto the last $(N-k+1)$ vectors of an Euclidean basis, obtaining

$$\lambda_k = \frac{1}{2} \left\langle \ln \left[\sum_i^{N-k+1} A_{ik}^2 \right] \right\rangle.$$

The average can be performed by noticing that $W_n := A_{1k}^2 + A_{2k}^2 + \cdots + A_{nk}^2$ (the index k in A_{ik} is irrelevant) has a χ^2 distribution. The use of standard relations for the χ^2 distribution (Abramowitz and Stegun, 1964) allows the determination of an explicit expression for the moments of W_n (Cohen and Newman, 1984),

$$\langle W_n^q \rangle = \frac{\sigma^{2q}}{2^{n/2}\Gamma(n/2)} \int_0^{+\infty} dw \, w^{n/2+q-1} e^{-w/2} \tag{C.1}$$

$$= \frac{(2\sigma^2)^q \Gamma(n/2+q)}{\Gamma(n/2)} = \exp[(\ln 2\sigma^2)q + \phi(n/2+q) - \phi(n/2)],$$

where $\phi(q) = \ln \Gamma(q)$ and $\Gamma(q)$ is the standard gamma function. An expression for $\langle \ln W_n \rangle$ is then obtained, by noticing that $\langle W_n^q \rangle = \langle \exp(q \ln W_n) \rangle$ and thereby expanding this equation around $q = 0$,

$$\langle \ln W_n \rangle = \frac{d}{dq} \langle W_n^q \rangle \Big|_{q=0} = \frac{d}{dq} \langle \ln W_n^q \rangle \Big|_{q=0}$$
$$= \ln 2\sigma^2 + \psi(n/2) \tag{C.2}$$

where $\psi(y) = \Gamma'(y)/\Gamma(y)$ is the digamma function. As a result,

$$\lambda_k = \ln \sigma + \frac{1}{2} \left[\ln 2 + \psi \left(\frac{N-k+1}{2} \right) \right].$$

C.2 Gaussian matrices: continuous time

Given the model defined in Section 8.2.4, one can express the volume defined in Eq. (8.39) as the determinant of a $k \times k$ matrix, obtained by projecting onto the first k vectors of the Euclidean basis,

$$\Lambda_k = \lim_{m \to \infty} \frac{m}{2} \left\langle \ln \det \left(\mathsf{P}_k e^{\mathsf{C}/\sqrt{m}} e^{\mathsf{C}/\sqrt{m}} \mathsf{P}_k \right) \right\rangle.$$

By now exploiting the smallness of $1/\sqrt{m}$, one can expand the exponentials in the aforementioned equation and make use of the standard relation

$$\ln \det(\mathsf{I} + \varepsilon \mathsf{A}) = \mathrm{Tr}\left[\ln (\mathsf{I} + \varepsilon \mathsf{A}) \right] = \sum_{i=1}^{\infty} (-1)^{i+1} \frac{\mathrm{Tr}\mathsf{A}^i}{i} \varepsilon^i.$$

As a result, it is found that

$$\Lambda_k = \lim_{m \to \infty} \frac{m}{2} \left\langle \mathrm{Tr} \left\{ \mathsf{P}_k \left[\frac{\mathsf{C}^\mathsf{T} + \mathsf{C}}{\sqrt{m}} + \frac{(\mathsf{C}^\mathsf{T} + \mathsf{C})^2}{2m} \right. \right. \right.$$
$$\left. \left. \left. - \frac{(\mathsf{C}^\mathsf{T} + \mathsf{C})\mathsf{P}_k(\mathsf{C}^\mathsf{T} + \mathsf{C})}{2m} \right] \mathsf{P}_k \right\} + o(1/m) \right\rangle, \tag{C.3}$$

where we have used that $\mathrm{Tr}(\mathsf{C}^\mathsf{T}\mathsf{C}) = \mathrm{Tr}(\mathsf{C}\mathsf{C}^\mathsf{T})$. Finally, by recalling that the entries of the matrix C have zero average, one obtains

$$\Lambda_k = \left\langle \sum_{i=1}^{k} \sum_{j=k+1}^{N} (C_{ji} + C_{ij})(C_{ij} + C_{ji}) \right\rangle = \frac{\sigma^2}{2} \sum_{i=1}^{k} (N-k),$$

which implies Eq. (8.40).

Symbolic encoding

In this appendix we briefly introduce the basic elements of a powerful approach that helps to characterise chaotic dynamical systems and eventually to obtain accurate estimates of its Lyapunov exponents. The idea is to partition the phase space into a collection $\mathcal{P}$ of disjoint elements $\{\mathcal{B}_i\}$ (the atoms) and thereby encode a generic trajectory $\{\mathbf{U}_n\}$ as a sequence of symbols $\{s_n\}$, where $s_n = \mathcal{B}_i$ if $\mathbf{U}_n \in \mathcal{B}_i$. The procedure is faithful only if the partition $\mathcal{P}$ is *generating*; i.e. an infinitely long trajectory is encoded by one and only one sequence of symbols.

In maps of the interval, a generating partition can be constructed by splitting the interval itself into subsets, where the map behaves monotonously (Collet and Eckmann, 1980). For instance the logistic map (A.2) $U' = aU(1 - U)$ has a maximum in $U = 1/2$, and its dynamics can be thereby encoded as a sequence of binary symbols, which are selected depending on whether the phase point belongs to the interval $[0, 1/2)$ or $[1/2, 1]$.

In two-dimensional spaces, the problem of constructing a generating partition is much harder. No rigorous results are, in fact, available, but there is compelling evidence that a method proposed by Grassberger and Kantz (1985) for the Hénon map works for generic dissipative models. It makes use of the homoclinic tangencies (i.e. the points where stable and unstable manifolds are mutually tangent). In practice, the two-dimensional plane is split into two parts by the polygonal line obtained by connecting the so-called *primary* tangencies (approximately, those characterised by a minimal value of the sum of the curvature of the two manifolds). As shown by Giovannini and Politi (1992) and Hansen (1992), the final result is not unique: a given dynamical system can be characterised by equivalent but different symbolic descriptions.

The approach can be extended to symplectic systems by complementing the use of homoclinic tangencies with that of suitable symmetry lines, which allow the partitioning of the ordered regions where no tangencies are present (Christiansen and Politi, 1997).

By definition, any trajectory of a map $\mathbf{F}(\mathbf{U})$ is encoded as a suitable symbolic sequence, but the converse is not generally true; typically, there exist infinitely many sequences that cannot be generated by a given mapping $\mathbf{F}$. This information is implicitly contained in the value of the topological entropy.

Markov partitions represent an important subclass of generating partitions. A partition $\mathcal{P}$ is said to be Markov if each atom $\mathcal{B}_i$ is mapped exactly onto the union of one or more atoms. As a result, whenever a finite Markov partion is available, the topological entropy can be exactly determined (see Chapter 5). In general, however, no finite Markov partition is typically available. In one-dimensional spaces, useful information can be extracted with the help of the kneading theory (Collet and Eckmann, 1980; Milnor and Thurston, 1988). In two-dimensional spaces, the leading tool is the so-called pruning front (Cvitanović et al., 1988).

Bibliography

Abrahams, E. (ed.). 2010. *50 years of Anderson localization*. World Scientific Publishing, Hackensack, NJ.

Abramowitz, M., and Stegun, I. A. 1964. *Handbook of mathematical functions with formulas, graphs, and mathematical tables*. National Bureau of Standards Applied Mathematics Series, vol. 55. For sale by the Superintendent of Documents, U.S. Government Printing Office, Washington, DC.

Acebrón, J. A., Bonilla, L. L., Pérez Vicente, C. J., Ritort, F., and Spigler, R. 2005. The Kuramoto model: a simple paradigm for synchronization phenomena. *Rev. Mod. Phys.* **77**: 137–185.

Ahlers, V., Zillmer, R., and Pikovsky, A. 2001. Lyapunov exponents in disordered chaotic systems: avoided crossing and level statistics. *Phys. Rev. E* **63**: 036213.

Ames, W. F. 1992. *Numerical methods for partial differential equations*. 3rd edn. Academic Press, Boston, MA.

Anderson, P. W. 1958. Absence of diffusion in certain random lattices. *Phys. Rev.* **109**: 1492–1505.

Arecchi, F., Giacomelli, G., Lapucci, A., and Meucci, R. 1992. Two-dimensional representation of a delayed dynamical system. *Phys. Rev. A* **45**: R4225–R4228.

Arnold, L. 1998. *Random dynamical systems*. Berlin: Springer-Verlag.

Arnold, L., and Imkeller, P. 1995. Furstenberg-Khas'minskii formulas for Lyapunov exponents via anticipative calculus. *Stochastics Stochastics Rep.* **54**: 127–168.

Arnold, L., Papanicolaou, G., and Wihstutz, V. 1986. Asymptotic analysis of the Lyapunov exponent and rotation number of the random oscillator and applications. *SIAM J. Appl. Math.* **46**: 427–450.

Artuso, R., Casati, G., and Guarneri, I. 1997. Numerical study on ergodic properties of triangular billiards. *Phys. Rev. E* **55**: 6384–6390.

Artuso, R., Guarneri, I., and Rebuzzini, L. 2000. Spectral properties and anomalous transport in a polygonal billiard. *Chaos* **10**: 189–194.

Ashwin, P., and Breakspear, M. 2001. Anisotropic properties of riddled basins. *Phys. Lett. A* **280**: 139–145.

Ashwin, P., Buescu, J., and Stewart, I. 1994. Bubbling of attractors and synchronisation of chaotic oscillators. *Phys. Lett. A* **193**: 126–139.

Aston, Ph. J., and Dellnitz, M. 1995. Symmetry breaking bifurcations of chaotic attractors. *Int. J. Bifurcat. Chaos* **5**: 1643–1676.

Aston, Ph. J., and Dellnitz, M. 1999. The computation of Lyapunov exponents via spatial integration with application to blowout bifurcations. *Comput. Methods Appl. Mech. Engrg.* **170**: 223–237.

Aston, Ph. J., and Laing, C. R. 2000. Symmetry and chaos in the complex Ginzburg-Landau equation. II. Translational symmetries. *Physica D* **135**: 79–97.

Aston, Ph. J., and Melbourne, I. 2006. Lyapunov exponents of symmetric attractors. *Nonlinearity* **19**: 2455–2466.

Aurell, E., Boffetta, G., Crisanti, A., Paladin, G., and Vulpiani, A. 1996. Growth of noninfinitesimal perturbations in turbulence. *Phys. Rev. Lett.* **77**: 1262–1265.

Aurell, E., Boffetta, G., Crisanti, A., Paladin, G., and Vulpiani, A. 1997. Predictability in the large: an extension of the concept of Lyapunov exponent. *J. Phys. A – Math. Gen.* **30**: 1–26.

Badii, R., and Politi, A. 1997. *Complexity: hierarchical structures and scaling in physics.* Cambridge University Press, Cambridge.

Bagnoli, F., Rechtman, R., and Ruffo, S. 1992. Damage spreading and Lyapunov exponents in cellular automata. *Phys. Lett. A* **172**: 34–38.

Barabási, A.-L, and Stanley, H. E. 1995. *Fractal concepts in surface growth.* Cambridge University Press, Cambridge.

Baranyai, A., Evans, D. J., and Cohen, E. G. D. 1993. Field-dependent conductivity and diffusion in a two-dimensional Lorentz gas. *J. Stat. Phys.* **70**: 1085–1098.

Barreira, L., and Pesin, Y. 2007. *Nonuniform hyperbolicity: dynamics of systems with nonzero Lyapunov exponents.* Encyclopedia of mathematics and its applications, vol. 115. Cambridge University Press, Cambridge.

Baxendale, P. H., and Goukasian, L. 2002. Lyapunov exponents for small random perturbations of Hamiltonian systems. *Ann. Probab.* **30**: 101–134.

Beck, C., and Schlögl, F. 1995. *Thermodynamics of chaotic systems: an introduction.* Cambridge University Press, Cambridge.

Beenakker, C. W. J. 1997. Random-matrix theory of quantum transport. *Rev. Mod. Phys.* **69**: 731–808.

Benettin, G., Galgani, L., Giorgilli, A., and Strelcyn, J.-M. 1980a. Lyapunov characteristic exponents for smooth dynamical systems and for Hamiltonian systems; a method for computing all of them. Part I: theory. *Meccanica* **15**: 9–20.

Benettin, G., Galgani, L., Giorgilli, A., and Strelcyn, J.-M. 1980b. Lyapunov characteristic exponents for smooth dynamical systems and for Hamiltonian systems; a method for computing all of them. Part II: numerical application. *Meccanica* **15**: 21–30.

Benzi, R., Paladin, G., Parisi, G., and Vulpiani, A. 1985. Characterization of intermittency in chaotic systems. *J. Phys. A – Math. Gen.* **18**: 2157.

Berlekamp, E. R., Conway, J. H., and Guy, R. K. 1982. *Winning ways for your mathematical plays. Vol. 2: Games in particular.* Academic Press, London and New York.

Bettencourt, J. H., López, C., and Hernández-García, E. 2013. Characterization of coherent structures in three-dimensional turbulent flows using the finite-size Lyapunov exponent. *J. Phys. A – Math. Theor.* **46**: 254022.

Beyn, W.-J., and Lust, A. 2009. A hybrid method for computing Lyapunov exponents. *Numer. Math.* **113**: 357–375.

Biktashev, V. N. 2005. Causodynamics of autowave patterns. *Phys. Rev. Lett.* **95**: 084501.

Bliokh, K. Y., Bliokh, Yu. P., Freilikher, V., Savel'ev, S., and Nori, F. 2008. Colloquium: Unusual resonators: plasmonics, metamaterials, and random media. *Rev. Mod. Phys.* **80**: 1201–1213.

Bochi, J., and Viana, M. 2005. The Lyapunov exponents of generic volume-preserving and symplectic maps. *Ann. Math.* **161**: 1423–1485.

Borland, R. E. 1963. The nature of the electronic states in disordered one-dimensional systems. *P. Roy. Soc. Lond. A Mat.* **274**: 529–545.

Bougerol, Ph., and Lacroix, J. 1985. *Products of random matrices with applications to Schrödinger operators*. Progress in Probability and Statistics, vol. 8. Birkhäuser, Boston, MA.

Bourgain, J. 2005. *Green's function estimates for lattice Schrödinger operators and applications*. Princeton University Press, Princeton, NJ.

Bridges, T. J., and Reich, S. 2001. Computing Lyapunov exponents on a Stiefel manifold. *Physica D* **156**: 219–238.

Broomhead, D. S., Jones, R., and King, G. P. 1987. Topological dimension and local coordinates from time series data. *J. Phys. A – Math. Gen.* **20**: L563–L569.

Brown, R., Bryant, P., and Abarbanel, H. D. I. 1991. Computing the Lyapunov spectrum of a dynamical system from an observed time series. *Phys. Rev. A* **43**: 2787–2806.

Bryant, P., Brown, R., and Abarbanel, H. D. I. 1990. Lyapunov exponents from observed time series. *Phys. Rev. Lett.* **65**: 1523–1526.

Butcher, J. C. 2008. *Numerical methods for ordinary differential equations*. 2nd edn. John Wiley & Sons, Chichester.

Campanino, M., and Klein, A. 1990. Anomalies in the one-dimensional Anderson model at weak disorder. *Comm. Math. Phys.* **130**: 441–456.

Carroll, T. L., and Pecora, L. M. 1991. Synchronizing chaotic circuits. *IEEE Trans. Circ. and Systems* **38**: 453–456.

Casetti, L., Livi, R., and Pettini, M. 1995. Gaussian model for chaotic instability of Hamiltonian flows. *Phys. Rev. Lett.* **74**: 375–378.

Cecconi, F., and Politi, A. 1999. An analytic estimate of the maximum Lyapunov exponent in products of tridiagonal random matrices. *J. Phys. A – Math. Gen.* **32**: 7603–7621.

Cencini, M., and Torcini, A. 2001. Linear and nonlinear information flow in spatially extended systems. *Phys. Rev. E* **63**: 056201.

Cencini, M., and Vulpiani, A. 2013. Finite size Lyapunov exponent: review on applications. *J. Phys. A – Math. and Theor.* **46**: 254019.

Cencini, M., Falcioni, M., Vergni, D., and Vulpiani, A. 1999. Macroscopic chaos in globally coupled maps. *Physica D* **130**: 58–72.

Cessac, B., Doyon, B., Quoy, M., and Samuelides, M. 1994. Mean-field equations, bifurcation map and route to chaos in discrete time neural networks. *Physica D* **74**: 24–44.

Chernov, N., and Markarian, R. 2006. *Chaotic billiards*. American Mathematical Society, Providence, RI.

Chernov, N. I., Eyink, G. L., Lebowitz, J. L., and Sinai, Ya. G. 1993. Derivation of Ohm's law in a deterministic mechanical model. *Phys. Rev. Lett.* **70**: 2209–2212.

Chirikov, B. V., and Vecheslavov, V. V. 1989. Chaotic dynamics of comet Halley. *Astronomy & Astrophysics* **221**: 146–154.

Christiansen, F., and Politi, A. 1997. Guidelines for the construction of a generating partition in the standard map. *Physica D* **109**: 32–41.

Christiansen, F., and Rugh, H. H. 1997. Computing Lyapunov spectra with continuous Gram-Schmidt orthonormalization. *Nonlinearity* **10**: 1063–1072.

Cipriani, P., and Politi, A. 2004. An open-system approach for the characterization of spatio-temporal chaos. *J. Stat. Phys.* **114**: 205–228.

Cohen, J. E., and Newman, Ch. M. 1984. The stability of large random matrices and their products. *Ann. Probab.* **12**: 283–310.

Cole, J. D. 1951. On a quasi-linear parabolic equation occurring in aerodynamics. *Quart. Appl. Math.* **9**: 225–236.

Collet, P., and Eckmann, J.-P. 1980. *Iterated maps on the interval as dynamical systems.* Birkhäuser, Boston, MA.

Collet, P., and Eckmann, J.-P. 1999. Extensive properties of the complex Ginzburg-Landau equation. *Comm. Math. Phys.* **200**: 699–722.

Cook, J., and Derrida, B. 1990. Lyapunov exponents of large, sparse random matrices and the problem of directed polymers with complex random weights. *J. Stat. Phys.* **61**: 961–986.

Cooper, F., Khare, A., and Sukhatme, U. 1995. Supersymmetry and quantum mechanics. *Phys. Rep.* **251**: 267–385.

Corazza, M., Kalnay, E., Patil, D. J., Yang, S. C., Morss, R., Cai, M., Szunyogh, I., Hunt, B. R., and Yorke, J. A. 2003. Use of the breeding technique to estimate the structure of the analysis "errors of the day". *Nonl. Processes in Geophysics* **10**: 233–243.

Crauel, H., Debussche, A., and Flandoli, F. 1997. Random attractors. *J. Dynam. Differential Equations* **9**: 307–341.

Crisanti, A., Paladin, G., and Vulpiani, A. 1993. *Products of random matrices in statistical physics.* Springer-Verlag, Berlin.

Crutchfield, J. P., and Kaneko, K. 1988. Are attractors relevant to turbulence? *Phys. Rev. Lett.* **60**: 2715–2718.

Curato, G., and Politi, A. 2013. Onset of chaotic dynamics in neural networks. *Phys. Rev. E* **88**: 042908.

Cvitanović, P., Gunaratne, G. H., and Procaccia, I. 1988. Topological and metric properties of Hénon-type strange attractors. *Phys. Rev. A* **38**: 1503–1520.

Cvitanović, P., Artuso, R., Mainieri, R., Tanner, G., and Vattay, G. 2013. *Chaos: classical and quantum.* Niels Bohr Institute, Copenhagen. www.chaosbook.org.

Dahlqvist, P. 1997. The Lyapunov exponent in the Sinai billiard in the small scatterer limit. *Nonlinearity* **10**: 159–173.

Daido, H. 1984. Coupling sensitivity of chaos: a new universal property of chaotic dynamical systems. *Progr. Theoret. Phys. Suppl.* **79**: 75–95.

Daido, H. 1985. Coupling sensitivity of chaos and the Lyapunov dimension: the case of coupled two-dimensional maps. *Phys. Lett. A* **110**: 5–9.

Daido, H. 1987. Coupling sensitivity of chaos: theory and further numerical evidence. *Phys. Lett. A* **121**: 60–66.

D'Alessandro, G., Grassberger, P., Isola, S., and Politi, A. 1990. On the topology of the Hénon map. *J. Phys. A – Math. Gen.* **23**: 5285–5294.

Darrigol, O. 2002. Stability and instability in nineteenth-century fluid mechanics. *Rev. Histoire Math.* **8**: 5–65.

Deissler, R. J., and Kaneko, K. 1987. Velocity-dependent Lyapunov exponents as a measure of chaos for open-flow systems. *Phys. Lett. A* **119**: 397–402.

Delfini, L., Denisov, S., Lepri, S., Livi, R., Mohanty, P. K., and Politi, A. 2007. Energy diffusion in hard-point systems. *Eur. Phys. J – Spec. Top.* **146**: 21–35.

Dellago, Ch., and Posch, H. A. 1995. Lyapunov exponents of systems with elastic hard collisions. *Phys. Rev. E* **52**: 2401–2406.

Dellago, Ch., and Posch, H. A. 1997. Lyapunov spectrum and the conjugate pairing rule for a thermostatted random Lorentz gas: numerical simulations. *Phys. Rev. Lett.* **78**: 211–214.

Dellnitz, M., and Hohmann, A. 1997. A subdivision algorithm for the computation of unstable manifolds and global attractors. *Numer. Math.* **75**: 293–317.

Derrida, B., and Gardner, E. 1984. Lyapounov exponent of the one-dimensional Anderson model: weak disorder expansions. *J. Physique* **45**: 1283–1295.

Derrida, B., and Hilhorst, H. J. 1983. Singular behaviour of certain infinite products of random 2×2 matrices. *J. Phys. A – Math. Gen.* **16**: 2641–2654.

Derrida, B., and Spohn, H. 1988. Polymers on disordered trees, spin glasses, and traveling waves. *J. Stat. Phys.* **51**: 817–840.

Derrida, B., Mecheri, K., and Pichard, J. L. 1987. Lyapounov exponents of products of random matrices: weak disorder expansion. Application to localisation. *J. Physique* **48**: 733.

Deutsch, J. M., and Paladin, G. 1989. Product of random matrices in a microcanonical ensemble. *Phys. Rev. Lett.* **62**: 695–699.

di Bernardo, M., Budd, C. J., Champneys, A.R., and Kowalczyk, P. 2008. *Piecewise-smooth dynamical systems: theory and applications*. Springer-Verlag, London.

Dieci, L., and Van Vleck, E. S. 1995. Computation of a few Lyapunov exponents for continuous and discrete dynamical systems. *Appl. Numer. Math.* **17**: 275–291.

Dieci, L., and Van Vleck, E.S. 2005. On the error in computing Lyapunov exponents by QR methods. *Numer. Math.* **101**: 619–642.

Dieci, L., Russell, R. D., and Van Vleck, E. S. 1997. On the computation of Lyapunov exponents for continuous dynamical systems. *SIAM J. Numer. Anal.* **34**: 402–423.

Dorfman, J. R., and van Beijeren, H. 1997. Dynamical systems theory and transport coefficients: a survey with applications to Lorentz gases. *Physica A* **240**: 12–42.

d'Ovidio, F., Fernández, V., Hernández-García, E., and López, C. 2004. Mixing structures in the Mediterranean Sea from finite-size Lyapunov exponents. *Geophys. Res. Lett.* **31**: L17203.

Dressler, U. 1988. Symmetry property of the Lyapunov spectra of a class of dissipative dynamical systems with viscous damping. *Phys. Rev. A* **38**: 2103–2109.

Dressler, U., and Farmer, J. D. 1992. Generalized Lyapunov exponents corresponding to higher derivatives. *Physica D* **59**: 365–377.

Eckhardt, B., and Yao, D. 1993. Local Lyapunov exponents in chaotic systems. *Physica D* **65**: 100–108.

Eckmann, J.-P., and Ruelle, D. 1985. Ergodic theory of chaos and strange attractors. *Rev. Mod. Phys.* **57**: 617–656.

Eckmann, J.-P., and Wayne, C. E. 1988. Liapunov spectra for infinite chains of nonlinear oscillators. *J. Stat. Phys.* **50**: 853–878.

Eckmann, J.-P., Forster, Ch., Posch, H. A., and Zabey, E. 2005. Lyapunov modes in hard-disk systems. *J. Stat. Phys.* **118**: 813–847.

Ershov, S. V., and Potapov, A. B. 1998. On the concept of stationary Lyapunov basis. *Physica D* **118**: 167–198.

Evans, D. J., and Morris, G. 2008. *Statistical mechanics of nonequilibrium liquids*. 2nd edn. Cambridge University Press, Cambridge. Books Online.

Evans, D. J., Cohen, E. G. D., and Morriss, G. P. 1990. Viscosity of a simple fluid from its maximal Lyapunov exponents. *Phys. Rev. A* **42**: 5990–5997.

Evers, F., and Mirlin, A. D. 2008. Anderson transitions. *Rev. Mod. Phys.* **80**: 1355–1417.

Family, F., and Vicsek, T. 1985. Scaling of the active zone in the Eden process on percolation networks and the ballistic deposition mode l. *J. Phys. A – Math. Gen.* **18**: L75.

Farmer, J. D. 1981. Spectral broadening of period-doubling bifurcation sequences. *Phys. Rev. Lett.* **47**: 179.

Farmer, J. D. 1982. Chaotic attractors of an infinite-dimensional dynamical system. *Physica D: Nonlinear Phenomena* **4**: 366–393.

Farmer, J. D., and Sidorowich, J. J. 1987. Predicting chaotic time series. *Phys. Rev. Lett.* **59**: 845–848.

Feigel'man, M. V., and Tsvelik, A. M. 1982. Hidden supersymmetry of stochastic dissipative dynamics. *Sov. Phys. JETP* **56**: 823.

Feudel, U., Kuznetsov, S., and Pikovsky, A. 2006. *Strange nonchaotic attractors: dynamics between order and chaos in quasiperiodically forced systems*. World Scientific, Hackensack, NJ.

Feynman, R. P., and Hibbs, A. R. 2010. *Quantum mechanics and path integrals*. Dover, Mineola, NY.

Fischer, R. A. 1937. The wave of advance of advantageous genes. *Ann. Eugenics* **7**: 353–369.

Francisco, G., and Matsas, G. E. A. 1988. Qualitative and numerical study of Bianchi IX models. *General Relativity and Gravitation* **20**: 1047–1054.

Friedman, B., Oono, Y., and Kubo, I. 1984. Universal behavior of Sinai billiard systems in the small-scatterer limit. *Phys. Rev. Lett.* **52**: 709–712.

Froeschlé, C., Lega, E., and Gonczi, R. 1997. Fast Lyapunov indicators: application to asteroidal motion. *Celest. Mech. Dyn. Astr.* **67**: 41–62.

Furstenberg, H., and Kesten, H. 1960. Products of random matrices. *Ann. Math. Statist.* **31**: 457–469.

Gardiner, C. 2009. *Stochastic methods: a handbook for the natural and social sciences*. Springer-Verlag, Berlin.

Gaspard, P. 2005. *Chaos, scattering and statistical mechanics*. Cambridge University Press, Cambridge.

Gaspard, P., and Nicolis, G. 1990. Transport properties, Lyapunov exponents, and entropy per unit time. *Phys. Rev. Lett.* **65**: 1693–1696.

Geist, K., Parlitz, U., and Lauterborn, W. 1990. Comparison of different methods for computing Lyapunov exponents. *Progr. Theoret. Phys.* **83**: 875–893.

Gertsenshtein, M. E., and Vasiljev, V. B. 1959. Waveguides with random inhomogeneities and Brownian motion in the Lobachevsky plane. *Theor. Prob. Appl.* **4**: 391–398.

Giacomelli, G., Hegger, R., Politi, A., and Vassalli, M. 2000. Convective Lyapunov exponents and propagation of correlations. *Phys. Rev. Lett.* **85**: 3616–3619.

Ginelli, F., Livi, R., Politi, A., and Torcini, A. 2003. Relationship between directed percolation and the synchronization transition in spatially extended systems. *Phys. Rev. E* **67**: 046217.

Ginelli, F., Poggi, P., Turchi, A., Chaté, H., Livi, R., and Politi, A. 2007. Characterizing dynamics with covariant Lyapunov vectors. *Phys. Rev. Lett.* **99**: 130601.

Ginelli, F., Takeuchi, K., Chaté, H., Politi, A., and Torcini, A. 2011. Chaos in the Hamiltonian mean-field model. *Phys. Rev. E* **84**: 066211.

Ginelli, F., Chaté, C., Livi, R., and Politi, A. 2013. Covariant Lyapunov vectors. *J. Phys. A – Math. Theor.* **46**: 254005.

Giovannini, F., and Politi, A. 1992. Generating partitions in Hénon-type maps. *Phys. Lett. A* **161**: 332–336.

Girko, V. L. 1984. The circular law. *Teor. Veroyatnost. i Primenen.* **29**: 669–679.

Goldhirsch, I., Sulem, P.-L., and Orszag, S. A. 1987. Stability and Lyapunov stability of dynamical systems: a differential approach and a numerical method. *Physica D* **27**: 311–337.

Goldobin, D. S., and Pikovsky, A. 2004. Synchronization of periodic self-oscillations by common noise. *Radiophys. Quantum El.* **47**: 910–915.

Goldobin, D. S., and Pikovsky, A. 2006. Antireliability of noise-driven neurons. *Phys. Rev. E* **73**: 061906.

Goldobin, D. S., Teramae, J., Nakao, H., and Ermentrout, G. B. 2010. Dynamics of limit-cycle oscillators subject to general noise. *Phys. Rev. Lett.* **105**: 154101.

Golub, G. H., and Van Loan, Ch. F. 1996. *Matrix computations*. Johns Hopkins University Press, Baltimore, MD.

Gorin, Th., Prosen, T., Seligman, Th. H., and Znidaric, M. 2006. Dynamics of Loschmidt echoes and fidelity decay. *Phys. Rep.* **435**: 33–156.

Goussev, A., Jalabert, R. A., Pastawski, H. M., and Wisniacki, D. A. 2012. Loschmidt echo. *Scholarpedia* **7**(8): 11687.

Gozzi, E., and Reuter, M. 1994. Lyapunov exponents, path-integrals and forms. *Chaos, solitons & fractals* **4**: 1117–1139.

Graham, R. 1988. Lyapunov exponents and supersymmetry of stochastic dynamical systems. *Europhys. Lett.* **5**: 101–106.

Grassberger, P., and Kantz, H. 1985. Generating partitions for the dissipative Hénon map. *Phys. Lett. A* **113**: 235–238.

Grassberger, P., and Procaccia, I. 1983. On the characterization of strange attractors. *Phys. Rev. Lett.* **50**: 346–349.

Grassberger, P., Badii, R., and Politi, A. 1988. Scaling laws for invariant measures on hyperbolic and nonhyperbolic attractors. *J. Stat. Phys.* **51**: 135–178.

Grebogi, C., Ott, E., and Yorke, J. A. 1983. Crises, sudden changes in chaotic attractors, and transient chaos. *Physica D* **7**: 181–200.

Gredeskul, S. A., and Freilikher, V. D. 1990. Localization and wave propagation in randomly layered media. *Physics-Uspekhi* **33**: 134–146.

Gupalo, D., Kaganovich, A. S., and Cohen, E. G. D. 1994. Symmetry of Lyapunov spectrum. *J. Stat. Phys.* **74**: 1145–1159.

Gutkin, E. 1986. Billiards in polygons. *Physica D* **19**: 311–333.

Haake, F. 2010. *Quantum signatures of chaos*. Springer-Verlag, Berlin.

Habib, S., and Ryne, R. D. 1995. Symplectic calculation of Lyapunov exponents. *Phys. Rev. Lett.* **74**: 70–73.

Hairer, E., Lubich, Ch., and Wanner, G. 2010. *Geometric numerical integration: structure-preserving algorithms for ordinary differential equations*. Springer, Heidelberg.

Hale, J. K. 1969. *Ordinary differential equations*. Wiley-Interscience, New York.

Haller, G. 2001. Distinguished material surfaces and coherent structures in three-dimensional fluid flows. *Physica D* **149**: 248–277.

Haller, G. 2002. Lagrangian coherent structures from approximate velocity data. *Phys. Fluids* **14**: 1851–1861.

Haller, G., and Yuan, G. 2000. Lagrangian coherent structures and mixing in two-dimensional turbulence. *Physica D* **147**: 352–370.

Halpin-Healy, T., and Zhang, Y.-Ch. 1995. Kinetic roughening phenomena, stochastic growth, directed polymers and all that. Aspects of multidisciplinary statistical mechanics. *Phys. Rep.* **254**: 215–414.

Hansen, K. T. 1992. Remarks on the symbolic dynamics for the Hénon map. *Phys. Lett. A* **165**: 100–104.

Harmer, G. P., and Abbott, D. 1999. Parrondo's paradox. *Statist. Sci.*, **14**: 206–213.

Hartung, F., Krisztin, T., Walther, H.-O., and Wu, J. 2006. Functional differential equations with state-dependent delays: theory and applications. In Cañada, A., Drábek, P., and Fonda, A. (eds.), *Handook of differential equations*, vol. 3. Amer. Math. Soc., Providence, RI, 435–546.

Heiligenthal, S., Dahms, Th., Yanchuk, S., Jüngling, Th., Flunkert, V., Kanter, I., Schöll, E., and Kinzel, W. 2011. Strong and weak chaos in nonlinear networks with time-delayed couplings. *Phys. Rev. Lett.* **107**: 234102.

Hénon, M. 1982. On the numerical computation of Poincaré maps. *Physica D* **5**: 412–414.

Herrmann, H. J. 1990. Damage spreading. *Physica A* **168**: 516–528.

Hoover, W. G. 1991. *Computational statistical mechanics*. Elsevier Science, Amsterdam.

Hopf, E. 1950. The partial differential equation $u_t + uu_x = \mu u_{xx}$. *Comm. Pure Appl. Math.* **3**: 201–230.

Horsthemke, W., and Bach, A. 1975. Onsager-Machlup function for one-dimensional nonlinear diffusion processes. *Z. Phys. B Cond. Mat.* **22**: 189–192.

Ilachinski, A. 2001. *Cellular automata: a discrete universe*. World Scientific, River Edge, NJ.

Inagaki, S., and Konishi, T. 1993. Dynamical stability of a simple model similar to self-gravitating systems. *Publ. Astron. Soc. Japan* **45**: 733–735.

Isopi, M., and Newman, Ch. M. 1992. The triangle law for Lyapunov exponents of large random matrices. *Comm. Math. Phys.* **143**: 591–598.

Izrailev, F. M., Ruffo, S., and Tessieri, L. 1998. Classical representation of the one-dimensional Anderson model. *J. Phys. A – Math. Gen.* **31**: 5263–5270.

Izrailev, F. M., Krokhin, A. A., and Makarov, N. M. 2012. Anomalous localization in low-dimensional systems with correlated disorder. *Phys. Rep.* **512**: 125–254.

Jalabert, R., and Pastawski, H. 2001. Environment-independent decoherence rate in classically chaotic systems. *Phys. Rev. Lett.* **86**: 2490–2493.

Johnson, R. A., Palmer, K. J., and Sell, G. R. 1987. Ergodic properties of linear dynamical systems. *SIAM J. Math. Anal.* **18**: 1–33.

Kaneko, K. 1985. Spatiotemporal intermittency in coupled map lattices. *Progr. Theor. Phys.* **74**: 1033–1044.

Kantz, H., and Grassberger, P. 1985. Repellers, semi-attractors and long-lived chaotic transients. *Physica D* **17**: 75–86.

Kantz, H., and Schreiber, Th. 2004. *Nonlinear time series analysis*. Cambridge University Press, Cambridge.

Kantz, H., Radons, G., and Yang, H. 2013. The problem of spurious Lyapunov exponents in time series analysis and its solution by covariant Lyapunov vectors. *J. Phys. A – Math. Theor.* **46**: 254009

Kaplan, J. L., and Yorke, J. A. 1979. Chaotic behavior of multidimensional difference equations. In Walter, H. O., and Peitgen, H.-O. (eds.), *Functional differential equations and approximation of fixed points*. Springer-Verlag, Berlin, 204–227.

Kardar, M., Parisi, G., and Zhang, Y.-Ch. 1986. Dynamic scaling of growing interfaces. *Phys. Rev. Lett.* **56**: 889–892.

Kargin, V. 2014. On the largest Lyapunov exponent for products of Gaussian matrices. *J. Stat. Phys.* **157**: 70–83.

Karrasch, D., and Haller, G. 2013. Do finite-size Lyapunov exponents detect coherent structures? *Chaos* **23**: 043126.

Katok, A., and Hasselblatt, B. 1995. *Introduction to the modern theory of dynamical systems*. Cambridge University Press, Cambridge.

Kauffman, S. A. 1969. Metabolic stability and epigenesis in randomly constructed genetic nets. *J. Theor. Biol.* **22**: 437–467.

Kenfack J., A., Politi, A., and Torcini, A. 2013. Convective Lyapunov spectra. *J. Phys. A* **46**: 254013.

Khasminskii, R. 2012. *Stochastic stability of differential equations*. Springer, Heidelberg. With contributions by G. N. Milstein and M. B. Nevelson.

Kockelkoren, J. 2002. *Dynamique hors d'équilibre et universalité en présence d'une quantité conservée*. Ph.D. thesis, Université Denis Diderot, Paris 7, CEA.

Kolmogorov, A. N., and Tikhomirov, V. M. 1961. ε-entropy and ε-capacity of sets in functional spaces. *Amer. Math. Soc. Transl. Ser. 2* **17**: 277–364.

Kolmogorov, N., Petrovsky, I., and Piscounov, N. 1937. A study of the diffusion equation with increase in the amount of substance, and its application to a biological problem. *Bull. Univ. Moscow, Ser. Int.* **A1**: 1.

Korabel, N., and Barkai, E. 2010. Separation of trajectories and its relation to entropy for intermittent systems with a zero Lyapunov exponent. *Phys. Rev. E* **82**: 016209.

Kostelich, E. J., Kan, I., Grebogi, C., Ott, E., and Yorke, J. A. 1997. Unstable dimension variability: a source of nonhyperbolicity in chaotic systems. *Physica D* **109**: 81–90.

Kramer, B., and MacKinnon, A. 1993. Localization: theory and experiment. *Rep. Prog. Phys.* **56**: 1469.

Kramer, B., MacKinnon, A., Ohtsuki, T, and Slevin, K. 1987. Finite size scaling analysis of the Anderson transition. In Abrahams, E. (ed.), *50 years of Anderson localization*. World Scientific, Singapore, 347–360.

Krug, J., and Meakin, P. 1990. Universal finite-size effects in the rate of growth processes. *J. Phys. A – Math. Gen.* **23**: L987.

Kruis, H. V., Panja, D., and van Beijeren, H. 2006. Systematic density expansion of the Lyapunov exponents for a two-dimensional random Lorentz gas. *J. Stat. Phys.* **124**: 823–842.

Krylov, N. S. 1979. *Works on the foundations of statistical physics*. Princeton University Press, Princeton, N.J. Translated by A. B. Migdal, Ya. G. Sinai and Yu. L. Zeeman. With a preface by A. S. Wightman.

Kunze, M. 2000. Lyapunov exponents for non-smooth dynamical systems. In Kunze, M. (ed.), *Non-smooth dynamical systems*. Springer, Berlin and Heidelberg, 63–140.

Kuptsov, P. V., and Kuznetsov, S. P. 2009. Violation of hyperbolicity in a diffusive medium with local hyperbolic attractor. *Phys. Rev. E* **80**: 016205.

Kuptsov, P. V., and Parlitz, U. 2012. Theory and computation of covariant Lyapunov vectors. *J. Nonl. Sci.* **22**: 727–762.

Kuptsov, P. V., and Politi, A. 2011. Large-deviation approach to space-time chaos. *Phys. Rev. Lett.* **107**: 114101.

Kuramoto, Y. 1975. Self-entrainment of a population of coupled nonlinear oscillators. In Araki, H. (ed.), *International symposium on mathematical problems in theoretical physics*. Springer, New York, 420.

Kuramoto, Y. 1984. *Chemical oscillations, waves and turbulence*. Springer, Berlin.

Kuznetsov, S. P., and Pikovsky, A. 1986. Universality and scaling of period-doubling bifurcations in dissipative distributed medium. *Physica D* **19**: 384–396.

Laffargue, T., Lam, Kh.-D. N.-Th., Kurchan, J., and Tailleur, J. 2013. Large deviations of Lyapunov exponents. *J. Phys. A – Math. Theor.* **46**: 254002.

Lai, Y.-Ch, and Tél, T. 2011. *Transient chaos: complex dynamics on finite time scales*. Springer, New York.

Lam, Kh.-D. N.-Th., and Kurchan, J. 2014. Stochastic perturbation of integrable systems: a window to weakly chaotic systems. *J. Stat. Phys.* **156**: 619–646.

Landau, L. D., and Lifshitz, E. M. 1958. *Quantum mechanics: non-relativistic theory. Course of theoretical physics, vol. 3*. Pergamon, Londo and Paris.

Laskar, J. 1989. A numerical experiment on the chaotic behaviour of the solar system. *Nature* **338**: 237–238.

Laskar, J., and Gastineau, M. 2009. Existence of collisional trajectories of Mercury, Mars and Venus with the Earth. *Nature* **459**: 817–819.

Latz, A., van Beijeren, H., and Dorfman, J. R. 1997. Lyapunov spectrum and the conjugate pairing rule for a thermostatted random Lorentz gas: kinetic theory. *Phys. Rev. Lett.* **78**: 207–210.

Ledrappier, F. 1981. Some relations between dimension and Lyapunov exponents. *Commun. Math. Phys.* **81**: 229–238.

Leimkuhler, B., and Reich, S. 2004. *Simulating Hamiltonian dynamics*. Cambridge University Press, Cambridge.

Leine, R. I. 2010. The historical development of classical stability concepts: Lagrange, Poisson and Lyapunov stability. *Nonlinear Dynam.* **59**: 173–182.

Leonov, G. A., and Kuznetsov, N. V. 2007. Time-varying lineraization and the Perron effects. *Int. J. Bifurcat. Chaos* **17**: 1079.

Lepri, S., Giacomelli, G., Politi, A., and Arecchi, F.T. 1994. High-dimensional chaos in delayed dynamical systems. *Physica D* **70**: 235–249.

Lepri, S., Politi, A., and Torcini, A. 1996. Chronotopic Lyapunov analysis. I: a detailed characterization of 1D systems. *J. Stat. Phys.* **82**: 1429.

Lepri, S., Politi, A., and Torcini, A. 1997. Chronotopic Lyapunov analysis. II: toward a unified approach. *J. Stat. Phys.* **88**: 31.

Lepri, S., Livi, R., and Politi, R. 2003. Thermal conduction in classical low-dimensional lattices. *Phys. Rep.* **377**: 1–80.

Letz, T., and Kantz, H. 2000. Characterization of sensitivity to finite perturbations. *Phys. Rev. E* **61**: 2533–2538.

Lifshits, I. M., Gredeskul, S. A., and Pastur, L. A. 1988. *Introduction to the theory of disordered systems*. John Wiley & Sons, New York.

Livi, R., Politi, A., and Ruffo, S. 1986. Distribution of characteristic exponents in the thermodynamic limit. *J. Phys. A – Math. Gen* **19**: 2033–2040.

Livi, R., Politi, A., Ruffo, S., and Vulpiani, A. 1987. Liapunov exponents in high-dimensional symplectic dynamics. *J. Stat. Phys.* **46**: 147–160.

Livi, R., Politi, A., and Ruffo, S. 1992. Scaling-law for the maximal Lyapunov exponent. *J. Phys. A – Math. Gen.* **25**: 4813–4826.

Lorenz, E. N. 1963. Deterministic nonperiodic flow. *J. Atmos. Sci.* **20**: 130–141.

Luccioli, S., Olmi, S., Politi, A., and Torcini, A. 2012. Collective dynamics in sparse networks. *Phys. Rev. Lett.* **109**: 138103.

Lyapunov, A. M. 1992. *The general problem of the stability of motion*. Taylor & Francis, Ltd., London. Translated from Edouard Davaux's French translation (1907) of the 1892 Russian original and edited by A. T. Fuller, with an introduction and preface by Fuller, a biography of Lyapunov by V. I. Smirnov, and a bibliography of Lyapunov's works compiled by J. F. Barrett, Lyapunov centenary issue. Reprint of *Internat. J. Control* **55** (1992), no. 3.

MacKinnon, A., and Kramer, B. 1981. One-parameter scaling of localization length and conductance in disordered systems. *Phys. Rev. Lett.* **47**: 1546–1549.

MacKinnon, A., and Kramer, B. 1983. The scaling theory of electrons in disordered solids: additional numerical results. *Z. Phys. B Cond. Mat.* **53**: 1–13.

Mainen, Z. F., and Sejnowski, T. J. 1995. Reliability of spike timing in neocortical neurons. *Science* **268**: 1503.

Mainieri, R. 1992. Cycle expansion for the Lyapunov exponent of a product of random matrices. *Chaos* **2**: 91–97.

Mallick, K., and Peyneau, P.-E. 2006. Phase diagram of the random frequency oscillator: the case of Ornstein–Uhlenbeck noise. *Physica D* **221**: 72–83.

Manneville, P. 1985. Liapounov exponents for the Kuramoto-Sivashinsky model. In *Macroscopic modelling of turbulent flows (Nice, 1984)*. Springer, Berlin, 319–326.

Marčenko, V. A., and Pastur, L. A. 1967a. Distribution of eigenvalues in certain sets of random matrices. *Mat. Sb. (N.S.)* **72**: 507–536.

Marčenko, V. A., and Pastur, L. A. 1967b. The spectrum of random matrices. *Teor. Funkciĭ Funkcional. Anal. i Priložen. Vyp.* **4**: 122–145.

Marinari, E., Pagnani, P., and Parisi, G. 2000. Critical exponents of the KPZ equation via multi-surface coding numerical simulations. *J. Phys. A – Math. Gen.* **33**: 8181.

Markoš, P. 1993. Weak disorder expansion of Lyapunov exponents of products of random matrices: a degenerate theory. *J. Stat. Phys.* **70**: 899–919.

Martin, B. 2007. Damage spreading and μ-sensitivity on cellular automata. *Ergodic Theory Dynam. Systems* **27**: 545–565.

McNamara, S., and Mareschal, M. 2001. Origin of the hydrodynamic Lyapunov modes. *Phys. Rev. E* **64**: 051103.

Mehta, M. L. 2004. *Random matrices*. Elsevier/Academic Press, Amsterdam.

Mello, P. A., and Robledo, A. 1993. Strongly coupled Ising chain under a weak random field. *Physica A* **199**: 363–386.

Milnor, J., and Thurston, W. 1988. On iterated maps of the interval. In *Dynamical systems (College Park, MD, 1986–87)*. Springer, Berlin.

Monteforte, M., and Wolf, F. 2010. Dynamical entropy production in spiking neuron networks in the balanced state. *Phys. Rev. Lett.* **105**: 268104.

Morton, K. W., and Mayers, D. F. 2005. *Numerical solution of partial differential equations: an introduction*. Cambridge University Press, Cambridge.

Motter, A. E. 2003. Relativistic chaos is coordinate invariant. *Phys. Rev. Lett.* **91**: 231101.

Motter, A. E., and Saa, A. 2009. Relativistic invariance of Lyapunov exponents in bounded and unbounded systems. *Phys. Rev. Lett.* **102**: 184101.

Müller, P. C. 1995. Calculation of Lyapunov exponents for dynamic systems with discontinuities. *Chaos Solitons Fractals* **5**: 1671–1681.

Murray, C. D., and Dermott, S. F. 1999. *Solar system dynamics*. Cambridge University Press, Cambridge.

Newman, Ch. M. 1986a. The distribution of Lyapunov exponents: exact results for random matrices. *Comm. Math. Phys.* **103**: 121–126.

Newman, Ch. M. 1986b. Lyapunov exponents for some products of random matrices: exact expressions and asymptotic distributions. In *Random matrices and their applications (Brunswick, Maine, 1984)*. Providence, RI, 121–141.

Olmi, S., Politi, A., and Torcini, A. 2012. Stability of the splay state in networks of pulse-coupled neurons. *J. Math. Neurosci.* **2**: 12

Oseledets, V. 2008. Oseledets theorem. *Scholarpedia* **3**(1): 1846.

Oseledets, V. I. 1968. A multiplicative ergodic theorem. Lyapunov characteristic numbers for dynamical systems. *Trans. Moscow Math. Soc.* **19**: 197–231.

Paladin, G., and Vulpiani, A. 1986. Scaling law and asymptotic distribution of Lyapunov exponents in conservative dynamical systems with many degrees of freedom. *J. Phys. A – Math. Gen.* **19**: 1881–1888.

Paoli, P., Politi, A., and Badii, R. 1989. Long-range order in the scaling behaviour of hyperbolic dynamical systems. *Physica D* **36**: 263–286.

Parks, P. C. 1992. A. M. Lyapunov's stability theory–100 years on. *IMA J. Math. Control Inform.* **9**: 275–303.

Parrondo, J.-M. R., and Dins, L. 2004. Brownian motion and gambling: from ratchets to paradoxical games. *Contemp. Phys.* **45**(2): 147–157.

Pastur, L., and Figotin, A. 1992. *Spectra of random and almost-periodic operators.* Springer-Verlag, Berlin.

Patil, D. J., Hunt, B. R., Kalnay, E., Yorke, J. A., and Ott, E. 2001. Local low dimensionality of atmospheric dynamics. *Phys. Rev. Lett.* **86**: 5878–5881.

Pazó, D., López, J. M., and Politi, A. 2013. Universal scaling of Lyapunov-exponent fluctuations in space-time chaos. *Phys. Rev. E* **87**: 062909.

Peacock, Th., and Haller, G. 2013. Lagrangian coherent structures: the hidden skeleton of fluid flows. *Phys. Today* **66**: 41–47.

Pecora, L. M., and Carroll, T. L. 1991. Driving systems with chaotic signals. *Phys. Rev. A* **44**: 2374–2383.

Pecora, L. M., and Carroll, T. L. 1998. Master stability functions for synchronized coupled systems. *Phys. Rev. Lett.* **80**: 2109–2112.

Pecora, L. M., and Carroll, T. L. 1999. Master stability functions for synchronized coupled systems. *Int. J. Bifurcat. Chaos* **9**: 2315–2320.

Pesin, Ya. B. 1977. Characteristic Lyapunov exponents and smooth ergodic theory. *Russ. Math. Surv.* **32**: 55.

Pettini, M. 2007. *Geometry and topology in Hamiltonian dynamics and statistical mechanics.* Springer, New York.

Pichard, J. L., and Sarma, G. 1981a. Finite-size scaling approach to Anderson localisation. *J. Phys. C – Solid State* **14**: L127.

Pichard, J. L., and Sarma, G. 1981b. Finite-size scaling approach to Anderson localisation. II. Quantitative analysis and new results. *J. Phys. C – Solid State* **14**: L617.

Pierrehumbert, R. T., and Yang, H. 1993. Global chaotic mixing on isentropic surfaces. *J. Atmos. Sci.* **50**: 2462–2480.

Pikovsky, A. 1984a. On the interaction of strange attractors. *Z. Physik B* **55**: 149–154.

Pikovsky, A. 1984b. Synchronization and stochastization of nonlinear oscillations by external noise. In Sagdeev, R. Z. (ed.), *Nonlinear and turbulent processes in physics*, vol. 3. Harwood Acad, Chur.

Pikovsky, A. 1984c. Synchronization and stochastization of the ensemble of autogenerators by external noise. *Radiophys. Quantum Electron.* **27**: 576–581.

Pikovsky, A. 1989. Spatial development of chaos in nonlinear media. *Phys. Lett. A* **137**: 121–127.

Pikovsky, A. 1991. Statistical properties of dynamically generated anomalous diffusion. *Phys. Rev. A* **43**: 3146–3148.

Pikovsky, A. 1993. Local Lyapunov exponents for spatiotemporal chaos. *Chaos*, 3: 225–232.

Pikovsky, A., and Feudel, U. 1995. Characterizing strange nonchaotic attractors. *Chaos* **5**: 253–260.

Pikovsky, A., and Grassberger, P. 1991. Symmetry breaking bifurcation for coupled chaotic attractors. *J. Phys. A: Math., Gen.* **24**: 4587–4597.

Pikovsky, A., and Politi, A. 1998. Dynamic localization of Lyapunov vectors in spacetime chaos. *Nonlinearity* **11**: 1049–1062.

Pikovsky, A., and Politi, A. 2001. Dynamic localization of Lyapunov vectors in Hamiltonian lattices. *Phys. Rev. E* **63**: 036207.

Pikovsky, A., Osipov, G., Rosenblum, M., Zaks, M., and Kurths, J. 1997a. Attractor-repeller collision and eyelet intermittency at the transition to phase synchronization. *Phys. Rev. Lett.* **79**: 47–50.

Pikovsky, A., Zaks, M., Rosenblum, M., Osipov, G., and Kurths, J. 1997b. Phase synchronization of chaotic oscillations in terms of periodic orbits. *Chaos* **7**: 680–687.

Pikovsky, A., Rosenblum, M. G., Osipov, G. V., and J., Kurths. 1997c. Phase synchronization of chaotic oscillators by external driving. *Physica D* **104**: 219–238.

Pikovsky, A., Rosenblum, M., and Kurths, J. 2001. *Synchronization: a universal concept in nonlinear sciences*. Cambridge University Press, Cambridge.

Pinsky, M. A. 1986. Instability of the harmonic oscillator with small noise. *SIAM J. Appl. Math.* **46**: 451–463.

Pinsky, M. A., and Wihstutz, V. 1988. Lyapunov exponents of nilpotent Itô systems. *Stochastics* **25**: 43–57.

Pinsky, M. A., and Wihstutz, V. 1992. Lyapunov exponents and rotation numbers of linear systems with real noise. In *Probability theory (Singapore, 1989)*. de Gruyter, Berlin, 109–119.

Pires, C. J. A., Saa, A., and Venegeroles, R. 2011. Lyapunov statistics and mixing rates for intermittent systems. *Phys. Rev. E* **84**: 066210.

Politi, A. 2014a. *Probability density of the Lyapunov vector orientation*. Unpublished manuscript.

Politi, A. 2014b. Stochastic fluctuations in deterministic systems. In Vulpiani, A., et al. (eds.), *Large deviations in physics*. Springer, Berlin and Heidelberg, 243–261.

Politi, A., and Torcini, A. 1992. Periodic orbits in coupled Hénon maps: Lyapunov and multifractal analysis. *Chaos* **2**: 293–300.

Politi, A., and Torcini, A. 1994. Linear and non-linear mechanisms of information propagation. *Europhys. Lett.* **28**: 545.

Politi, A., and Torcini, A. 2010. Stable chaos. In *Nonlinear dynamics and chaos: advances and perspectives*. Springer, Berlin, 103–129.

Politi, A., and Witt, A. 1999. Fractal dimension of space-time chaos. *Phys. Rev. Lett.* **82**: 3034–3037.

Politi, A., Livi, R., Oppo, G.-L., and Kapral, R. 1993. Unpredictable behavior of stable systems. *Europhys. Lett.* **22**: 571.

Politi, A., Torcini, A., and Lepri, S. 1998. Lyapunov exponents from node-counting arguments. *J. Phys. IV France* **8**(Pr6), Pr6–263–Pr6–270.

Politi, A., Ginelli, F., Yanchuk, S., and Maistrenko, Y. 2006. From synchronization to Lyapunov exponents and back. *Physica D* **224**: 90–101.

Pollicott, M. 2010. Maximal Lyapunov exponents for random matrix products. *Invent. Math.* **181**: 209–226.

Pomeau, Y., Pumir, A., and Pelcé, P. 1984. Intrinsic stochasticity with many degrees of freedom. *J. Stat. Phys.* **37**: 39–49.

Popovych, O. V., Maistrenko, Yu. L., and Tass, P. A. 2005. Phase chaos in coupled oscillators. *Phys. Rev. E* **71**: 065201.

Posch, H. A., and Hirschl, R. 2000. Simulation of billiards and of hard body fluids. In *Hard ball systems and the Lorentz gas*. Encyclopaedia Math. Sci., vol. 101. Springer, Berlin, 279–314.

Pyragas, K. 1996. Weak and strong synchronization of chaos. *Phys. Rev. E* **54**: 4508–4511.

Pyragas, K. 1997. Conditional Lyapunov exponents from time series. *Phys. Rev. E* **56**: 5183–5188.

Quarteroni, A., and Valli, A. 1994. *Numerical approximation of partial differential equations*. Springer-Verlag, Berlin.

Radons, G. 2005. Disordered dynamical systems. In Radons, G. Just, W., and Hussler, P. (eds.) Collective dynamics of nonlinear and disordered systems, Springer-Verlag, Berlin 271–299.

Ramasubramanian, K., and Sriram, M. S. 2000. A comparative study of computation of Lyapunov spectra with different algorithms. *Physica D* **139**: 72–86.

Rangarajan, G., Habib, S., and Ryne, R. D. 1998. Lyapunov exponents without rescaling and reorthogonalization. *Phys. Rev. Lett.* **80**: 3747–3750.

Risken, H. 1989. *The Fokker-Planck equation: methods of solution and applications*. 2nd edn. Springer-Verlag, Berlin.

Romeiras, F. J., Bondeson, A., Ott, E., M., Antonsen T., and Grebogi, C. 1987. Quasiperiodically forced dynamical systems with strange nonchaotic attractors. *Physica D* **26**: 277–294.

Rössler, O. E. 1979. An equation for hyperchaos. *Phys. Lett. A* **71**: 155–157.

Ruelle, D. 1979. Ergodic theory of differentiable dynamical systems. *Inst. Hautes Études Sci. Publ. Math.* **50**: 27–58.

Ruelle, D. 1982. Large volume limit of the distribution of characteristic exponents in turbulence. *Comm. Math. Phys.* **87**: 287–302.

Ruelle, D. 1985. Rotation numbers for diffeomorphisms and flows. *Ann. Inst. H. Poincaré Phys. Théor.* **42**: 109–115.

Ruffo, S. 1994. Hamiltonian dynamics and phase transition. In *Transport, chaos and plasma physics*. World Scientific, Singapore, 114.

Rulkov, N. F., Sushchik, M. M., Tsimring, L. S., and Abarbanel, H. D. I. 1995. Generalized synchronization of chaos in directionally coupled chaotic systems. *Phys. Rev. E* **51**: 980–994.

Sano, M. M., and Kitahara, K. 2001. Thermal conduction in a chain of colliding harmonic oscillators revisited. *Phys. Rev. E* **64**: 056111.

Sauer, T. D., Tempkin, J. A., and Yorke, J. A. 1998. Spurious Lyapunov exponents in attractor reconstruction. *Phys. Rev. Lett.* **81**: 4341–4344.

Schmalfuß, B. 1997. The random attractor of the stochastic Lorenz system. *Z. Angew. Math. Phys.* **48**: 951–975.

Shadden, Sh. C., Lekien, F., and Marsden, J. E. 2005. Definition and properties of Lagrangian coherent structures from finite-time Lyapunov exponents in two-dimensional aperiodic flows. *Physica D* **212**: 271–304.

Shcherbakov, P. S. 1992. Alexander Mikhailovitch Lyapunov: on the centenary of his doctoral dissertation on stability of motion. *Automatica J. IFAC* **28**: 865–871.

Shepelyansky, D. L. 1983. Some statistical properties of simple classically stochastic quantum systems. *Physica D* **8**: 208–222.

Shibata, T., Chawanya, T., and Kaneko, K. 1999. Noiseless collective motion out of noisy chaos. *Phys. Rev. Lett.* **82**: 4424–4427.

Shimada, I., and Nagashima, T. 1979. A numerical approach to ergodic problem of dissipative dynamical systems. *Prog. Theor. Phys.* **61**: 1605–1616.

Sinai, Ya. 2009. Kolmogorov-Sinai entropy. *Scholarpedia* **4**(3): 2034.

Sinai, Ya. G. 1996. A remark concerning the thermodynamical limit of the Lyapunov spectrum. *Int. J. Bifurcat. Chaos* **6**: 1137–1142.

Skokos, Ch., Bountis, T. C., and Antonopoulos, Ch. 2007. Geometrical properties of local dynamics in Hamiltonian systems: the generalized alignment index (GALI) method. *Physica D* **231**: 30–54.

Slevin, K., and Ohtsuki, T. 1999. Corrections to scaling at the Anderson transition. *Phys. Rev. Lett.* **82**: 382–385.

Smirnov, V. I. 1992. Biography of A. M. Lyapunov. *International Journal of Control*, **55**: 775–784. Translated by J. F. Barrett from *A M Lyapunov: Izbrannie Trudi*, Izdat. Akad. Nauk SSSR, 1948.

Sommerer, J. C. 1994. Fractal tracer distributions in complicated surface flows: an application of random maps to fluid dynamics. *Physica D* **76**: 85–98.

Sompolinsky, H., Crisanti, A., and Sommers, H.-J. 1988. Chaos in random neural networks. *Phys. Rev. Lett.* **61**: 259–262.

Stöckmann, H.-J. 1999. *Quantum chaos: an introduction*. Cambridge University Press, Cambridge.

Stratonovich, R. L. 1967. *Topics in the theory of random noise*. Taylor & Francis.

Straube, A. V., and Pikovsky, A. 2011. Pattern formation induced by time-dependent advection. *Math. Model. Nat. Phenom.* **6**: 138–148.

Sussman, G. J., and Wisdom, J. 1992. Chaotic evolution of the solar system. *Science* **257**: 56–62.

Tailleur, J., and Kurchan, J. 2007. Probing rare physical trajectories with Lyapunov weighted dynamics. *Nat. Phys.* **3**: 203–22207.

Takens, F. 1981. Detecting strange attractors in turbulence. In *Dynamical systems and turbulence*, edited by D. A. Rand and L.-S. Young. Springer, London, 366–381.

Takeuchi, K. A., and Chaté, H. 2013. Collective Lyapunov modes. *J. Phys. A* **46**: 254007.

Takeuchi, K. A., Chaté, H., Ginelli, F., Politi, A., and Torcini, A. 2011a. Extensive and subextensive chaos in globally coupled dynamical systems. *Phys. Rev. Lett.* **107**: 124101.

Takeuchi, K. A., Yang, H.-L., Ginelli, F., Radons, G., and Chaté, H. 2011b. Hyperbolic decoupling of tangent space and effective dimension of dissipative systems. *Phys. Rev. E* **84**: 046214.

Tanase-Nicola, S., and Kurchan, J. 2003. Statistical-mechanical formulation of Lyapunov exponents. *J. Phys. A – Math. Gen.* **36**: 10299.

Taylor, T. J. 1993. On the existence of higher order Lyapunov exponents. *Nonlinearity* **6**: 369.

Teramae, J., and Tanaka, D. 2004. Robustness of the noise-induced phase synchronization in a general class of limit cycle oscillators. *Phys. Rev. Lett.* **93**: 204103.

Torcini, A., Grassberger, P., and Politi, A. 1995. Error propagation in extended chaotic systems. *J. Phys. A – Math. Gen.* **28**: 4533–4541.

Toth, Z. and Kalnay, E. 1997. Ensemble forecasting at NCEP and the breeding method. *Weather Rev.* **125**: 3297–3319.

Vallejos, R. O., and Anteneodo, C. 2012. Generalized Lyapunov exponents of the random harmonic oscillator: cumulant expansion approach. *Phys. Rev. E* **85**: 021124.

van Beijeren, H., and Dorfman, J. R. 1995. Lyapunov exponents and Kolmogorov-Sinai entropy for the Lorentz gas at low densities. *Phys. Rev. Lett.* **74**: 4412–4415.

van Saarloos, W. 1988. Front propagation into unstable states: marginal stability as a dynamical mechanism for velocity selection. *Phys. Rev. A* **37**: 211–229.

van Saarloos, W. 1989. Front propagation into unstable states. II. Linear versus nonlinear marginal stability and rate of convergence. *Phys. Rev. A* **39**: 6367–6390.

Vanneste, J. 2010. Estimating generalized Lyapunov exponents for products of random matrices. *Phys. Rev. E* **81**: 036701.

Venegerolcs, R. 2012. Thermodynamic phase transitions for Pomeau-Manneville maps. *Phys. Rev. E* **86**: 021114.

Viana, M. 2014. *Lectures on Lyapunov exponents.* Cambridge University Press, Cambridge.

von Bremen, H. F., Udwadia, F. E., and Proskurowski, W. 1997. An efficient QR based method for the computation of Lyapunov exponents. *Physica D* **101**: 1–16.

Walters, P. 1982. *An introduction to ergodic theory.* Springer-Verlag, Berlin.

Wigner, E. P. 1967. Random matrices in physics. *SIAM Review* **9**: 1–23.

Wolfe, C. L., and Samelson, R. M. 2007. An efficient method for recovering Lyapunov vectors from singular vectors. *Tellus A* **59**: 355–366.

Wolfram, S. 1986. *Theory and applications of cellular automata: including selected papers, 1983–1986.* World Scientific, Singapore.

Yanchuk, S., and Wolfrum, M. 2010. A multiple time scale approach to the stability of external cavity modes in the Lang-Kobayashi system using the limit of large delay. *SIAM J. Appl. Dyn. Syst.* **9**: 519–535.

Yang, H.-L., and Radons, G. 2013. Hydrodynamic Lyapunov modes and effective degrees of freedom of extended systems. *J. Phys. A* **46**: 254015.

Young, L.-S. 1982. Dimension, entropy, and Lyapunov exponents. *Ergod. Theor. Dyn. Syst.* **2**: 109–124.

Yu, L., Ott, E., and Chen, Q. 1990. Transition to chaos for random dynamical systems. *Phys. Rev. Lett.* **65**: 2935–2938.

Zanon, N., and Derrida, B. 1988. Weak disorder expansion of Liapunov exponents in a degenerate case. *J. Stat. Phys.* **50**: 509–528.

Zaslavsky, G. M. 2007. *The physics of chaos in Hamiltonian systems*. Imperial College Press, London.

Zaslavsky, G. M., and Edelman, M. A. 2004. Fractional kinetics: from pseudochaotic dynamics to Maxwell's demon. *Physica D* **193**: 128–147.

Zhou, D., Sun, Y., Rangan, A.V., and Cai, D. 2010. Spectrum of Lyapunov exponents of non-smooth dynamical systems of integrate-and-fire type. *J. Comput. Neurosci.* **28**: 229–245.

Zillmer, R., and Pikovsky, A. 2003. Multiscaling of noise-induced parametric instability. *Phys. Rev. E* **67**: 061117.

Zillmer, R., and Pikovsky, A. 2005. Continuous approach for the random-field Ising chain. *Phys. Rev. E* **72**: 056108.

Zillmer, R., Ahlers, V., and Pikovsky, A. 2000. Scaling of Lyapunov exponents of coupled chaotic systems. *Phys. Rev. E* **61**: 332–341.

Zillmcr, R., Ahlers, V., and Pikovsky, A. 2002. Coupling sensitivity of localization length in one-dimensional disordered systems. *Europhys. Lett.* **60**: 889–895.

Zillmer, R., Livi, R., Politi, A., and Torcini, A. 2006. Desynchronization in diluted neural networks. *Phys. Rev. E* **74**: 036203.

Index